The U

LEPEN

DEVELOPMENTS IN FOOD PROTEINS—2

CONTENTS OF VOLUME 1

Volume 1

DEVELOPMENTS IN FOOD PROTEINS—2

Edited by

B. J. F. HUDSON

Senior Research Fellow, Department of Food Science, University of Reading, London Road, Reading, UK

APPLIED SCIENCE PUBLISHERS
LONDON and NEW YORK

APPLIED SCIENCE PUBLISHERS LTD
Ripple Road, Barking, Essex, England

Sole Distributor in the USA and Canada
ELSEVIER SCIENCE PUBLISHING CO., INC.,
52 Vanderbilt Avenue, New York, NY 10017, USA

Catalogued in the British Library as:

Developments in food proteins.—1–.
—London: Applied Science, 1982–.
— v.; 23 cm.— (Developments series)
Irregular.
ISSN 0263-4708 = Developments in food proteins
1. Food—Protein contents—Periodicals
I. Sr
641.1′2′05 TX553.P7
(B82-26157)

ISBN 0-85334-176-1

WITH 85 TABLES AND 63 ILLUSTRATIONS

Photoset in Malta by Interprint Ltd
Printed in Great Britain by Galliard (Printers) Ltd, Great Yarmouth

PREFACE

This volume extends the range of currently significant topics in food proteins that were reviewed in Volume 1. Again, I have tried to keep a parity between protein sources or commodities, treated from a general point of view, and specialised considerations such as manufacturing processes or techniques of investigation.

It is not easy, or perhaps even desirable, to keep these aspects apart. For instance, Chapter 4 on lupin seed proteins not only draws attention to a useful, though somewhat neglected, food source, but is also a timely case study of methodology: the classification of a typical complex protein in terms of fractions and sub-fractions, with data on their individual characteristics. Similarly, Chapter 7, which considers yeast protein, analyses the general problem of removal of nucleic acids from proteins.

Probably more protein of cereal origin than from any other source is consumed in food (see Chapter 1). This is recognised in a comprehensive treatment of trends in wheat protein technology. Soyabean protein is an increasingly important protein source and Chapter 2 is devoted to this subject. Another developing commodity, rapeseed, provides a further timely topic, as interest grows in the utilisation of its by-product protein (Chapter 3).

Fish remains one of the most attractive and nutritious of foods. The treatment of it in Chapter 6 as a protein source is particularly opportune in view of changes taking place in the composition of the catch and the form in which it is presented to the consumer.

Processing is represented by the application of cereal flour classification methods to legumes (Chapter 5), with valuable potential in terms of protein concentration.

Finally, no collection of topics representing the protein field is complete without reference to nutrition. Present thinking on the vexed question of the evaluation of biological quality is the subject of Chapter 8.

The theme that links the book together is the recurring importance to the food technologist of functionality in proteins. The subtle and very diverse variations in physical properties that are dependent on protein structure, conformation and environment, are highlighted, and the technological consequences discussed.

For so ably helping me to complete this second volume, I acknowledge with sincere gratitude the enthusiasm and diligence of my eight contributors, and the sustained encouragement of the publisher.

B. J. F. Hudson

CONTENTS

LIST OF CONTRIBUTORS

MICHAEL R. BOOTH
Scientific Adviser, RHM Cereals Limited, c/o RHM Research Limited, The Lord Rank Research Centre, Lincoln Road, High Wycombe, Buckinghamshire HP12 3QR, UK.

PAOLO CERLETTI
Professor of General Biochemistry, University of Milan, Via Celoria, 2-20133, Milano, Italy.

M. GUZMAN-JUAREZ
Research Food-Scientist, Department of Food Science, University of Reading, London Road, Reading RG1 5AQ, UK.

I. M. MACKIE
Head of Protein Chemistry Section, Ministry of Agriculture, Fisheries and Food, Torry Research Station, PO Box 31, 135 Abbey Road, Aberdeen AB9 8DG, UK.

A. M. PEARSON
Professor of Food Science and Human Nutrition, Michigan State University, East Lansing, Michigan 48827, USA.

J. DAVID SCHOFIELD
Principal Scientist, RHM Research Limited, The Lord Rank Research Centre, Lincoln Road, High Wycombe, Buckinghamshire HP12 3QR, UK.

FRANK W. SOSULSKI
Professor of Crop Science, University of Saskatchewan, Saskatoon, Canada S7N 0W0.

ANN F. WALKER
Lecturer, Department of Food Science, University of Reading, London Road, Reading RG1 5AQ, UK.

Chapter 1

WHEAT PROTEINS AND THEIR TECHNOLOGICAL SIGNIFICANCE

J. DAVID SCHOFIELD and MICHAEL R. BOOTH
RHM Research Limited, High Wycombe, UK

SUMMARY

Wheat is the world's largest cereal crop[1] *with annual production of approximately* 450 *million tonnes and it is a major component of the staple diet in many areas of the world. Its importance owes much to its almost unique ability to be baked into bread, a property governed largely by the physicochemical properties of its endosperm storage proteins, collectively termed gluten. These same proteins are also of considerable importance in determining the suitability of wheat for the production of paste products (pasta), such as spaghetti and macaroni, and other end uses. However, wheat proteins are not only important as components of unfractionated wheat flour; a number of commercial processes have been devised to separate flour into gluten and starch. Although gluten is quantitatively the minor product, the demand for it ensures a relatively high price and makes it a valuable commodity in which there is significant national and international trade. This demand for gluten again results from its peculiar physicochemical properties. In this chapter we describe the nature of flour/gluten proteins and show how these properties account for their technological importance.*

THE NATURE OF WHEAT PROTEINS

Classification of Wheat Proteins

The bulk of the proteins of wheat flour are derived from the starchy endosperm of the wheat caryopsis and to a large extent comprise the endosperm storage proteins. Other proteins may also be present that are remnants of the metabolic apparatus and structural elements of the developing endosperm, and there will also be proteins derived from contaminating fragments of the outer layers of the grain, the bran and the germ.

Wheat proteins have been classified traditionally according to their extractabilities in various solvents.[2] More modern protein separation techniques such as electrophoresis and chromatography have complemented rather than supplanted extraction techniques and have allowed a more detailed and precise chemical characterisation of these proteins.[3,4] Nevertheless, despite the battery of techniques now available to protein chemists, wheat proteins remain an extremely difficult system to study because of their insolubility in solvents normally employed for protein separation and analysis, and because of their polydispersity with regard to quaternary structure and subunit compositions.

The most commonly used extraction procedures for fractionating wheat proteins are based on the techniques originally devised by Osborne[2] but since modified by numerous workers. Five fractions may be produced by such sequential extraction schemes: albumins, extractable in saline and remaining soluble on dialysis against water; globulins, extractable in saline and precipitating on dialysis against water; gliadins (wheat prolamins), extractable in concentrated aqueous alcohol solutions; glutenins (wheat glutelins), extractable in dilute aqueous acid or alkali; and a residual fraction that is not extracted, 'residue' protein.

A number of factors, including wheat variety, affect the proportion of the total flour protein present in each fraction but the exact conditions used for extraction also affect the distribution of protein in different fractions.[4] In general, albumins and globulins each account for about 6–10 % of the total protein, and gliadins and glutenins each comprise about 35 %. 'Residue' protein accounts for the remainder and can be as little as 10% or as much as 40% of the total protein.[5,6] It is clear, however, that each fraction is heterogeneous and that in most cases considerable overlap occurs between fractions, particularly between gliadins and glutenins and between glutenins and 'residue' protein.[7]

The protein fraction of greatest technological significance is that known as gluten; this is the viscoelastic, proteinaceous material produced when a flour/water dough is washed with excess water to remove starch and other more minor components. The gross composition of gluten is somewhat variable and, although it is often described as a protein, implying purity and a fairly well defined composition, gluten, like flour proteins, is extremely heterogeneous. Usually only about 75–85% of its dry weight actually comprises protein, the remainder being residual starch, lipid, hemicelluloses, etc.

The protein component of gluten consists in large part of gliadins, glutenins and 'residue' protein, of which the gliadins confer extensibility while the glutenins and 'residue' protein confer elasticity; some albumins and globulins are also usually present, the level depending on the thoroughness of the washing-out process.

Amino Acid Composition

Examination of the amino acid compositions of wheat protein fractions illustrates that these classes comprise different types of protein/polypeptide subunits (Table 1). The compositions of the gliadin and glutenin fractions show many overall similarities although there are some notable differences, amongst these being the lower glycine, alanine, lysine, aspartic acid and threonine contents and the higher glutamic acid, proline and cystine contents of the gliadin fraction. Two important points to note concerning these fractions are that the glutamic acid contents of these proteins are extremely high and, moreover, that a very large proportion of the glutamic acid is in the amide form, glutamine; for some isolated gliadin subunits it appears that essentially all of the glutamic acid is in this form.[13] Much of the aspartic acid is also in the amide form and this, together with the relatively low contents of basic amino acids (lysine particularly, but also histidine and arginine), means that these proteins carry a very low charge. Indeed, Kasarda *et al.*[3] have indicated that some of the gliadins are amongst the least charged proteins known. As a result, ionic interactions between the gluten proteins may be quantitatively less important than they are in other proteins.

The high levels of glutamic and aspartic acid, and the fact that the bulk of these side chains are present in the amide form, has important consequences as far as interactions between gluten proteins is concerned. Synthetic homopolypeptides of glutamine have been reported to form β-structures that are stable under very harsh conditions.[14] The stability of these structures results from very extensive hydrogen bonding, for the amide groups of glutamine side chains can act both as hydrogen bond donors and acceptors (Fig. 1). The viscoelastic properties of flour/gluten proteins are impaired or lost if the amide groups are converted to carboxyl groups by mild acid hydrolysis[15] or to ester groups by methanolysis.[16] On the other hand, when flour or gluten is solvated with deuterated water (D_2O) rather than ordinary water (H_2O) the rheolo-

TABLE 1

AMINO ACID COMPOSITIONS OF WHEAT PROTEIN FRACTIONS, SOYA PROTEIN ISOLATE, CASEIN AND EGG WHITE (moles/1000)

	Albumin[a]	*Globulin*[a]	*Gliadin*[a]	*Glutenin*[a]	*Gluten*[b]	*Soya isolate*[c]	*Casein*[d]	*Egg white*[e]
Asp	69·8	60·4	28·4	36·5	29·5	119·9	58·7	97·7
Thr	27·9	48·3	24·4	34·5	27·5	47·1	42·2	47·3
Ser	51·6	110·8	61·3	69·5	60·8	74·6	72·2	84·5
Glu	140·0	51·0	344·5	288·6	327·2	111·8	174·6	122·2
Pro	85·0	36·4	162·4	119·0	147·7	78·0	127·1	37·6
Gly	48·1	95·0	30·7	75·3	54·5	66·5	31·2	56·5
Ala	72·6	61·1	32·7	43·8	36·5	63·1	41·4	82·9
Cys	65·2	134·0	33·2	26·1	22·2	11·2	7·4	30·4
Val	80·9	23·6	47·6	48·1	46·7	55·0	73·3	75·7
Met	0	3·6	12·5	14·2	13·8	12·9	26·2	33·3
Ile	36·7	13·7	43·5	37·1	40·4	56·6	63·8	52·2
Leu	95·2	90·1	69·3	65·3	67·9	93·9	85·1	82·2
Tyr	21·6	16·3	18·3	24·8	26·0	32·4	39·4	29·3
Phe	35·5	24·5	42·9	36·4	40·3	53·5	42·6	46·1
Lys	87·3	106·4	5·7	19·8	12·5	21·1	64·9	53·5
Hist	32·1	17·7	18·6	19·0	18·0	30·8	23·0	17·6
Arg	50·1	106·5	19·9	29·8	22·2	71·4	26·9	39·2
Trp	ND	ND	4·3	12·6	6·4	ND	ND	11·8

[a] Ref. 8; [b] Ref. 9; [c] Ref. 10; [d] Ref. 11; [e] Ref. 12.

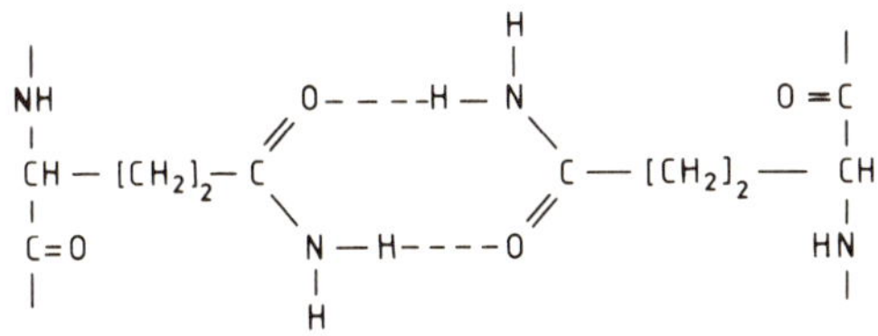

FIG. 1. Hydrogen bonding between glutamine residues in adjacent polypeptide chains.

gical properties of dough or gluten proteins are altered such that the protein matrix appears mechanically stronger[17,18] because of stronger interaction between the proteins due to the greater bond strength of deuterium bonds compared with hydrogen bonds. The observations suggest that the numerous weak hydrogen bonds are indeed very important in the interactions between wheat proteins. The presence of relatively high levels of amino acids with hydrophobic side chains also means that there is considerable potential for interactions of the hydrophobic type and recent models for the structure of glutenin complexes have stressed the importance of this type of interaction (see below).

Gliadins and glutenins both contain high levels of the amino acid proline, and this also has important consequences for the conformational structure of these proteins, particularly for the gliadins, which contain the highest levels of proline. The fact that the N–C bond that comprises part of the polypeptide backbone is part of a cyclic structure (Fig. 2) means that neither free rotation about the N–C bond nor hydrogen bonding to the nitrogen atom can occur. As a result, whenever proline occurs in a polypeptide chain, ordered structures such as α-helix and β-sheet are interrupted and the lack of freedom to rotate may also introduce rigidity into the conformational structure.

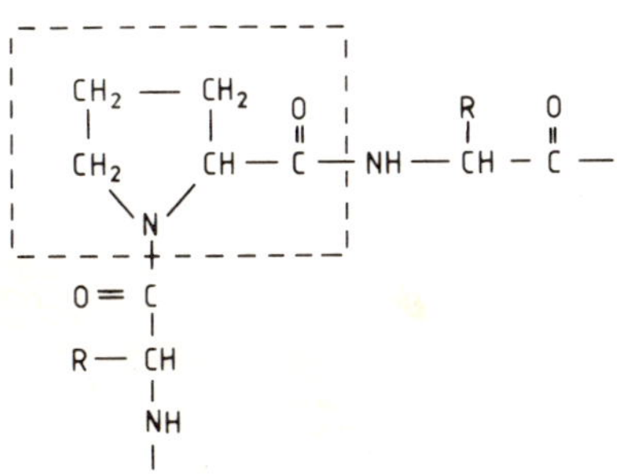

FIG. 2. The cyclic structure of the side chain of peptide-bound proline. The fact that the N–C bond of the polypeptide backbone is part of a cyclic structure eliminates free rotation about this bond.

The compositions of the albumin and globulin fractions are quite distinct from those of the gliadin and glutenin fractions; this is particularly so for the globulin fraction. Both albumins and globulins have a much more even distribution of amino acids and it is particularly noteworthy that glutamic acid is not nearly so predominant as in the gliadins and glutenins. Lysine is present at much higher levels in both albumins and globulins, as are aspartic acid, alanine, cystine, leucine and arginine; methionine levels appear to be particularly low in these fractions.

Gluten Proteins

The major protein components of gluten are gliadins, glutenins and 'residue' protein. These components are difficult to define in precise chemical and biological terms since much more work still remains to be done to determine their structure and biology. As discussed above, gliadins and glutenins have been defined operationally according to their extractability/solubility characteristics.

In the present review gliadins are regarded as being the monomolecular proteins that comprise the bulk of the proteins extractable in aqueous alcohol solution. Glutenins are regarded as being the protein complexes, linked by disulphide bonds, that are largely, but not completely, unextractable in aqueous alcohol solution but extractable in dilute acid or alkali solution.

Gliadins

Gliadins comprise a heterogeneous group of proteins. On starch gel or acrylamide gel electrophoresis in the presence of lactate buffers, approximately 20–25 bands are usually discernible although the band patterns for gliadins from different varieties of wheat show considerable variation (Fig. 3) and in all as many as about 45 different bands have been detected in the commonly occurring wheats.[19–22] However, greater resolution can be achieved on two-dimensional electrophoresis systems, and, whereas a particular variety may yield up to 25 gliadin components on single-dimension starch or acrylamide gels, almost 50 different components may be discernible on two-dimensional systems.[23] It seems likely that even this number may be a conservative estimate, and application of even higher resolution electrophoretic methods or combinations of chromatographic and electrophoretic separations may be expected to increase the number of identifiable gliadin subunits.

Gliadins are present in gluten or flour in a monomolecular form and the available evidence suggests that they have somewhat asymmetric,

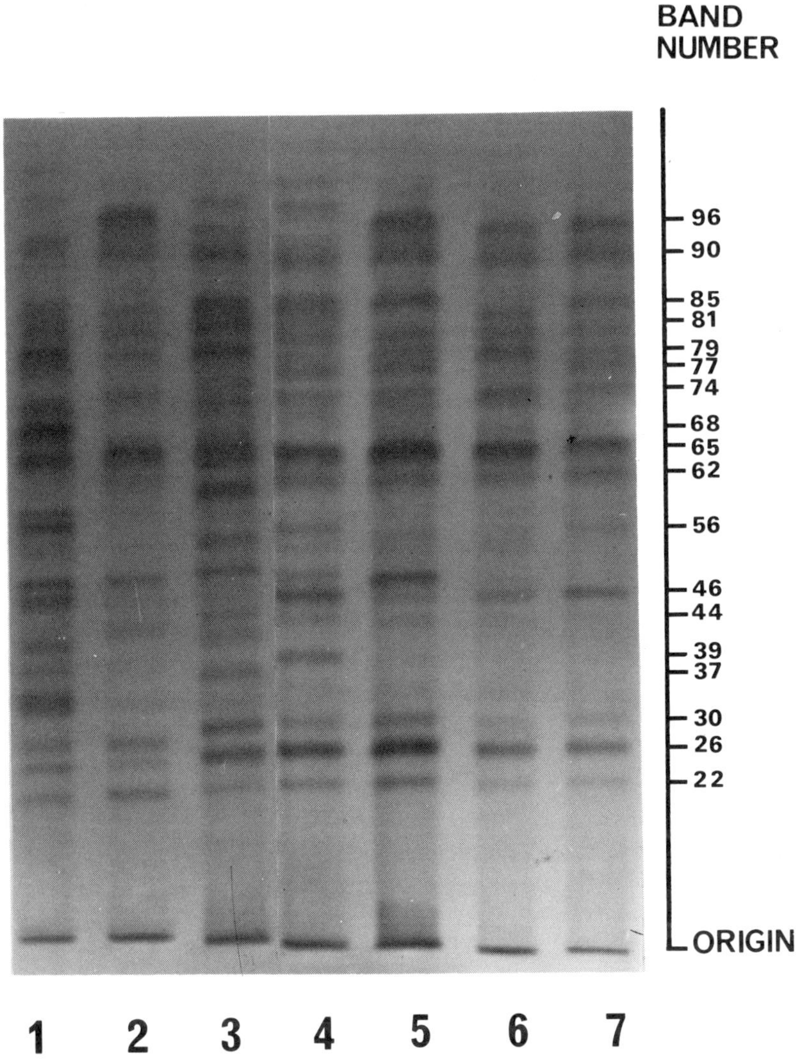

FIG. 3. Starch gel electrophoresis of gliadins extracted from different wheat varieties. Up to 25 bands may be discernible on single-dimension gels. The differences in gliadin components may be used to differentiate between varieties. Varieties present in this electropherogram are: 1, Neepawa (Canadian, hard red spring); 2, Broom (English, hard red spring); 3, Highbury (English, hard red spring); 4, Norman (English, soft red winter); 5, Avalon (English, hard red winter); 6, Rapier (English, soft red winter); 7, Maris Huntsman (English, hard red winter). Band numbering system as in Refs. 20, 21. (Photograph provided courtesy of Dr J. R. S. Ellis and Mr C. J. Boyce, RHM Research Ltd.)

compact structures that are stabilised by intramolecular disulphide bonds. In general, they contain substantial proportions of α-helical structure[24–27] (about 35% of the structure in some α-gliadins is α-helix[27]) and lesser amounts (10%) of β-sheet structure. The conformational structure of gliadins, although stabilised by disulphide bonds, does nevertheless exhibit some flexibility as evidenced by partial unfolding (i.e. loss of α-helix) at acid pH or in 3 M urea[24–27] or increased folding in other solvents.[28] Viscosity measurements on gliadins indicate that the molecules, although globular in nature, may not actually be spherical but rather may have more of an ellipsoid structure.[26,29]

Gliadins are also heterogeneous with respect to their molecular weights although not to the same extent as the glutenins (see below). The molecular weights of most gliadin species fall within the range 30 000 to 45 000.[3] The ω-gliadins, however, have much higher molecular weights and values of 75 500 and 79 000 have been determined for some components of this group using sedimentation equilibrium analysis.[13] The ω-gliadins are also unusual in that they are deficient, and in some cases completely lacking, in sulphur amino acids so that intramolecular disulphide bonding in these molecules is minimal or absent.[13] Little work has been reported on the conformational structure of the ω-gliadins although the extremely high content of proline residues may confer greater conformational stability than might be expected of a protein lacking disulphide bonds.[13] A further feature of ω-gliadins is that essentially all of the glutamic acid and aspartic acid side chains are in the amide forms[13] and this, together with their low contents of basic amino acids, accounts for their very low electrophoretic mobilities.

Glutenins

Glutenins are generally believed to exist as high molecular weight complexes comprising subunits linked together by disulphide bonds.[3,4,30] These disulphide-linked complexes are extremely polydisperse with respect to molecular size; molecular weights ranging from in the order of 100 000 to several millions have been deduced by column chromatography experiments.[5,31–33] Ewart considers glutenin molecules to be linear 'concatenations' in which neighbouring subunits are held together by one or more disulphide bonds[34] while others consider that three-dimensional aggregates may also exist involving lateral aggregation of the subunits.[3,4,35] It is clear from gel filtration chromatography, however, that there are no groups of complexes with discrete molecular weights (i.e. no discrete peaks) but rather there is a continuum of molecular weights from the smallest to the largest (Fig. 4).

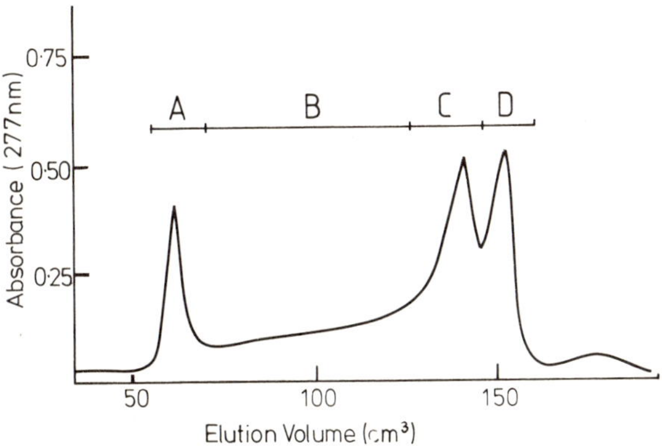

FIG. 4. Gel filtration chromatography of sodium dodecyl sulphate extracted flour proteins on Sepharose CL-4B in the presence of sodium dodecyl sulphate. The protein subunits in regions A and B are present in SDS extracts as disulphide-linked molecules and these are designated as glutenins. Glutenins in region A are at the void volume of the column (i.e. of such a size as to be excluded from the gel) and glutenins in region B are polydisperse with regard to mol. wt., there being no discrete peaks in this region. Polypeptides present in region C are monomolecular and are designated as gliadins while those in region D are of low molecular weight and comprise albumin and globulin species.

Disulphide bonding is generally recognised as being an important feature of the structure of glutenin molecules but less attention has been paid to the nature and importance of non-covalent bonding. However, there is now much debate about the contribution of non-covalent bonding to glutenin structure. Such bonds are undoubtedly of considerable importance in the interactions between glutenin subunits and between glutenins and gliadins and other wheat proteins, as indeed they are in other protein systems.

Considering the amino acid composition of glutenins and gliadins (see above) there is considerable potential for non-covalent bonding among wheat protein subunits. The presence of very large amounts of glutamine means that there is the potential for very extensive interaction through hydrogen bonds. Although these bonds are relatively weak, the occurrence of large numbers of them in wheat protein enhances their importance.

Another type of interaction that is considered to play an especially important role in the stabilisation of native conformation, and also in the aggregation and binding of proteins, is hydrophobic interaction. Such interactions occur because in an aqueous environment there is a favourable free energy change when non-polar hydrocarbon amino acid side chains, such as those of valine, leucine, isoleucine and phenylalanine, come together by appropriate folding of a polypeptide chain or by aggregation of two protein subunits having non-polar side chains at their surfaces; in this way minimal disruption of the strong attractive forces between water molecules occurs.[36,37] Jones and Carnegie[38] proposed some time ago that hydrophobic interactions may be important in the changes that wheat proteins undergo upon hydration. However, Kasarda *et al.*[3] have recently proposed a model for glutenin polymers in which the subunits are not covalently linked at all but are linked into extremely stable complexes mainly through hydrophobic interactions. The disulphide bonds that undoubtedly exist in glutenins are envisaged in this model as being intramolecular bonds stabilising the three-dimensional structure of each subunit.

The results of work on the effects of the sodium salts of fatty acids on glutenin solubility in unbuffered aqueous systems[39] have been quoted as providing evidence in support of Kasarda *et al.*'s hypothesis. However, relatively little work has been reported on the characterisation of the proteins solubilised by soaps nor on the mechanism of solubilisation. On the other hand, an aqueous solution of the detergent sodium dodecyl sulphate (SDS) is an excellent solvent for the extraction and gel filtration chromatography of wheat proteins,[32,33] the solubilising effect of SDS presumably being similar to that of the sodium soaps. Moreover, it is clear that the glutenins solubilised by SDS are still present as very high molecular weight complexes as determined by gel filtration chromatography (Refs. 32,33—see Fig. 4). These results suggest that the glutenin complexes are indeed linked by inter-subunit disulphide bonds.

Other recent work from our laboratory, involving the application of cross-polarisation carbon-13 nuclear magnetic resonance (NMR) techniques to study wheat protein structure,[40] has provided preliminary evidence to suggest that, while hydrophobic interactions between glutenins and gliadins are important, such interactions are perhaps of less importance amongst glutenin subunits than the model of Kasarda *et al.* would suggest. In these experiments, a gliadin-rich fraction gave very broad resonance peaks in the region of the NMR spectrum corresponding to hydrophobic amino acid side chains, which suggests that these side chains are not mobile and that they have a high level of interaction with each other; this would be consistent with a tightly folded structure for gliadins generally. A glutenin-enriched fraction, by contrast, exhibited sharp resonance peaks in the same region of the spectrum indicating that these side chains are extremely mobile and do not interact to anywhere near the same extent as in the gliadin fraction. However, it is noteworthy that a weighted summation of the spectra for the two fractions could not give a spectrum equivalent to that of the parent gluten. The implication of this finding is that hydrophobic interactions between glutenins and gliadins may well be important in gluten itself and may be important determinants of the rheological properties of this material.

The subunit composition of glutenins is complex, with polypeptides ranging in molecular weight from about 30 000 to about 140 000.[31-33,41] The subunits present may vary from one variety to another and many of the subunits, particularly those of high molecular weight, have been assigned to specific genes.[41-45] There is probably not the same degree of multiplicity found for gliadins although as many as 12 subunits have been identified in single-dimension SDS–polyacrylamide gel electrophoresis (PAGE) systems.[31] It is unclear at present whether or not there are any polypeptides that are common to both the disulphide-linked glutenin complexes and the monomolecular gliadins although sequence studies suggest that the bulk of the monomolecular gliadins are unlike the bulk of glutenin subunits.[46,47] Components with very similar mobilities in SDS–PAGE systems (mol. wt. in the region of 30 000–40 000) are found in both gliadins and glutenins prepared by gel filtration chromatography (Fig. 5), but identical mobility, even in high resolution, single-dimension, SDS–PAGE systems is not sufficient to conclude that these components are the same polypeptides. It is clear, however, that most of the glutenin subunits having molecular weights from about 40 000 to about 140 000 are not found in the monomolecular gliadin fraction.

Examination of the different molecular size classes of glutenin complexes suggests that those complexes of lowest molecular weight, although comprising the same polypeptides as those of high molecular weights, do in fact, exhibit different proportions of subunits.[31] Thus, glutenin complexes of lowest molecular weight have a higher proportion of subunits with molecular weights in the 30 000–50 000 range and a lower proportion of those subunits with molecular weights in the 90 000–140 000 range. Whether or not this observation indicates some sort of oligomeric structure comprising low molecular weight subunits, from which the very large glutenin complexes are constructed, with the high molecular weight subunits acting as bridging components, cannot be ascertained at present.

The lower molecular weight, disulphide-linked glutenin complexes have appreciable solubility in aqueous alcohol solutions although those of very high molecular weight do not. As a result, these lower molecular

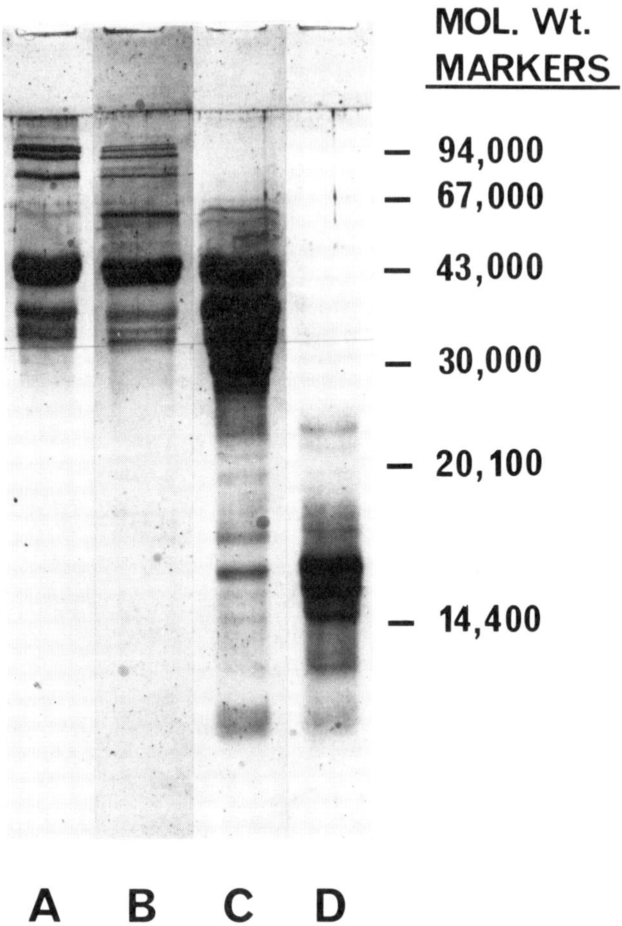

FIG. 5. Polyacrylamide gel electrophoresis in the presence of sodium dodecyl sulphate of the polypeptides present in different regions of the Sepharose CL-4B chromatograms (Ref. 33). Most of the polypeptides present in column regions A and B (glutenins) are identical but relative proportions of polypeptides of different molecular weight vary according to the size of the disulphide-linked molecules. Some polypeptides in region C (gliadins) have similar molecular weights to some of these in regions A and B but a large proportion of the polypeptides in region C are not present in regions A and B.

weight glutenins are extracted along with the monomolecular gliadin components and they have, in fact, been designated as high molecular weight gliadins by some workers.[41,48] However, sequence data indicate that the disulphide-bonded and monomolecular components are different classes of protein.[46,47] The solubility properties of the lower molecular weight glutenins, plus the fact that the subunits of the glutenins are also soluble to a large extent in aqueous alcohol after reduction of disulphide bonds,[47,49] have been contributory factors in the confusion and controversy over the definition of gliadins and glutenins.

'Residue' Protein

The definition of 'residue' protein is even less satisfactory than definitions for gliadins and glutenins. Operationally it is the protein remaining insoluble after extraction of the albumin, globulin, gliadin and glutenin fractions, although procedures for obtaining such fractions vary widely; some workers carry out sequential extractions of flour with salt, alcohol and acid[6] whereas others extract flours directly with acetic acid.[50] Despite the lack of an exact chemical/biological definition of 'residue' protein, this fraction may be important since it has been shown that there is a positive correlation between breadmaking quality and the amount of 'residue' protein.[6]

The reasons for the insolubility of 'residue' protein and for its possible importance in breadmaking are unclear. It has been suggested that soluble glutenins may be linear polymers whereas the bulk of 'residue' protein may comprise glutenin molecules that are aggregated in three dimensions to form very large extended complexes.[4,51] Ewart, however, considers it unnecessary to postulate three-dimensional complexes to account for the insolubility of glutenins in 'residue' protein but considers that the glutenins of 'residue' protein may also be linear polymers that are insoluble solely on account of their very high molecular weights.[34]

TECHNOLOGICAL SIGNIFICANCE OF THE GLUTEN PROTEINS OF WHEAT

Breadmaking

Wheat flour has many food uses, but the most important in the Western World is in the production of bread. For instance, in the UK in 1980/81, 67% of flour produced was bread flour.[52] This ability of wheat flour to be turned into bread has been recognised since the times of the ancient Egyptians, and with the exception of rye, wheat is the only cereal that can be used successfully for this purpose. This almost unique property of wheat flour is attributed to the wheat proteins.

The Importance of Protein—Quantity and Quality

Finney demonstrated a strong positive relationship between the protein content of a series of North American flours and the volume of bread baked from them by a traditional fermentation method (Ref. 53—see also Fig. 10 below). The level of protein required in a wheat flour to produce satisfactory bread varies considerably according to the type of bread, the process employed for manufacture and the required keeping quality. For instance, in France, the protein content of the flour ($N \times 5{\cdot}7$, 14% m.b.) used to produce typical 'baguettes', the envy of many other countries, is normally between 9 and 10%. This product, however, stales rapidly. Most bread in the UK, some 75%, is produced under conditions in which the dough is intensively mixed: 5 watt h per 1b dough ($40\,kJ\,kg^{-1}$) must be imparted in under 5 min, typically 3 min. This technique, known as the Chorleywood Bread Process (CBP)[54] obviates the need for several hours of bulk fermentation although proving is still required. Good bread can be produced with a flour protein level of 11%, whereas traditional fermentation processes use flours of 12% protein or higher. The CBP thus allows the use of grists with less high protein, North American wheat and more home-grown grain, which in turn has helped to stabilise flour and hence bread prices. In the UK, flours of higher protein content are still used by the craft baker, and for speciality products such as morning goods, Vienna rolls, etc.

The kind of wheat available, has determined the type of bread produced in individual countries or areas, dependent upon whether grown locally or obtained through an established importing system. Thus France uses French wheat but the UK has traditionally imported grain from the Commonwealth countries, Canada and Australia, and continues to import 'third country' wheat mainly from Canada and the USA.

A large proportion (some 80–85%) of wheat flour protein comprises the viscoelastic complex gluten, described earlier, and it is this which is largely responsible for the dough properties and breadmaking ability of wheat flour. Thus a good positive relationship also exists between the quantity of gluten present in a flour and its breadmaking performance.[55] Indeed many millers, especially in parts of Europe, use this as one of their quality parameters for flour and a semi-automatic machine has been developed to extract and determine the gluten content.[56]

The prime importance of gluten proteins in the preparation of bread is the 'strength' they impart to the dough—i.e. its resistance to extension and its stability during mixing, although the dough must not be over-strong. 'Strong' doughs have the ability to retain gas cells in the structure. A fine silky crumb is the result of an even distribution of small gas cells that expand during proving and during the initial stages of baking. If the rheological properties of the gluten are such that the dough is too tough and insufficiently extensible, resistance to expansion of the gas cells will be too great and a poor volume loaf with a rather rubbery texture will result. Conversely, for doughs containing gluten that is too soft and extensible, the gas cells expand excessively and coalesce with each other and this bread will have a very open texture with coarse cell wall structure.

Flours of equivalent protein or gluten content may contain gluten that is too extensible, insufficiently extensible or with optimal rheological characteristics. Thus, not only is the protein and/or gluten content of a flour of importance in breadmaking but there are also qualitative characteristics in flour proteins/gluten that are crucial. Although more convenient and more generally applicable definitions of quality than a baking test are desirable (e.g. a rapid physical or chemical test), the complex nature of gluten and the incomplete knowledge of the relationship between fundamental protein structure, rheological parameters and product quality, only allow a definition of flour quality in terms of a baked product. In the final analysis, of course, baking performance is the criterion that is all-important and against which any other more quantitative or convenient test of flour/protein quality must be correlated.

Factors Influencing Flour Protein and Gluten Quality

Many factors influence the quality of flour protein, not least the variety of wheat. Other factors having an effect on gluten quality include the action of proteolytic enzymes (p. 51) and of heat. Heat damage to gluten normally arises when wheat has been harvested wet and excessive air temperatures have been used to dry the grain. A considerable toughening of the gluten occurs and in extreme cases it is not possible to extract any gluten at all from the dough. Heat denatured wheat is unacceptable for use in breadmaking since very poor loaf volume and texture arise (Fig. 6). The effect of heat on gluten will be discussed more fully later.

FIG. 6. Comparison of the baking performances of flours from normal (left) and heat-denatured (right) wheat.

The Molecular Basis of Baking Quality

The complexity of wheat proteins and in particular of the viscoelastic material, gluten, has been discussed above. Many studies have been conducted with the objective of elucidating the relationship between composition of structure of the proteins and their functionality, i.e. the ability to produce bread with satisfactory loaf volume and crumb texture characteristics.

The physical properties of gluten are somewhere between those of hydrated gliadin, which is soft and extensible, and those of glutenin, which is firm, elastic and resistant to extension.[57] There has been speculation that the relative proportions of gliadin and glutenin account for the overall gluten properties.[58] Thus, properties of gluten from poor breadmaking wheats would be explained by a higher proportion of gliadin and less glutenin. However, despite many attempts to demonstrate such a relationship there are still no firm data in its support. Difficulties over adequate definitions of gliadins and of glutenins and, hence, confusion concerning methods for their determination, have led to the abandonment of this approach.[59,60] It must be remembered that gliadin and glutenin are themselves not single proteins but comprise many species of varying molecular size. It is within these that the differences in quality arise.

Starch and polyacrylamide gel electrophoresis have proved to be powerful tools for demonstrating the multitude of components or subunits present in both gliadin and glutenin. Very little correlation between the individual components present and the bread baking performance of the original flour has been found until recently when Payne and his colleagues, using polyacrylamide gel electrophoresis in the presence of SDS, presented evidence to show that flour quality can be related to the presence of particular subunits of glutenin.[44,61] In the initial experiments[61] the presence of a particular glutenin subunit of molecular weight about 145 000, in varieties from a UK wheat breeding programme, was found to correlate with baking quality (Fig. 7) when this was measured

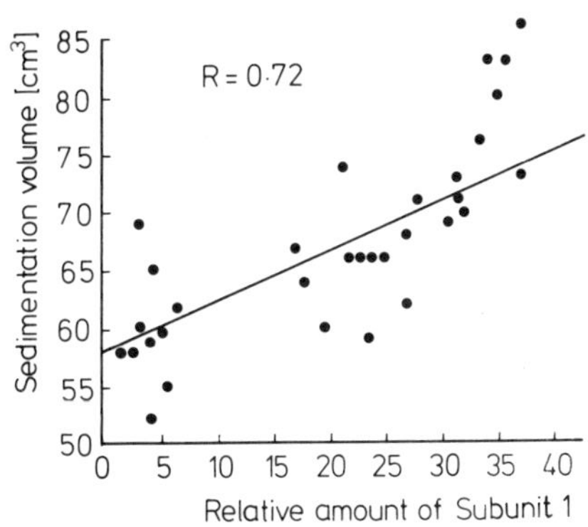

FIG. 7. Relationship between baking performance of wheat samples, as measured by the SDS sedimentation test, and the relative amount of glutenin subunit 1. (Reproduced courtesy of Dr P. I. Payne, Plant Breeding Institute, Cambridge, UK from Ref. 61.)

by the SDS sedimentation test.[62] However, it was recognised that varieties from other breeding programmes do not possess this particular subunit but are, nevertheless, of good baking quality. In subsequent studies correlations between flour quality and the presence of other glutenin subunits were identified although the biochemical mechanism by which some glutenin subunits confer better baking quality than others is not known.[44] Determination of which glutenin subunits are responsible for superior baking quality appears to offer the possibility of substantial advances in breeding for baking quality since such analyses should provide a rational basis for bringing together the appropriate subunits by conventional breeding.

The importance to baking quality of the disulphide-linked glutenin molecules is well recognised and there is evidence that flours of higher baking quality contain a greater proportion of glutenins with highest molecular weight than flours of poorer quality.[5, 33] Lee and MacRitchie fractionated flour proteins into different molecular size classes, by repeated extraction using urea and then alkali, and then determined the effect of these fractions on dough mixing properties.[63, 64] Consistent with the findings of Huebner and Wall,[5] fractions rich in high molecular weight proteins were observed to increase dough strength and mixing stability. More recently Ewart determined the intrinsic viscosities of glutenins from 25 different wheat varieties and observed that intrinsic viscosities correlate with specific loaf volumes.[65] These results again suggest that molecular size of glutenin molecules is an important factor in determining flour baking quality.

Extension of the traditional Osborne fractionation scheme (see above) to include 'residue' protein (i.e. that left insoluble after the final extraction with dilute acetic acid) enabled Orth and Bushuk[6] to obtain a reasonable positive correlation between loaf volume per unit flour protein and the proportion of 'residue' protein. There was a negative correlation of loaf volume with proportion of glutenin protein but there was no correlation between loaf volume and the proportion of gliadin protein. The technique has been simplified by Orth and O'Brien[50] to arrive at a predictive test of bread baking quality for plant breeders, again based upon the proportion of 'residue' protein.

Studies by Booth and Melvin,[66] which were aimed at ascertaining the cause of poor baking functionality of the high yielding European wheat variety Maris Huntsman, showed, by interchanging flour components with the quality CWRS wheat, that the problem was associated solely with the gluten component and not the starch and water-soluble material (Table 2). The gluten was fractionated into acid (lactic) soluble and insoluble material. A greater proportion of insoluble protein was found for CWRS (49% of the gluten protein as compared to 17% for Maris Huntsman), again suggesting the importance of this material. However, the quality of Maris Huntsman was not improved significantly by increasing the proportion of its insoluble protein to that found in CWRS, showing that qualitative factors in the protein fractions are important as well as their relative proportions. Interchanging both soluble and insoluble fractions from CWRS with those from Maris Huntsman indicated that both fractions from Maris Huntsman were defective in some way since incorporation of either fraction from CWRS resulted in significant improvement in baking performance (Table 3).

TABLE 2

INTERCHANGE OF GLUTEN, STARCH AND WATER SOLUBLE MATERIAL OF MARIS HUNTSMAN AND CWRS FLOUR AND ITS EFFECT ON BAKING PERFORMANCE

Component source				*Baking quality*	
Gluten	*Starch*	*Water solubles*	*Protein (% dry wt)*	*Loaf volume (cm³)*	*Texture score*
CWRS	CWRS	CWRS	15·6	184	5
CWRS	MH	MH	15·6	187	5
CWRS	MH	MH	11·2	169	4
CWRS	MH	CWRS	15·6	210	6
CWRS	CWRS	MH	15·1	180	6
MH	MH	MH	11·2	120	1
MH	CWRS	CWRS	11·2	119	1
MH	CWRS	CWRS	15·6	125	1
MH	CWRS	MH	11·2	109	1
MH	MH	CWRS	11·3	136	1

TABLE 3

EFFECT ON BAKING PERFORMANCE OF THE SEPARATE REPLACEMENT OF THE LACTIC ACID INSOLUBLE AND SOLUBLE GLUTEN COMPONENTS IN CWRS WITH MARIS HUNTSMAN COMPONENTS

Component source		*Baking quality*	
Insoluble gluten	*Soluble gluten*	*Loaf volume (cm³)*	*Texture score*
CWRS	CWRS	169	5
MH	CWRS	142	1
CWRS	MH	144	3
MH	MH	131	1

The definition of 'residue' protein is not very satisfactory and the proportion of the total flour or gluten protein present in this fraction depends upon the conditions used during extraction (e.g. sequence of extractions, number of extractions performed with each solvent, sample agitation technique, etc.). Indeed, in experiments in which gluten was extracted more exhaustively (four times with acetic acid) to leave an insoluble, 'residue' protein (which accounted for 10% of the total gluten protein), this 'residue' protein, when added to two base flours, had a slight dough-weakening effect and reduced loaf volume significantly.[67] The acid soluble gluten proteins, on the other hand, had a dough-strengthening effect and significantly increased loaf volume in two different types of baking test. These results suggest that the protein remaining insoluble after exhaustive extraction of gluten is not important to baking quality, but the proteins that are important to baking quality are all acid soluble.

The effects of reducing agents such as cysteine on both the rheological properties and baking quality of a dough also indicate the importance of disulphide bonding among glutenin subunits. Addition of large amounts of such reagents produces a very soft, extensible, non-elastic dough and the resultant bread has a poor, open crumb structure due to the gas cells coalescing during proving and the early stages of baking. Thiol compounds analogous to cysteine also occur naturally in wheat flour, being derived in large part from the outer layers of the grain, particularly the germ.[68,69] Of the thiol compounds detected, the only one having any marked rheological activity in doughs is glutathione,[70] and examination of the levels of glutathione in different varieties of wheat has indicated a negative correlation between glutathione content and some dough rheological properties (e.g. Farinogram mixing tolerance) and between glutathione content and flour quality tests such as the Pelshenke fermentation and Zeleny sedimentation procedures.[71-73]

High molecular weight components may also arise by non-covalent aggregation of subunits, and hydrophobic interactions have been cited as being of primary importance in the structure of glutenin.[3,39,51] Indeed, Kobrehel and Matignon have indicated a relationship between baking quality and the level of soap required to aid the solubilisation of the flour protein, particularly the higher molecular weight glutenins, less soap being required for varieties of poor breadmaking ability.[74] The inference drawn by the authors is that the protein of quality wheats is more strongly aggregated by hydrophobic bonds. This may be so, but to suggest that disulphide bonds are not implicated in breadmaking quality appears rather extreme. They obviously have a role to play in determining protein structure although may be only partly responsible for the differences in quality between varieties.

It has been noted that hydrogen bonding probably plays a key role in gluten structure because of the very high levels of amidated amino acid side chains, particularly glutamine. Recently MacRitchie has produced some evidence which suggests that the bread baking potential of a flour is positively correlated with the degree of amidation of the protein.[75] The difficulty in assessing this, however, lies in the relatively small range of amide content variation, the precision of the analytical technique, and

glutenin complexes of flour proteins, which he considers to be the most important fraction in determining dough viscoelasticity.[34, 93–95] This model tries to account not only for the nature of reactions occurring at the level of individual amino acid residues, specifically cysteine or cystine residues, but also for the macromolecular nature of, and interactions between, glutenin polymers.

In the most recent refinement of this model,[34] glutenin complexes are considered as linear concatenations of polypeptide subunits held together by one interchain disulphide bond per subunit. The individual subunits are folded and kinked because of high contents of proline residues; these structures are stabilised by secondary (non-covalent) forces but once these forces are overcome (e.g. on dough mixing) great capacity for extension ensues. However, because the folded, kinked forms of the subunits are thermodynamically more stable, the structures will tend to contract on removal of the stresses, to regain the lowest free energy state, and the glutenins will therefore exhibit elastic behaviour. At higher stresses the secondary forces holding adjacent glutenin polymers will be overcome and the molecular entanglements resulting from the complex conformations of the concatenations will be unravelled; the polymers will therefore slide past each other. The tendency of the extended polymers to contract is then frustrated and the proteins exhibit viscous flow; the monomolecular gliadins may act as plasticisers under such circumstances. Dough resistance would be at a maximum when the greatest number of subunits are in an extended state but before molecular slippage is significant.

A consequence of such a model might be that the degree of polymerisation of the glutenin molecules would be an important determinant of dough strength[95] although in a recent paper this is given less emphasis.[34] This is due to the increased probability of interaction between glutenin polymers through non-covalent bonds and molecular entanglements as the degree of polymerisation increases. The question of how particular glutenin subunits affect glutenin polymer structure such that different baking qualities are produced has not so far been explained.

Reactions involving SH and S–S groups during dough mixing are taken into account in Ewart's model. It is considered that sulphydryl–disulphide interchange may aid in stress relaxation in rested doughs or in yeasted doughs undergoing expansion due to the production of CO_2 during fermentation (i.e. at low shear rates). Interchange between strained S–S bonds and suitably juxtaposed SH groups is envisaged as occurring to relieve the stresses without necessarily causing depolymerisation of the concatenations. Nevertheless, stress relaxation is considered to take place predominantly through viscous flow (i.e. molecular slip) and S–S bond scission.

Dough breakdown is considered in Ewart's model to be due to mechanical scission of S–S bonds when the shear stress induced by dough mixing cannot be relieved rapidly enough by molecular slip or SH/S–S interchange. Scission of S–S bonds would produce thiyl radicals, which might either be converted to SH groups by hydrogen abstraction from water, or, according to Hoseney and his colleagues,[96] react with activated double bond compounds. Whatever the subsequent free radical reaction(s) the effect of S–S bond scission would be to reduce the average molecular size of the glutenin concatenations. The lower molecular weight polymers would have a lower probability of interacting with each other and would therefore be less rheologically effective.

Although it is outside the scope of this review to examine in detail the other components of flour, some mention should be made of interactions between flour proteins and non-protein components of flour or dough. Particularly important in this respect are the endogenous lipids of flour and fats added as ingredients in bread recipes. Flour lipids in breadmaking have been discussed in greater detail in recent reviews.[97–99] Both the endogenous flour lipids and added bakery fat are important in breadmaking: of particular significance are the endogenous free polar lipids, principally the glycolipids. Although the action of lipids becomes apparent at the baking stage the causes of these effects are almost certainly established at the dough mixing stage and other stages leading up to the oven stage. Interchange experiments have shown that differences in baking performance between flours are not due to differences in the flour lipids themselves but to differences in the flour proteins, which interact differently with the lipid.[100] However, little is known at present about the precise protein components and amino acid sequences or conformations that are involved in these interactions with lipids.

Quite clearly there is still a great deal to be learned about the modifications that occur in flour proteins during dough mixing, and their interactions with other flour/dough components, although considerable progress has been made in pinpointing the types of reaction that are important. The goal of explaining how flour proteins are modified on dough mixing will undoubtedly be aided by a more complete description of the structure of flour proteins.

Pasta Manufacture

Paste products (pasta) are amongst the oldest types of food made from wheat and include products such as macaroni, spaghetti, lasagna, and sundry shaped products (alphabet, wheels, twists, etc.). The raw material of choice for the production of pasta is a type of hard wheat known as durum (*Triticum durum*). Although durum wheat belongs to the same genus as bread wheat (*Triticum aestivum*) the two differ in that bread wheat is hexaploid whereas durum wheat is tetraploid.

In the manufacture of pasta, durum wheat is milled into a granular product termed semolina, semolina being preferred to flour mainly because it requires less water to form a dough, which in turn reduces processing (drying) time and costs. A stiff dough having a moisture content of about 30% is mixed, kneaded under vacuum to remove occluded air, extruded under a pressure of about 80–130 $kg\,cm^{-2}$ through a die to produce the required shape, and finally dried to a moisture content of about 12·5% or below.[101]

Factors Influencing Pasta Quality

Pasta made from good quality durum wheat is bright and yellow in the dry form and, when cooked, resists disintegration and has a firm texture. The pasta manufacturer also requires dry pasta to have good mechanical properties, i.e. to resist breaking and cracking (checking) during and after processing, for broken or cracked pasta cannot be sold and this may represent a serious commercial problem. Whereas the colour characteristics of pasta are determined largely through its content of carotenoid pigments and the activity of the enzyme lipoxygenase, which at high levels may cause bleaching of these pigments, the textural characteristics of the dry and the cooked product are largely governed by the proteins of durum wheat.

Both protein content and the qualitative nature of the protein are important in determining the mechanical properties of dry pasta and the quality of cooked pasta prepared from durum wheat semolina. Superior cooking quality of pasta is characterised by a high degree of firmness accompanied by a low level of compressibility and a high degree of recovery (elasticity), following compression. The relationship between these quality parameters and protein content has been demonstrated by examining the quality of cooked spaghetti prepared from semolinas covering a wide range of protein contents.[102–104] In each case, the quality of the cooked spaghetti was directly proportional to the protein content of the semolina. Protein content also has an important influence on the mechanical properties of dry pasta and a semolina of higher protein content yields dry pasta possessing better mechanical properties.[102] A semolina protein content of 10% has been quoted as being the minimum necessary for producing spaghetti with appropriate mechanical properties and of acceptable cooking quality, a protein content of about 12% generally being considered most appropriate.[102]

It is also well known that just as the breadmaking quality of hexaploid wheats depends on protein quality as well as quantity, so pasta cooking quality and mechanical properties depend upon qualitative factors associated with durum wheat proteins as well as protein content. Thus, some durum wheat varieties yield pasta of much better cooking quality than other varieties even though the protein contents may be exactly equivalent. Considering only durum wheats, the 'stronger' the gluten (i.e. the greater the resistance to extension under a given load) the better is the quality of the cooked spaghetti.[105] The relationship does not hold, however, if one also considers hexaploid bread wheats, some of which have 'stronger' glutens than durum wheats but which yield spaghetti of inferior cooking quality.[105] Gluten quality also affects the mechanical properties of dry pasta, although to a lesser extent than protein content. Pasta from semolina with 'strong' gluten yields dry pasta with greater resistance to cracking and breaking.[102]

Molecular Basis of Pasta Quality

The biochemical features of durum wheat protein that are responsible for superior pasta quality are not well defined. Evidence has been presented to suggest that a higher proportion of glutenins among the semolina proteins is associated with better cooking quality.[104,106,107] More recently it has been reported that the amount of 'residue' protein correlates with spaghetti cooking quality.[108,109] Somewhat paradoxically, however, fortification of semolina with semolina protein fractions soluble in dilute lactic acid had been observed previously to lead to an improvement in cooking quality, whereas fortification with the lactic acid insoluble protein did not improve cooking quality to the same extent.[110] Moreover, it has been reported that the proportion of acetic acid insoluble 'residue' protein may remain constant or even decrease as semolina protein content and cooking quality increase.[104] D'Egidio *et al.*[111] consider that firmness of cooked pasta is influenced mainly by protein content while stickiness is affected by protein composition as determined using solvent extraction techniques.

Correlations have also been sought between pasta cooking quality and semolina protein subunit compositions. As with bread wheat proteins, the subunits are complex but a relationship between cooking quality and the relative levels of two glutenin subunits of molecular weights 53 000 and 60 000 has been noted,[112] although the data are much less extensive than for bread wheats. In similar vein, it has been noted that a strong relationship exists between the presence or absence of two gliadin components, termed 45 and 42, determined electrophoretically, and certain properties of gluten considered to be indicative of pasta cooking quality.[113, 114] The main use of this observation is probably as an aid in breeding programmes but it is possible that the presence of component 45 might confer good cooking quality or that the presence of component 42 might confer poor cooking quality; the relationship between cooking quality and the presence or absence of these gliadin components, and the mechanisms that might be involved, remains to be defined.

Effect of Processing on Semolina Proteins

A number of studies have been conducted to discover the fate of semolina proteins during pasta manufacture and during cooking. A decrease in the proportion of semolina proteins extractable in dilute acetic acid occurs when a dough is mixed at pasta dough moisture levels (30–35 % moisture content) although at higher moisture levels similar to those of bread doughs (about 50 %) extractability increases. Despite the high level of mechanical work imparted to the dough during its extrusion through the die, there is no further change in acetic acid extractability.[115] Application of modified Osborne extraction techniques has indicated that the proportion of globulins decreases on dough mixing and insoluble 'residue' protein increases;[108, 116, 117] Wasik[108] also noted a decrease in the proportion of proteins in the soluble glutenin fraction.

We have also examined extractability in a dissociating solvent (2% SDS; 0·1 M Tris-HCl, pH 8·0) and have found extractability to decrease substantially during dough mixing and during extrusion from 94 % to 86 %.[117] Gel filtration chromatography indicated that there was a marked decrease in the proportion of extracted proteins that comprised glutenins whereas the proportion that comprised gliadin remained constant. Moreover, the glutenin species that were affected to the greatest extent were those of highest molecular weight that were excluded from the gel. Similar results in terms of the decrease in the proportion of glutenins in protein extracted using another dissociating solvent (acetic acid/urea/cetyltrimethylammonium bromide—AUC) were obtained by Wasik[108] when he examined proteins in semolina and finished pasta. In other studies it has been demonstrated that there is a decrease in the total number of sulphydryl groups during dough mixing and extrusion although no concomitant increase in the total number of disulphide bonds could be detected.[115]

The extractability and chromatographic data indicate that dough mixing results in the formation of glutenin complexes that are of too great a molecular size to be soluble even in highly dissociating solvents and that this insolubilisation is likely to involve disulphide bonds. It seems most probable that the change in extractability of the globulin fraction involves non-covalent binding of these proteins to other semolina components, either through an increase in the extent of such binding or through a change in the nature of the binding. Little change in the proteins has been observed during pasta drying.[115]

Scanning electron microscopy has also been used to study the alterations during pasta processing.[81,118] Semolina was observed to have a tight, compact, irregular structure with no starch granules visible. Upon dough mixing a more open structure evolves with some starch granules visible within a protein matrix. The structure of the protein continues to be modified during dough mixing and pressing such that by the time the dough is extruded the protein network is continuous, although not developed to the same extent as in bread doughs, with starch granules deeply embedded in it: longitudinal sections show aligned starch granules surrounded by a laminar protein network. The surface of freshly extruded pasta is covered with a substantially continuous protein film with few pores visible. The picture that emerges from this and the biochemical work is that the major changes that occur in durum semolina proteins during pasta manufacture do so during the dough mixing and extrusion stages.

Biscuit Manufacture

In the UK flour for biscuit manufacture is the single most important use of wheat flour after bread production and accounts for 11% of milling.[119] The flour, and the requirements of its protein, differ according to the type of biscuit.

For crackers, the requirements are similar to those for bread manufacture. A reasonably strong flour with protein content of about 11% (14% m.b.) is normal since formation of a gluten network is important to maintain gas cells between the laminations.[120]

For semi-sweet biscuits, a gluten quality very different from that used in bread and cracker production is necessary—indeed the quality is often found in high yielding soft wheat varieties. In general, flours which contain a fairly weak, extensible gluten with a low degree of elasticity perform best.[121] High elasticity results in contraction of the sheeted, rolled dough before cutting giving non-uniform products, particularly in terms of biscuit thickness. Treatment of the dough with sodium metabisulphite is normal to bring about the desired properties, and acts by breaking disulphide crosslinks in the protein, preventing proper gluten formation.[122,123] Level of protein in the flour appears unimportant.

Short-dough biscuits have a high fat content and the level and quality of protein in the flour appears unimportant.[124]

For wafers, the characteristics of the gluten in the flour are relatively unimportant. Crispness or hardness of the wafer does appear to relate to protein level, the crisper product being made from the higher protein flour. However, it is essential to avoid gluten separation in the thin batter before it is poured onto the sheet die, and, in practice, flours with lower protein content are used. Flour particle size is said to be the most important quality parameter.[125]

Cake Manufacture and Domestic Flour

These together account for approximately 12% of UK flour production[119] but discussion of these is basically outside the scope of this chapter since, except for strong domestic flours intended for home bread production, where the attributes discussed on p. 14 apply, the protein has no great technological significance.

Cake flours generally have low protein contents, although for those which will be required to carry a heavy fruit load, a higher protein level may be employed. Cake flours are usually milled from soft wheat which by nature often has a lower protein content. In most recipes structure of the product comes from the other proteinaceous ingredients, such as egg or milk.

Domestic flours are general purpose flours mostly employed for making cakes or pastries, which do not require high protein levels or specific protein qualities.

Commercial Gluten Manufacture and Applications

From the foregoing sections it can be readily appreciated that gluten is a highly important constituent of flour, necessary for good bread production. To exploit its unique properties many commercial plants have been set up throughout the world to separate the gluten from wheat flour. Total free world production of gluten is estimated at 120 000 tonnes per annum.[126] Although only a very small proportion of total wheat production (0·25%) is utilised in this manner, production has expanded regularly as new uses for the gluten and the starch develop, some far removed from the original use in bread products. The developments that have occurred in production processes, end-use products, and the protein qualities required for such, are discussed below.

Raw Materials

Most types of bread wheat (*T. aestivum*) have been cited at some time as being suitable starting materials for commercial separation of gluten; processes are available for production directly from wheat or from flour.

In theory, the ideal material should have a high protein content since the primary objective is to isolate this material, which has a much higher commercial value than starch. However, for the total process to be economic the production and sales pattern of the starch or starch-derived products such as syrups, must also be considered. In addition, in some countries high protein content wheats are more expensive either because they need to be imported and may be subject to levy, or, if home-grown, command a premium. This applies particularly in the EEC where it is more normal to utilise home-grown wheat with lower protein content. Occasionally, air classification is used to produce a protein-rich fraction. Where high protein wheat is available at little extra cost, as in the USA or Canada, this wheat is employed. Hard and soft textured wheat, spring and winter, have all been employed as raw materials.

Most washing-out processes, probably all those in commercial operation, use a flour rather than wheat. The flour can be either straight-run, that is, the total flour made during the milling operation, or it can be a low grade flour. The latter, used commonly in the USA and referred to as 'second clears', is a poorer quality material, which remains after removal of the whiter patent flour. It sells at a discount and can be attractive in some gluten separation plants despite its higher fibre and soluble solids content. Most important to the gluten/starch processor is a consistent raw material since the plant operation can then be optimised. A variable wheat or flour has an effect on the efficiency of the washing-out process, operation of the driers and, inevitably, on the quality of the final products.

The other raw material to consider is the water. Although any potable source can be utilised, problems often arise when the water is soft; the gluten tends to disintegrate, to be soft and slimy, and difficulty is experienced in removing the starch satisfactorily. Hard water is therefore preferred, although the problems with soft water can be alleviated to a certain extent by using salt, which toughens the gluten and prevents break-up.

Processes
Techniques for preparation of gluten can be divided into six types.

(*a*) *Dough*. These involve mixing the flour and water into a dough in which the amount of water is equivalent to about 60% of the weight of flour. The dough may be rested for up to 30 min to allow hydration of the gluten, depending on the quality of the flour. It is then tumbled in large rotating drums through which water flows. Starch forms a suspension in the wash water whilst the gluten coalesces to form a viscoelastic mass. The two are subsequently separated on screens, and the starch recovered by centrifugation. Alternatively, the dough is kneaded in a trough with reciprocating rolls or screws. All products are normally further washed to purify before drying. This process, often referred to as the Martin process, is one of the oldest (1835) and is the most widely used method, particularly in USA and Canada, Europe and Australia. A flow diagram is presented in Fig. 9 and further details are given by

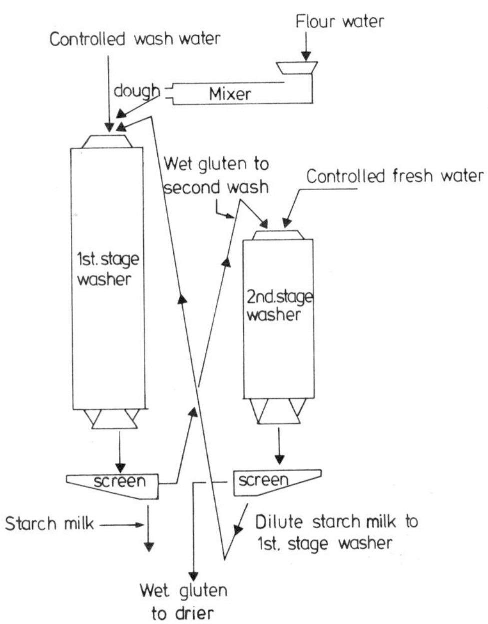

FIG. 9. Schematic representation of a possible layout for a gluten/starch separation plant employing a continuous form of the Martin dough process. (Reproduced courtesy of Mr A. Jackson, Tenstar Products Ltd, Kent, UK from Ref. 128.)

Knight[127] and by Jackson.[128] One draw-back of the process is that large quantities of water are required to obtain the end products (7–8 tonnes per tonne flour) and this results in a dilute aqueous effluent from which it is often uneconomic to recover the dissolved water soluble components of the flour.

(*b*) *Batter.* These resemble dough processes since in both the gluten is brought together in a network before any starch is separated. However, a batter with a water addition equivalent to between 70 % and 200 % of the flour weight is prepared rather than a dough. The constituents are allowed to hydrate and the addition of further water plus a work input using a cutting pump results in separation of gluten curds or strands which can be removed from the starch on vibrating screens.[127,128] Like dough processes, batter processes use much fresh water and incur resultant effluent penalties: they are known to operate in Australia and Europe.

(*c*) *Aqueous dispersion.* These are more recent developments and differ from the foregoing two techniques in that some of the starch is separated from a flour–water dispersion before formation of a gluten network is allowed to proceed. In one such technique,[129,130] a modification of the 'Fesca' process,[131] differences in sedimentation rate are exploited to separate components rapidly before full hydration and development of the gluten can occur. A flour slurry is continuously centrifuged to give a reasonably pure prime starch, and a fraction containing most of the protein plus remaining starch. This latter fraction has a protein content of approximately 20–40 % (cf. a minimum level of approximately 70 % usually obtained by dough and batter techniques). A process has been described of late that allows solubilisation of the starch component using α-amylase, and subsequent recovery of the gluten.[132]

A development of the 'Fesca' process by Raisio/Alfa Laval has received much attention, not least because of the claim of significant reductions in usage of water ($\sim$2 tonnes per tonne flour) and pollution-free processing due to a closed system.[133,134] Flour and water are made into a thick batter which is homogenised to give a uniform suspension without gluten development. Centrifugation of this, as in the modified 'Fesca' process, separates the heavy 'A' or prime starch, leaving a protein-rich fraction. Gluten in the latter develops and agglomerates when worked by shearing in a pin mill, after which it can be removed on screens. Washing of products is carried out. The quality of the gluten is claimed to be very good. In addition to the original Finnish plant, one has been built in the USA and another is under construction in the UK.

The Raisio/Alfa Laval process is often referred to as a whole wheat wet fractionation process. This is not strictly true since only flour is mixed with water, not whole grain or whole ground wheat. However, the total process does incorporate a simplified dry milling stage which substantially removes bran and germ, but produces a coarser flour than that produced by conventional milling. This has the advantage of lower energy requirements and a material with better flow characteristics together with less damage to the starch.

Another batter process has been developed in Holland by KSH.[135,136] A batter formed from recycled wash water and flour is introduced directly into a series of hydrocylones. The 'A' starch is washed out directly with countercurrent fresh water. Because of the intense fluid shear in the hydrocyclones a rapid agglomeration of the gluten occurs after passage through a few hydrocyclones. The gluten curds can be washed and separated on a rotating dough washer with screens. Again one of its main advantages is its low usage of water.

(*d*) *Chemical dispersion*. In these processes the gluten is dispersed in weak acids, such as acetic acid, but problems have been encountered with glutens from flours of high ash content. Straight dispersion of wheat protein directly from flour in dilute ammonium hydroxide and without separate washing-out stages has also been proposed.[137] The flour/ammonium hydroxide slurry is subjected to high shear and then centrifuged. This yields a compacted starch fraction and a stream containing dispersed gluten plus solubles which is spray-dried. It is understood that a similar technique for dispersion in ammonia solutions is used commercially in Japan although the product appears rather dark grey.

(*e*) *Wet milling*. These differ from the preceding approaches as they start directly from wheat rather than flour, with the wheat included in the wet processing. A technique has been patented by the Pillsbury Company, the main objective being to provide a product having good doughing and baking characteristics,[138,139] but this could be considered the starting point for gluten and starch separation. Essentially this is a hydromilling process in which wheat grains are macerated with water; the bran can be screened off, and the starch and gluten can be separated.

The Far-Mar-Co process is also a wet milling technique but with a much lower water requirement. Grain is tempered or steeped with water to between 15% and saturation and then it is flaked in a roller mill. Further water is added and mixed with the flakes to form a dough from which the bran, germ and starch can be separated under high pressure water spray, leaving behind the gluten.[140-142]

Advantages claimed for these wet milling techniques include substantially lower damage to the starch with resultant increase in prime starch yields and more complete recovery of the outer layers of the wheat grain, particularly the aleurone. However, these outer layers contain more water soluble components which could lead to increased effluent disposal problems. No true wet milling processes are known to be in commercial operation.

(*f*) *Non-aqueous separation.* These have been considered and although such processes appear attractive, they do not seem to have been exploited commercially as yet. Starch and gluten have different buoyant densities and the difference can be exploited to effect separation in non-aqueous liquids. The potential advantage with such processes is the avoidance of hydration of the gluten and loss of functionality on drying. For instance, fluorinated hydrocarbons of appropriate density are available, and Finley has investigated their use at a laboratory and pilot scale.[143-145] These solvents have low boiling points and heats of vaporisation, thus reducing energy requirements. Water pollution problems are also overcome.

It is essential to detach all protein from the starch granules during the milling process. The flour must therefore have fine particle size, otherwise overall recovery of the separate components is impaired and impure components are produced.

Results obtained more recently in another laboratory,[146] however, have indicated that a less satisfactory separation of starch and gluten was obtained compared to that achieved by Finley.

Gluten Drying

Gluten deteriorates rapidly if kept in the wet state. Proteolytic enzymes, both those occurring naturally in the wheat grain and ones derived from bacterial activity, lead to considerable degradation of the protein which becomes extremely soft and extensible. It is then unsuitable for most major uses, which rely on retention of the unique cohesive, viscoelastic properties. Refrigeration has been used, particularly where product application is adjacent to the washing-out plant, but this is not normal. Retardation of gluten spoilage, by storage in brine solution, has also been practised, but extra transport costs are involved; gluten is also considerably toughened by salt and in many applications this must be washed away before use.

Most commercial gluten is preserved by processing in pneumatic driers that operate at fairly high temperatures. Like many proteins, hydrated gluten (about 65% moisture content) is very sensitive to heat, and great care has to be taken to avoid devitalisation, i.e. loss of functionality due to heat denaturation. Gluten is less susceptible to heat denaturation at lower moisture contents and attempts are often made to reduce the overall moisture in the wet gluten by blending it with a larger quantity of previously dried gluten. This can be carried out external to the drier proper, particularly for flash driers, to produce a friable material consisting of discrete fragments. Alternatively, the gluten is introduced into a ring drier by a pelletiser which 'fires' small pellets of wet material into a hot air stream that is carrying almost dry, recycling gluten.[128] At the same time a rotating disintegrator in the air stream fragments the gluten further, and the wet material is coated with the recycling product. A certain amount of moisture is thus transferred from the wet gluten to the dry, and the coated material should not stick to itself or the walls of the ring. The disadvantage of this system and the external coating in flash driers is that dried product can go through several rewetting/drying cycles with the risk of increased heat damage.

Spray-drying is also employed to dehydrate gluten but is confined to protein concentrates (products from Fesca-type processes for instance) and material from chemical dispersion. It is not suitable for gum gluten. Temperature is again critical.

Quality of Commercial Dried Gluten

The quality of dried gluten can really only be defined in terms of its performance in a particular end use. Some users have developed small scale tests based on their performance requirements, especially where the end product is bread. More general criteria of quality are rather unsophisticated and include assessment of protein content, extent and rate of rehydration, particle size, colour, physical properties, ether-extractable (free) fat level and solubility in acetic acid. The ranges of values encountered for such parameters are presented in Table 4.

At first sight it would seem logical to conclude that a gluten of higher protein content might be more economical in use but some buyers believe such protein levels are only achieved by sacrificing quality due to extra stretching and working of the gluten. A higher rate and degree of

TABLE 4
VALUES FOR SOME GENERAL CRITERIA OF COMMERCIAL GLUTEN QUALITY

Moisture	4–10%
Protein (N × 5·7)—d.b.	75–85%
Water absorption	150–200%
Rehydration time	5 s–10 min
Ash	0·6–1·2%
Free fat	0·5–1·5%
Acetic acid solubility	80% minimum

rehydration are often of great importance, particularly in short-time processes such as modern high speed breadmaking processes (e.g. the Chorleywood process) in which it is important that the gluten is almost immediately functional and can interact with the natural flour protein. Conversely, in some applications which require pumping of slurries or batters, delayed rehydration may be advantageous to prevent clumping of the gluten until a later stage. In general, gluten that is denatured least will rehydrate most quickly and to the greatest extent.

Particle size should be similar to that of the materials with which the gluten is to be mixed; in general, fine particles rehydrate more rapidly than coarse. Storage stability may be affected by free fat level and some users specify a maximum level to reduce this risk. Assessment of gluten physical properties may simply involve manipulation with the fingers and a subjective appraisal of the rehydrated mass or more objective instrumental methods to measure the degree of, and resistance to, extension at specified loads. Conventional dough testing instruments such as the Brabender Extensograph and Farinograph and the Chopin Alveograph are also widely applied to doughs prepared from standard base flours plus the gluten under examination. Gluten colour may also be important. Solubility in dilute acetic acid is used as an index of loss of functionality caused by the drying process; at least 80% solubility is required.

Bread baking tests. The biggest outlet for gluten is in baking and the ultimate test of quality in this application is therefore some form of breadmaking performance assessment. No test is universally acceptable but generally the methods are based upon addition of gluten to a standard flour followed by the breadmaking procedure relevant to the interests of the investigators. Any test must be highly standardised. The level of gluten addition may vary from as little as 1% to more than 5%. Quality of gluten is assessed by its effect upon loaf volume and crumb texture.

The authors have experience of a somewhat different test[66,147] developed along similar lines to that used in the fractionation/reconstitution studies of Finney.[148] However, instead of reconstituting all components of a flour, freeze-dried starch and water soluble material, prepared by washing out from a standard flour (in our case milled from Canadian Western Red Spring wheat) are recombined with the gluten under examination. Baking follows either a traditional 3 h bulk fermentation procedure, in which case the reconstituted flour protein content is adjusted to 14% at 14% moisture, or a high energy (Chorleywood) technique, for which the flour protein content is adjusted to 11%. The bread is subsequently evaluated for volume and texture. Alternatively, reconstituted flours can be prepared covering a range of gluten protein levels (8–16%) for each gluten and a factor termed the 'protein response' determined, i.e. the increment of loaf volume per unit increase in flour protein content.[147] Figure 10 illustrates how protein level affects loaf volume for two glutens of different quality. This technique is very

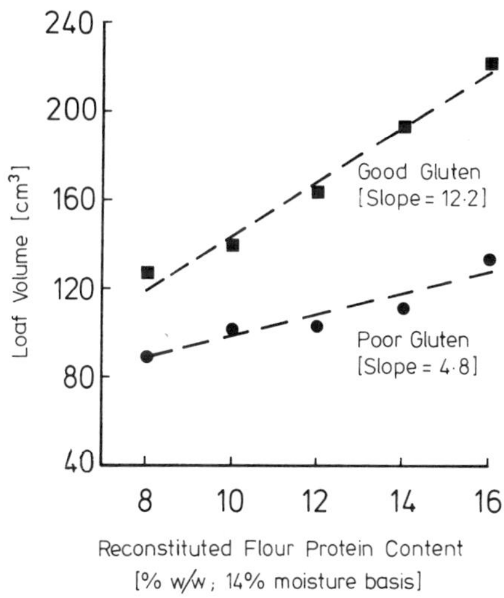

FIG. 10. Effect of protein level on the loaf volume of bread baked from glutens of differing functionality. Assessment of baking performance was made using a reconstitution, bulk fermentation baking procedure (Ref. 147). The gradient of each line is termed the 'protein response'.

sensitive to quality differences in gluten since the only effective protein is that arising from the added gluten. The reconstituted flour does not contain any native, unfractionated protein such as is present in most fortification techniques.

A number of glutens from most major manufacturing countries have been examined by this technique. Figure 11 illustrates the differences in

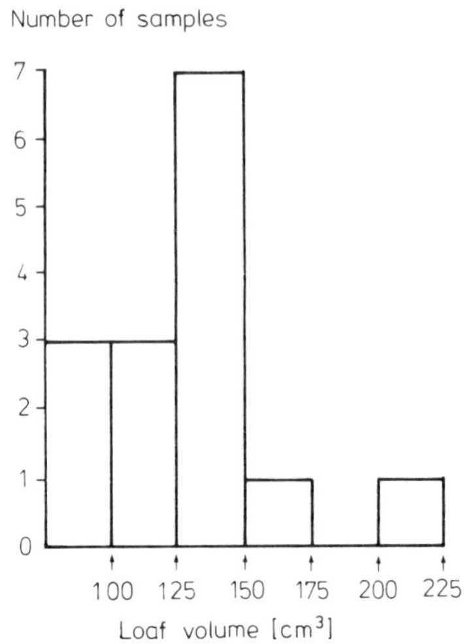

FIG. 11. Comparison of the baking qualities (loaf volumes) of a number of commercially produced glutens (Ref. 153). The glutens were compared in a bulk fermentation baking test using flours reconstituted at 14% protein (14% moisture basis).

functionality between the samples when baked at a flour protein content of 14%. It is clear that although all were termed 'vital' the quality varied widely as measured by breadmaking potential. Most produced smaller loaves than freeze-dried gluten, suggesting all had suffered a measure of protein denaturation during the critical drying process. The user must decide whether this loss of functionality is important in his product at the level of incorporation employed.

Work in this laboratory has also shown that a small scale pin-mixer[149] can be used to assess gluten quality.[147] In this test, the gluten under examination is mixed with water at two levels of water addition, 150% and 200% of the gluten weight. Recordings of the torque transmitted by the rehydrating gluten mass from the rotating pins to the static pins in the mixing bowl are drawn on a chart recorder. The areas beneath the curves are measured from the start of mixing to the points at which maximum torque is achieved. A plot of log (protein response) v. log (area (200% WA–150% WA)) gives a good straight line relationship (Fig. 12). Based on this, a larger scale procedure using the Farinograph has also been described recently.[150]

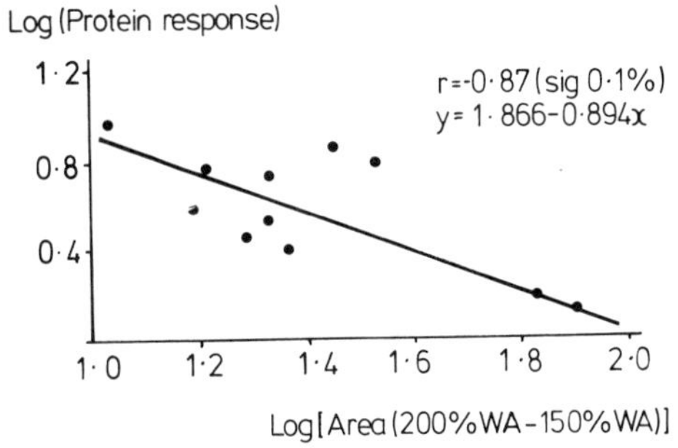

FIG. 12. Relationship between 'protein response' and micro-pin mixing data for a number of commercial glutens (Ref. 147). The logarithm of 'protein response' is plotted versus the logarithm of the difference between the areas beneath micro-pin mixer curves measured from the start of mixing to the point of minimum mobility (maximum viscosity) at water additions of 150% and 200% of gluten weight.

Nature of the Effect of Heat on Gluten

The denaturing effect of heat on gluten is well known. Pence *et al.*[151] demonstrated that the loss of functionality is essentially a first order reaction having an activation energy of approximately 146 kJ mol^{-1}. However, since gluten is a complex mixture of gliadins and glutenins it is unlikely that denaturation arises through a single step. Indeed, Pence and co-workers demonstrated the greater sensitivity to heat of the whole complex as compared with gliadin.

More recently, Jeanjean and colleagues showed that heating gluten leads to decreased compressibility and increased firmness (i.e. a tougher gluten). Formation of new bonds occurs which results in the insolubilisation of proteins previously extractable in aqueous ethanol.[152] It was suggested therefore that the net state of aggregation of the gluten is increased on heat denaturation. Similar studies in our laboratory[153] also indicated the involvement of disulphide bond formation in the loss of functionality on heating. Extraction of proteins in a denaturing solvent from glutens heated at different temperatures, and examination of the extracted proteins by gel filtration chromatography, indicated that both glutenins and gliadins became less soluble upon heating and that the glutenins were affected at lower temperatures than the gliadins (Fig. 13). Examination of the proteins soluble in aqueous propan-1-ol by starch and polyacrylamide gel electrophoresis showed that only the ω-gliadins

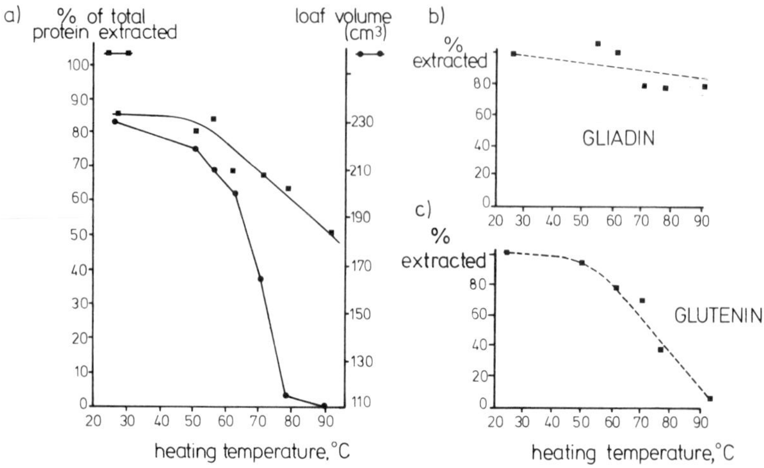

FIG. 13. Effect of heating wet gluten at different temperatures on baking quality and extractability of gluten proteins in a dissociating solvent. (a) Effect on baking quality and extractability of total gluten protein. (b) Effect on gliadin extractability. (c) Effect on glutenin extractability. The extractant was 0·04 M cetyltrimethylammonium bromide/4·5 M urea/0·5 M sodium citrate, pH 5·5 (CUC) (Ref. 153).

were unaffected by heat (Fig. 14). The other gliadins, and the alcohol-extractable glutenins of low molecular weight were rendered much less soluble in the alcohol extractant after heating. The importance of the lack of involvement of ω-gliadins is that they are deficient in sulphur amino acids, particularly cyst(e)ine.[13]

The results suggest that an important effect of heat therefore is to cause conformational changes in glutenins that allow new intermolecular disulphide bonds to be formed giving increased molecular size of the gluten aggregates and decreased extractability and baking performance. Similar changes occur in the cyst(e)ine-containing gliadins but at higher temperatures since they are less heat sensitive. The new disulphide bonds stabilise the proteins in the denatured state on cooling, rendering the process irreversible.

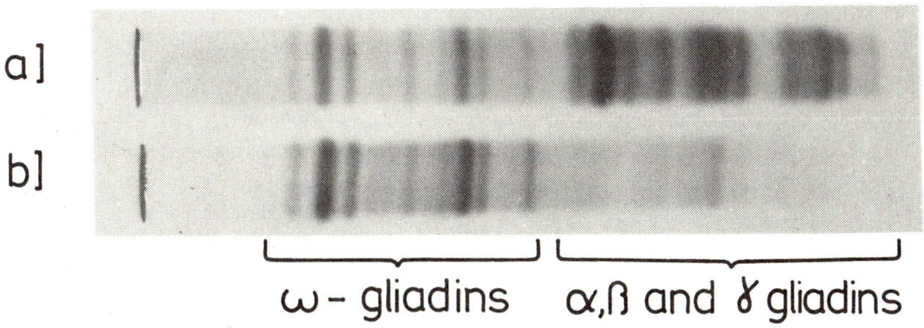

FIG. 14. Starch gel electrophoresis of proteins extracted from native and heat-denatured gluten using aqueous 50% (v/v) propan-1-ol (Ref. 153). (a) Native gluten; (b) gluten heated at 100 °C.

Modification of Gluten Functionality

Most commercial gluten is sold without further modification. However, problems have been encountered in certain applications and a number of processes and products have been described to improve the performance. These have been related chiefly to bread manufacture in the USA. For instance, in the continuous sponge-dough process, gluten and a portion of the flour are mixed with water, yeast and fat to a sponge which is fermented before addition of the remaining flour, salt and sugar. The main problem is in rehydration of the gluten. Initial surface hydration is rapid and small lumps form which, due to the cohesive nature of gluten, tend to coalesce. This hinders further hydration and also distribution of the gluten throughout the sponge to give a uniform product. To overcome this disadvantageous tendency towards particle cohesion, a product has been developed in which the gluten is coated with small quantities of certain non-ionic hydrophilic emulsifiers.[154] Improved loaves in terms of volume and crumb texture are reported. An alternative method, claimed to be cheaper, has been described[155] in which the gluten is mixed with a finely divided edible additive rich in protein (e.g. soya flour, milk whey). The added protein presumably binds to the gluten at accessible sites, thereby preventing such extensive hydrogen bonding between the gluten proteins and thus lessening the degree of cohesion. Delayed functionality to minimise gluten cohesion has also been achieved by producing either a coarser particle size or a semi-denatured material, although the latter presumably will have some influence on final performance.

Gluten treated with a hydrophilic emulsifier, according to another patent,[156] also improves the tolerance to processing changes during food manufacture, bread production being a prime example. The emulsifying agent is melted, but at a temperature sufficiently low to prevent gluten denaturation, mixed in with wet, gum gluten at approximately a 15–30% level (dry gluten basis) and dried as usual. The emulsifier-treated gluten is said to possess improved handling characteristics and to produce bread with better volume and a softer, finer crumb texture. Bread with improved volume and texture can also be obtained by using a gluten that has been modified by mixing with xanthan gum at a preferred level of 2·5% (wet basis).[157] The mode of action is unknown but the suggestion is that the effect is due to a proportion of the gliadin complexing with the xanthan gum to impart the desired properties to the product.

As described previously, rate of rehydration is critical for many applications. Gluten that is not hydrated is not functional. It is claimed that incorporation into the gluten of a non-toxic chelating agent, sodium citrate being particularly successful, significantly improves hydration properties.[158] It is not clear how this agent affects gluten rehydration.

Overall functionality of most modified glutens will still depend upon the temperature profile the gluten has encountered during drying. Quality lost by heat denaturation is rarely recovered, although Stenvert and colleagues have suggested that reducing agents such as sodium metabisulphite may be beneficial.[150]

Applications of Commercial Gluten

The market for commercial vital gluten has grown dramatically over the last 10–15 years and projections indicate that this trend is likely to continue. Approximately 40% of total free world production (120 000 tonnes per annum) is utilised in the USA[126] and it is instructive to examine growth for that country. Between 1967 and 1977 consumption increased from 15 000 to 45 000 tonnes. By 1985 a figure of 62 000 tonnes is projected.[159] At present only about 45% of this requirement is home-produced, the majority being imported, mainly from Australia and Canada.

An analysis of usage indicates the primary application of gluten, not surprisingly, to be baked goods, accounting for 69% of usage in USA and 75% in Australia.[126,160] Similar proportions are likely for other western countries. Breakfast cereals are responsible for 10–15% whilst a similar proportion is utilised in the meat analogue/processed meat/pet food area. This latter grouping is by far the most important in Japan with insignificant bakery and breakfast cereal usage. The remaining gluten is utilised in miscellaneous industrial and novel food processes.

(*a*) *Baked goods.* Vital commercial gluten possesses the same unique viscoelastic properties that are found in unfractionated flour proteins and which are essential in a dough to ensure gas retention and to produce bread with a fine, even crumb texture. Commercial gluten can, therefore, supplement natural flour protein.

There are many reasons why this may be necessary. For the miller with little or no access to the high protein wheats that are grown in Canada, Australia or the USA, who relies almost solely on the home-grown crop, gluten can be extremely valuable to fortify his flour, especially in years which have produced wheats with below average protein content. France is a particular example, where up to 1% gluten is regularly added to breadmaking flours.

Australian millers have also been known to fortify with gluten, depending upon the relative prices of home-grown high protein wheat and commercial gluten. For gluten-producing companies associated with flour milling businesses, flour fortification can be used to control the price of gluten on the open market. When there is a surplus of gluten and its price drops, the value can be maintained by taking the excess off the market and using it in bread flours. However, recently the demand has been high, and most Australian gluten is currently exported.

Gluten can also be used to advantage by the baker when his requirements include both a flour of average protein content and one of strong, high protein flour. By buying just one standard bread flour he can add the required quantity of gluten to produce the high protein flour necessary for his speciality products, such as morning goods and 'Vienna' products. This reduces the requirement for flour bins and thus reduces capital costs. A further advantage is that because the protein added causes an increase in the amount of water that must be incorporated into the dough, bread is produced that is softer and has a longer shelf-life.

Increasing the 'hinge' strength of hamburger buns is another application of gluten in a baked product. Hamburger buns are cut horizontally, almost in half, before use, but leaving a small portion, the hinge, on one side uncut. Some users of these buns find it important that there is sufficient strength in the crumb structure for the bun to be opened without the hinge tearing. Gluten brings this increased strength to the product and is widely used in the USA for this purpose.

In addition gluten is widely used to increase the tolerance of a dough to physical abuse during processing. Tolerance is difficult to define and it is even more difficult to develop test methods to demonstrate it. In general, tolerant flours/doughs are ones that will produce consistently good quality products despite minor variations in plant operating conditions such as mixing time and energy input, dough handling particularly during dividing and moulding, proof time, etc. Flours with a higher protein content are usually more tolerant: vital gluten can be added by the baker should he experience problems with a particular batch of flour.

Low density bread is another area which consumes much gluten in some countries. The high protein level of the dough enables increased expansion of the product during proof: suitable processing will retain a fine crumb structure. Thus loaves of conventional volume can be produced from lighter dough pieces and each slice cut will contain less dry weight material, particularly starch. Such products have found certain favour in the slimming market, and for those with preference for a less dense structure. It is possible to produce rolls solely from gluten, but these have a very open texture. Products are available containing high levels of gluten but retaining some flour.

Perhaps the fastest growing application for vital gluten in the bakery trade is in the production of high fibre bread. In the UK, sales of brown bread doubled in the 10-year period to 1981 and now account for 18% of total bread sales. However, incorporation of wheat bran and germ, or other sources of dietary fibre, has a deleterious effect on the loaf characteristics of volume and crumb texture. The exact mechanism is not clear; in addition to dilution of the proportion of gluten protein in the flour, it is likely the material physically disrupts the gluten network formed in the dough. The wheat grain aleurone layer often remains attached to the bran after milling. This layer is rich in protein but it is essentially non-gluten in character and may interact to affect the gluten network and hence dough structure. Germ constituents, particularly in wholemeal flours, also influence the structure. For this reason, flours for wholemeal bread production are often prepared from wheats with higher protein content than those for white bread manufacture. Protein content can be raised advantageously by incorporating vital gluten in the recipe for high fibre bread production. This improves both the volume of the bread and the crumb texture which otherwise would be quite solid and have low resilience. The gluten thus enables the dough to 'carry' the fibre.

(*b*) *Breakfast cereals.* Gluten is used in certain breakfast cereals for its binding ability. It also improves the nutritional properties by increasing the protein content. The low level of lysine in gluten is not too important since most breakfast cereals are consumed with milk which supplements this essential amino acid. It is reported that one major breakfast cereal manufacturer in the USA uses about 5500 tonnes of vital gluten per year in a single product.[159]

(*c*) *Breading and batter mixes and pasta.* Many precooked foods on the market are coated with breadcrumbs or batter, e.g. fish and fish fingers, chicken pieces. Most are stored frozen, and there may be a tendency for the coating to come away from the product. A dusting of the product with gluten prior to application of the coating can improve the adhesion. Dry egg albumen is also very effective but considerably more expensive.[161]

Gluten can be used directly in batter mixes to improve batter characteristics, including the adhesive properties. The gluten, because of its film-forming properties, encases the natural juices in the product thereby retaining the flavour.[162]

The importance of the native wheat protein in durum wheat to pasta quality has been described on p. 27. To improve the mechanical properties of the dried product, commercial gluten may be incorporated at the doughing stage. Pasta made from low quality durum wheat or hard bread wheats tends to stick together on cooking. This is related to release of amylose from the product, resulting from incompleteness of the protein network. Addition of vital wheat gluten will improve this structure and reduce stickiness, but the gluten must be of such quality that it is functional almost immediately.[162]

At present, the quantities of vital wheat gluten used in these applications is not thought to be very large.

(*d*) *Meat analogues, processed meat and pet foods.* The properties of gluten when mixed with water (i.e. the formation of an insoluble visco-elastic material with strong cohesive properties), the capacity for this mass to be stretched into thin films in a three-dimensional network, and its thermosetting ability giving rise to a firm, clean cutting resilient gel, make it ideal for use in reformulated meat products. After the traditional choice cuts of meat have been removed from a carcass, there remains much meat of poorer quality, although not necessarily poor from a nutritional view point. These pieces may be small, may be surrounded by fat and may contain gristle. Removal of some of the fat and flaking does not leave an appetising product. However, reforming with the aid of gluten binds the various pieces of meat tissue, starch and fat together to give such products as meat chunks, steaks, turkey rolls and sausages. Gluten contributes to the 'bite' of the final cooked product because of its thermosetting properties. A recent paper has shown that it is far superior to other types of protein in its ability to bind meat pieces.[163] Gluten also contributes to a reduction in cooking losses since it has strong affinity both for water and for fat, with which it binds readily. For similar reasons gluten is used widely in pet food products, most of which incorporate meat that would otherwise be classed as waste, or low grade materials such as offals. It is also a relatively cheap source of protein in these products and it is estimated that in the USA, in 1975, about 8% of gluten usage was for pet foods.[160]

Wheat gluten is also used quite extensively in textured vegetable protein products, for instance in combination with soya protein where it can be extruded into 'meat' chunks or slices. The relative proportions of soya protein and gluten can vary considerably, either being the major component.

(*e*) *Novel and industrial uses.* In addition to fried gluten entrees, which are popular in the Far East, a novel process from Japan suggests that it is possible to produce a crab meat analogue from gluten.[164] The wet gluten is blended with a reducing agent such as sodium bisulphite and a raising or gassing agent, e.g. sodium bicarbonate. The reducing agent will also bleach the carotenoid pigments in the gluten producing a white fibrous material. The blended material is then gelled by heating it in water, which is brought to boiling point. Stirring speed is critical to produce the right textured product: slow speeds result in formation of thick and heavy fibrous material, fast speeds giving a shredded product. The appearance and likeness of the product to crab meat is enhanced by preparing a second batch with red dye and mixing a proportion of this with the undyed batch.

Chemical or enzymic treatment of gluten can modify its properties and give the opportunity for use in several other food products, including cheese, fish or meat and whippable foams.[162]

It is clear that vital wheat gluten has an important role to play in the food industry both in the traditional area of baked products, and in the convenience and fast food markets. Its adhesive properties have been used in the past in industrial products such as glue, and new uses are being constantly sought.

SIGNIFICANCE OF NON-GLUTEN PROTEINS

Albumins and Globulins

The albumin/globulin fraction of wheat flour protein comprises proteins with enzymic functions, and other proteins with unknown activities. Some of the enzymes are of considerable significance in bakery technology, particularly in breadmaking.

The importance of enzymic proteins will be discussed separately below. Regarding the importance in breadmaking of the albumin/globulin proteins without recognised enzymic activity, the situation is not well understood. Fractionation and reconstitution techniques have demonstrated that the water-extractable fraction of flour (including non-protein components) is necessary in breadmaking although it is not responsible for the differences in baking quality between wheat varieties.[165]

Since flour itself contains salts, a proportion of the globulin fraction is extracted when flour is extracted with water; dialysis of the water extract leads to precipitation of the globulins and also separates low molecular weight diffusible materials from the retained high molecular weight solutes. Examination of the baking quality of flours reconstituted from these sub-fractions indicated that neither heat coagulable albumins, nor globulins precipitated on dialysis, played any part in flour baking quality.[165] The diffusible fraction could be replaced by yeast food, which comprises mainly salts, and the uncoagulated fraction after boiling apparently comprised mainly hemicellulose-type material. However, the fact that only a proportion of the globulins is extracted by water, and the fact that the albumin fraction was prepared by boiling, which would certainly have denatured the protein, and might not have precipitated all of the protein present, means that the question of whether or not the albumin/globulin fraction is important in breadmaking has not been answered unequivocally.

Pence *et al.*[166] obtained evidence to suggest that the albumin fraction might well be important. Their experiments indicated that the albumins were easily denatured and that their effectiveness was easily altered by this. Furthermore, a significant correlation has been observed for 32 flours of widely varying baking quality between the ratio of albumins to globulins and protein quality, defined as the coefficient of regression of loaf volume on flour protein when a flour's protein was increased by fortifying with its own gluten and water solubles or decreased by dilution with starch.[167] However, the results of this experiment are difficult to extrapolate to the situation where differences in baking quality are due to natural differences in protein content since it has been found that for a Canadian wheat, Neepawa, the ratio of albumins to globulins decreases as the ratio of glutenin to gliadin increases and as protein content increases.[168]

The role of the salt- and water-extractable proteins in breadmaking is, therefore, far from clear. The water-extractable fraction would also include low molecular weight thiol peptides such as glutathione. Many workers hold the view that glutathione may cause modification of the high molecular weight glutenin complexes during dough mixing by catalysing sulphydryl–disulphide interchange reactions. There is evidence that the content of glutathione in flours may be significantly correlated with variations in flour quality assessment parameters observed for different varieties (see above). There is also considerable circumstantial evidence that the removal of glutathione by oxidation to the dipeptide form may be the mechanism by which improvers such as bromate, iodate, dehydroascorbate and azodicarbonamide exert their effects.[73,169] The identification of components in the water- or salt-extractable fractions of flour and the definition of the role of these components in baking technology are still problems to which the answers are incomplete.

Enzymic Proteins

Although enzymic proteins are present at relatively low levels in quantitative terms compared with the storage proteins, some can have a profound effect on the performance of wheat flours during food manufacture, particularly breadmaking. A brief account of the effects of certain endogenous enzymes is presented below.

Amylases

In traditional bulk fermentation breadmaking procedures the presence of sufficient levels of amylase activity is important in terms of yeast activity. At dough temperatures amylases (both α- and β-) catalyse the hydrolysis of damaged starch granules to give various breakdown products, including maltose, that are used by the yeast as fermentation substrates. Thus, amylase activity is important for gas production in bulk fermentation systems. However, damaged starch granules also absorb water and this water is released after digestion of the starch by amylases leading to a softer, more fluid dough. It is, therefore, important that amylase activity is not excessive so that the dough handling characteristics are not adversely affected.[170–172]

In short-time breadmaking procedures, where the bulk fermentation step is eliminated, it is not clear how important the role of amylase activity is in producing yeast-fermentable substrates by starch hydrolysis. In the Chorleywood Bread Process, for example, the time available is of the order of 60 min (time from addition of water to the flour to baking in the oven) but whether sufficient substrate is already available in the flour itself to support the level of yeast fermentation required, or whether amylolytic degradation of starch must also contribute to the production of adequate levels of substrate, is not clear. Amylase activity is, nevertheless, important in these systems, for during proof starch is converted to dextrins and maltose, which affect crust colour; during the early stages of baking, starch begins to gelatinise before inactivation of the amylase occurs, leading to the formation of dextrins, which improve loaf volume. Therefore, there should be sufficient amylase activity, otherwise loaf volume and crust colour are adversely affected. On the other hand, if amylase activity is excessive, too much starch may be converted to dextrins, leading to a dough that is sticky and difficult to process and to bread having a sticky crumb that is difficult to slice.

It is clear, therefore, that there is a fine balance between having sufficient but not excessive levels of amylase activity in a flour.

Proteases

Endogenous proteases in normal, sound wheat usually cause few problems in technological applications of flour. If proteolytic enzymes are active they have a gluten-softening effect. Where there is excessive activity doughs become sticky and difficult to handle. The dough has excessive flow characteristics and poor gas retention and as a result it tends to flatten in the tin at the proof stage and the baked product has low volume and a coarse crumb of poor colour.[172] Glutens isolated from such doughs soften quickly on standing under water and in extreme cases become almost liquid.[173]

Two situations arise when high levels of protease activity are encountered. During wet harvest years, sprouting of the grain in the ear may be prevalent. In this case, the grain germinates prematurely and proteases are released from the aleurone to mobilise the amino acid reserves in the endosperm storage proteins. Amylases are also produced under these circumstances and Redman has demonstrated a significant correlation between proteolytic and amylolytic enzyme activity in such wheats.[174] On the other hand, wheats grown in Eastern Europe and the USSR sometimes have high levels of protease activity due to wheat bug damage (*Eurygaster* and *Aelia* spp.). This proteolytic activity is not strictly endogenous but rather arises as a result of the injection of digestive enzymes into the grain to mobilise the endosperm reserves which are then used as food by these insects.[175] Redman has also obtained evidence to suggest that proteolytic activity, arising as a result of bacterial or mould contamination, may also cause gluten softening.[176]

Exogenous proteases may be added to certain flours to lower mixing time, to improve machinability and handling and to reduce 'buckiness' in the dough when the flour is from wheats whose gluten possesses excessive strength.

Lipoxygenases

Enzymes are present in wheat that catalyse the oxidation of unsaturated fatty acids, particularly linoleic acid, to lipid peroxides via lipid hydroperoxide intermediates. Their presence may be either advantageous or disadvantageous depending upon the particular process being considered. The two technologically important effects of these enzymes are that they catalyse bleaching of pigments and that they may catalyse reactions that lead to rheological changes in dough proteins which increase tolerance to overmixing.

In pasta production, where the colour of the dried product is an important criterion of quality, bleaching of the yellow carotenoid pigment during processing must be minimised; thus low lipoxygenase activities are desirable in durum wheats. However, the level of available substrates is also an important factor in determining the degree of pigment destruction. By contrast, bleaching of pigments is considered to be an advantage in white bread production and the addition of lipoxygenase activity is well known. This extra activity is usually added in the form of 'enzyme-active' soya bean flour but in some countries, France in particular, in the form of horse bean flour.

The significance of endogenous wheat lipoxygenase in terms of dough rheology is difficult to assess, although there is a good deal of evidence to indicate that exogenous soya lipoxygenase does have beneficial effects. The presence of soya flour during dough mixing increases the 'strength' of the dough and thus leads to some improvement in loaf volume.[177,178] However, the magnitude of these changes is small compared with the changes brought about by oxidants such as bromate and azodicarbonamide. It has been claimed that the particular advantage of soya flour lies in its effects on increasing tolerance to overmixing;[177,178] most oxidative improvers have the effect of decreasing tolerance. The effects of lipoxygenase and of oxidative improvers on mixing tolerance are apparently additive,[177,178] and thus the more substantial benefits of the oxidative improvers on loaf volume as well as the benefits of soya lipoxygenase on mixing tolerance may be obtained simultaneously.

Enzymes Involved in Ascorbic Acid Interconversions

L-Ascorbic acid has been used for many years as an improver in breadmaking in a similar way to bromate, although there is still disagreement as to the mechanisms involved. It is generally considered, however, that it is not ascorbic acid itself but the oxidised form, dehydro-L-ascorbic acid, that is the active component in the improver reaction since it is more potent than ascorbic acid and is, in fact, equivalent to bromate on a weight-for-weight basis.[179] The specificity of the improver reaction (several ascorbic acid analogues are not improvers) has suggested the involvement of enzymically catalysed reactions.[180–183] The picture is complicated, however, in that in chemical dough development, as opposed to conventional bulk fermentation procedures, dehydroascorbic acid apparently cannot replace ascorbic acid and presumably, therefore, the effects of ascorbic acid in these two dough systems are different.[184]

The conversion of ascorbic acid to dehydroascorbic acid is the reaction over which there is, perhaps, the greatest controversy. Some early work indicated that in water extracts of flour this reaction occurred by both enzymic and non-enzymic reactions.[181] Subsequent studies have confirmed the presence of the enzyme, ascorbic acid oxidase, in extracts of wheat flour,[185–187] but whether or not the enzyme is active in doughs, and the relative rates of enzymic and non-enzymic oxidation reactions, are not clear.

Oxidation of ascorbic acid to dehydroascorbic acid does nevertheless occur in doughs. Elkassabany *et al.*[188] showed that ascorbic acid and the analogue, D-isoascorbic acid, were converted to the corresponding oxidised dehydro compounds in yeasted flour/water doughs although in doughs mixed without yeast the oxidation appeared to be relatively slow. However, the low recoveries of ascorbic acid plus dehydroascorbic acid from unyeasted doughs makes interpretation of the effects of yeast in this experiment difficult. Nevertheless the fact that oxidation of the analogue occurs in the dough system does suggest that the oxidation of ascorbic acid may be largely a non-enzymic reaction since the work of Sanstedt and Hites[181] indicated that D-isoascorbic acid was a much poorer substrate for ascorbic acid oxidase than ascorbic acid itself.

The actual improver reaction in conventional dough systems involves not ascorbic acid itself but dehydroascorbic acid. It appears that this improver reaction is enzymically catalysed, involving reduced glutathione as co-substrate.[181] The presence of the enzyme which catalyses this reaction, glutathione dehydrogenase, has been demonstrated in flour extracts[181] and the overall result of the reduction of dehydroascorbate and the concomitant co-oxidation of glutathione is to remove the rheologically deleterious thiol peptide from the dough system. The competing reactions are illustrated in Fig. 15.

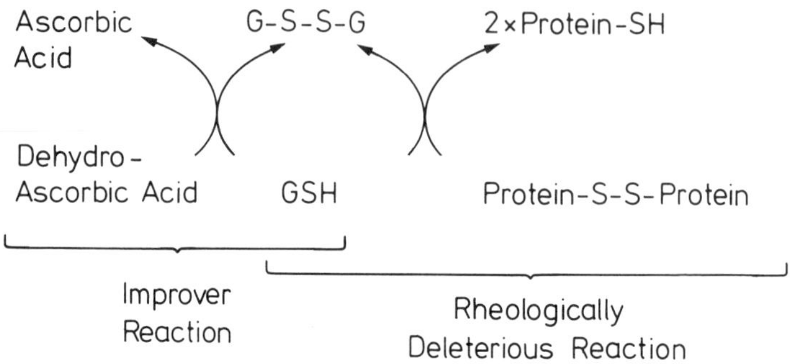

FIG. 15. Postulated competing reactions in flour/water doughs involving interchain protein disulphide bonds, reduced glutathione and dehydroascorbic acid.

Protein Disulphide Isomerase

There is considerable evidence that low molecular weight thiol compounds, such as reduced glutathione, can react with protein disulphide groups in dough systems and that low molecular weight disulphides, such as oxidised glutathione, can react with protein sulphydryl groups. There is also some evidence that similar reactions may occur between protein sulphydryl and disulphide groups. Such reactions are envisaged as being of importance in dough development; the oxidation of low molecular weight thiol compounds, so as to prevent their reaction with protein disulphides, is likewise widely regarded as being the mechanism by which improvers such as bromate, dehydroascorbic acid and azodicarbonamide produce their effects.

Protein disulphide isomerase activity, catalysing the interchange of protein disulphide and sulphydryl groups, has been identified in a number of plant as well as animal tissues: such enzymes require low molecular weight thiols, such as glutathione, as co-factors. An enzyme of this type has been identified in wheat germ,[189, 190] but to date there have been no reports of a similar enzyme activity in the endosperm of the wheat grain; a similar enzyme is also present in soya and horse bean flours which are often added to bread doughs.[190] It is not clear whether the enzyme is present in white flour or not, although presumably it is present in wholemeal flour. It will be of interest to determine the activity of this enzyme in wheat endosperm and to ascertain whether or not it plays a significant role in dough development, particularly in the presence of oxidising improver reagents, such as bromate or iodate, which will remove low molecular weight thiol co-factors that the enzyme requires for its activity.

ANTINUTRITIONAL EFFECTS OF WHEAT PROTEINS

Although it has been used for thousands of years in many areas of the world to supply much of the daily requirement of protein, carbohydrate and other nutrients, wheat, in common with many other foodstuffs of both plant and animal origin (e.g. soya, milk, etc.), contains materials, particularly in the protein fraction, that may produce antinutritional effects or adverse reactions in a small proportion of the population. Perhaps the best known condition involving wheat proteins is coeliac disease (gluten-induced enteropathy) although a few studies have implicated wheat proteins in the aetiology of schizophrenia. The antinutritional effects of proteinaceous inhibitors of animal α-amylases have also received some attention.

Wheat flour and wheat gluten are used to exploit their technological properties in such a wide range of foods, in addition to those most readily identified by the public as containing wheat flour (bread, cakes, biscuits, etc.), that the selection of appropriate foods can be a real problem for those people unfortunate enough to suffer a reaction against wheat proteins. These adverse reactions, therefore, are clearly relevant to the technological exploitation of wheat proteins, particularly when foods are being fabricated that, without specialist knowledge, one could not expect most people to recognise as containing wheat flour or gluten.

People with coeliac disease suffer symptoms of diarrhoea, abdominal cramps, vomiting, flatulence, fatty stools and weight loss or failure to grow. The incidence of the disease is put at 1 in 2000–3000 in the UK and USA but at 1 in 300 in certain parts of Eire, although the basis for this variation is unclear. The symptoms are associated with the ingestion of wheat-derived products (rye, barley and oats are also active) and it is now believed that it is in the protein fraction that the active factor resides.

There are a number of theories to explain the disease ranging from the lack of a peptidase, causing incomplete digestion of gluten, to an abnormal immune response to gluten. But at present the basic defect is not known. Most of the symptoms are due to the malabsorption of nutrients from the intestine, which is brought about because of damage to the mucosa, the absorptive surface of the small intestine. The complexity of wheat proteins and their variability from one wheat variety to another has contributed towards making the task of identifying the toxic factor extremely difficult.

Digestion of gluten with proteases, fractionation of the peptides and assessment of the toxicity of the sub-fractions has been one approach used to determine the nature of the peptide responsible for triggering the disease.[191] With this knowledge it might then be possible to devise some method, perhaps enzymic or chemical, whereby the food could be treated to destroy this sequence and, thus, to detoxify the food.

Flour proteins, as explained earlier, are very complex mixtures of polypeptides and digestion with proteases increases the level of heterogeneity enormously and makes separation and identification of the toxic factors very difficult. Hekkens and his colleagues[192] have sought to simplify matters by separating the gliadin fraction into its component polypeptides following on from the demonstration that the gliadin fraction was the most toxic. In particular it has been shown that the α-gliadin fraction contains toxic polypeptides.[192] Other gliadin species show differing degrees of toxicity, ω-gliadins being the least toxic.[193] This approach of identifying the toxic gliadin species is a promising one since, if these polypeptides can be identified, there is a possibility of being able to breed wheats lacking those particular species. Indeed, Kasarda *et al.*[194] considered that the α-gliadin fraction was the only toxic fraction, and in experiments in which they fed to coeliac patients certain wheat types lacking the chromosomes carrying the genes for these α-gliadins, no toxic effects were noted.

Although considerable progress has been made, a solution to the problem of coeliac disease remains distant.

Epidemiological and clinical trial evidence has led Dohan to suggest tentatively that there may also be a causal relationship between the prevalence of certain forms of schizophrenia and the consumption of wheat.[195] It is surprising, in view of the relatively high incidence of schizophrenia, that there has not been more work carried out to test the validity of this hypothesis. Despite this some evidence already exists to suggest a possible mechanism for the effect of wheat proteins in individuals predisposed to the disease.[196] Digestion of gluten and gliadin with pepsin was reported to produce peptides with similar activity to certain peptides, termed opioid peptides or endorphins, which are produced in the brain, and which have neurological effects similar to those of morphine and related compounds. These peptides are extremely potent, exerting their effects at concentrations in the nanomolar range and there is evidence of their involvement in schizophrenia. Peptic digests of two other food proteins, bovine serum albumin and casein, exhibited similar activities. Further work in this area seems justified.

Wheat is also known to contain proteinaceous inhibitors of mammalian α-amylases and suggestions have been made that these may have important nutritional consequences, particularly in animal feeds, in which wheat may form a large proportion of a relatively invariant diet. The inhibitors may comprise up to 66% of the albumin fraction of wheat proteins,[197] i.e. up to 1% of the weight of flour; their most likely physiological role seems to be in conferring resistance to insect attack.[197]

Technologically, the close association between the albumin fraction (and hence the α-amylase inhibitors) and starch granules may be important in determining the milling hardness of wheat. It has been suggested that these proteins adhere strongly to starch granules and may act to 'cement' the storage protein matrix to the starch granules.[198] Nutritionally, the presence of the amylase inhibitors may reduce the availability of dietary starch. However, since inhibitory activity is destroyed by pepsin, monogastric animals have some capacity for inactivation. In fact, Keys found that inclusion of as much as 70% of wheat into diets fed to young pigs caused no loss of starch availability[199] and it has been reported that pelleting of feeds leads to partial starch gelatinisation and thus increases its susceptibility to amylolysis.[200]

The amylase inhibitors are relatively heat stable and they persist in human food products such as bread, pasta and wheat-based breakfast cereals.[197] It has been proposed that the inhibitors might be of therapeutic use in diabetes mellitus and obesity on the basis of experiments in which inhibitors were fed with raw starch to human volunteers with the result that glucose availability from dietary starch was reduced.[201] However, the much diminished effects of the inhibitors when fed with gelatinised starch rather than raw starch, and the fact that chronic ingestion of these inhibitors was observed to induce histological changes in some tissues of chickens, throws doubt on the usefulness of this approach.[197]

THE SIGNIFICANCE OF PROTEINS IN THE TECHNOLOGY OF WHEAT SEED PRODUCTION AND GRAIN TRADING

The fact that wheat varieties differ substantially in their suitability for particular end uses (e.g. breadmaking) or in terms of their agronomic characteristics (e.g. yield potential, disease resistance), has resulted in an increasing requirement in recent years on the part of seed producers to segregate harvested wheat on the basis of variety and on the part of grain users (e.g. flour millers) to recognise which varieties of wheat are present.

Traditionally, the most widely used method for differentiating one variety from another has been based upon the morphological characteristics of the ear. However, after harvesting, morphological distinctness between grains of different varieties is lacking and more definitive techniques were required. This requirement was fulfilled in the mid-1970s by the successful exploitation of electrophoretic techniques developed earlier, in which the gliadin components of single grains could be separated and compared.

These procedures were based on electrophoretic separation of wheat gliadins in starch gels in the presence of an aluminium lactate buffer.[19,20] By starch gel electrophoresis (SGE) as many as 43 components with different mobilities have been distinguished in all the varieties examined, although a maximum of 25 are present in any single variety; different varieties have different complements of the various gliadin components (Fig. 3). The complement of gliadin components is a genotypic character and as such has been shown to be independent of area of cultivation, protein level, heating of the grain (as in artificial drying) and of harvest time, at least up to three weeks before normal harvest ripeness, although in some instances differences in relative band intensities do occur.[21] Recently reported results also suggest a relationship between the presence of certain gliadin bands and both grain and flour quality parameters.[202]

In a few instances the gliadin band patterns for two different varieties cannot be distinguished using SGE (e.g. the good breadmaking varieties Sappo and Sicco), and in such cases additional procedures must be employed. Grain morphological characteristics or other features, such as the phenol test or endosperm hardness,[19] may be useful in this context; alternatively examination of a different endosperm protein fraction, requiring a different type of electrophoresis technique, polyacrylamide gel electrophoresis in the presence of the detergent sodium dodecyl sulphate (SDS–PAGE) may be used.[203] This technique separates proteins according to molecular weights and in variety identification work it is the variation in the molecular weights of certain glutenin subunits that is exploited. However, the total number of glutenin subunits in all varieties, as resolved by SDS–PAGE, is considerably smaller than the total number of gliadin components resolved by SGE and the intervarietal variability of glutenin subunit patterns analysed by SDS–PAGE is not so great as that of gliadin components analysed by SGE. Thus, SDS–PAGE is very much a 'reserve' method in varietal identification work in comparison with SGE.

Despite the success of SGE in this area the technique does have some drawbacks, one of the most notable being the difficulty in preparing reproducible, highly uniform gels that perform consistently in terms of the resolution achieved. To overcome these drawbacks there has been a move in the past 2 or 3 years towards the use of polyacrylamide gels as the support medium, but still using lactate buffers.[22,204–206] Reproducible, uniform polyacrylamide gels are much more easily prepared than are starch gels and resolution of gliadin components may be at least equivalent to, and in some cases superior to, that achieved using starch gels. Further increase in the resolution of some components may be achieved by using polyacrylamide gels with a gradient of acrylamide concentration from one end of the gel to the other (lowest concentration at the origin) but the compression of bands in other zones may be disadvantageous.[204]

It can be expected that future developments in this area will be in reducing the actual time taken for electrophoresis and the replacement of visual inspection and interpretation of the stained gels by computer-assisted image analysis procedures.

REFERENCES

1. KENT. N. L. (1975). *The Technology of Cereals with Special Reference to Wheat*, 2nd Edition, Pergamon Press, Oxford.
2. OSBORNE, T. B. (1907). *The Proteins of the Wheat Kernel*, Carnegie Institution, Washington, D. C.
3. KASARDA, D. D., BERNARDIN, J. E. and NIMMO, C. C. (1977). *Adv. Cereal Sci. Technol.*, **1**, 158.
4. WALL, J. S. (1979). in *Recent Advances in the Biochemistry of Cereals*, Editors D. L. Laidman and R. W. Wyn-Jones, Phytochem. Soc. Eur. Symp. Ser. No. 16, Academic Press, London, New York and San Francisco, p. 275.
5. HUEBNER, F. R. and WALL, J. S. (1976). *Cereal Chem.*, **53**, 258.

6. ORTH, R. A. and BUSHUK, W. (1972). *Cereal Chem.*, **49**, 268.
7. BIETZ, J. A. and WALL, J. S. (1975). *Cereal Chem.*, **52**, 145.
8. KENT, N. L. (1970). in *Proteins as Human Food*, Editor R. A. Lawrie, Butterworth & Co. (Publ.) Ltd, London, p. 280.
9. *Amino Acid Content of Foods and Biological Data on Proteins*, FAO (1970).
10. BRESSANI, R., VITERI, F., ELIAS, L. G., de ZAGHI, S., ALVARADO, J. and ODELL, A. D. (1967). *J. Nutr.*, **93**, 349.
11. GRANT, R. A. (1980). in *Applied Protein Chemistry*, Editor R. A. Grant, Applied Science Publishers, London, p. 224.
12. LUNVEN, P., le CLEMENT de ST. MARC, C., CARNOVALE, E. and FRATONI, A. (1973). *Brit. J. Nutr.*, **30**, 189.
13. BOOTH, M. R. and EWART, J. A. D. (1969). *Biochim. Biophys. Acta*, **181**, 226.
14. KRULL, L. H., WALL, J. S., ZOBEL, H. and DIMLER, R. J. (1965). *Biochemistry*, **4**, 626.
15. HOLME, J. and BRIGGS, D. R. (1959). *Cereal Chem.*, **36**, 321.
16. NIELSEN, H. C., BECKWITH, A. C. and WALL, J. S. (1968). *Cereal Chem.*, **45**, 37.
17. KRETOVICH, V. L. and VAKAR, A. B. (1964). *Proc. Acad. Sci. USSR, Biochem. Section*, **155**, 71.
18. TKACHUK, R. and HLYNKA, I. (1968). *Cereal Chem.*, **45**, 80.
19. WRIGLEY, C. W. and SHEPHERD, K. W. (1974). *Aust. J. Exp. Agric. Anim. Husb.*, **14**, 796.
20. AUTRAN, J. –C. and BOURDET, A. (1975). *Ann. Amelior Plantes*, **25**, 277.
21. ELLIS, J. R. S. and BEMINSTER, C. H. (1977). *J. Natn. Inst. Agric. Bot.*, **14**, 221.
22. BUSHUK, W. and ZILLMAN, R. R. (1978). *Can. J. Plant Sci.*, **58**, 505.
23. WRIGLEY, C. W. and SHEPHERD, K. W. (1973). *Ann. N.Y. Acad. Sci.*, **209**, 154.
24. WU, Y. V. and CLUSKEY, J. E. (1965). *Arch. Biochem. Biophys.*, **112**, 32.
25. CLUSKEY, J. E. and WU, Y. V. (1966). *Cereal Chem.*, **43**, 119.
26. WU. Y. V., CLUSKEY, J. E. and SEXSON, K. R. (1967). *Biochim. Biophys. Acta*, **133**, 83.
27. KASARDA, D. D., BERNARDIN, J. E. and GAFFIELD, W. (1968). *Biochemistry*, **7**, 3950.
28. CLUSKEY, J. E. and WU, Y. V. (1971). *Cereal Chem.*, **48**, 203.
29. BECKWITH, A. C., NIELSEN, H. C., WALL, J. S. and HUEBNER, F. R. (1966). *Cereal Chem.*, **43**, 14.
30. HUEBNER, F. R., BIETZ, J. A. and WALL, J. S. (1977). in *Protein Crosslinking—Biochemical and Molecular Aspects*, Editor M. Friedman, Adv. Exp. Med. Biol., Vol. 86A, Plenum Press, New York and London, p. 67.
31. PAYNE, P. I. and CORFIELD, K. G. (1979). *Planta*, **145**, 83.
32. HUEBNER, F. R. and WALL, J. S. (1980). *J. Agric. Food Chem.*, **28**, 433.
33. BOTTOMLEY, R. C., KEARNS, H. F. and SCHOFIELD, J. D. (1982). *J. Sci. Food Agric.*, **33**, 481.
34. EWART, J. A. D. (1979). *J. Sci. Food Agric.* **30**, 482.
35. KAHN, K. and BUSHUK, W. (1978). *Baker's Digest*, **52**, 14.
36. KAUZMANN, W. (1959). *Adv. Protein Chem.* **14**, 1.

37. TANFORD, C. (1980). *The Hydrophobic Effect: Formation of Micelles and Biological Membranes*, Wiley-Interscience, Johy Wiley & Sons, New York.
38. JONES, I. K. and CARNEGIE, P. R. (1971). *J. Sci. Food Agric.*, **22**, 358.
39. KOBREHEL, K. and BUSHUK, W. (1977). *Cereal Chem.*, **54**, 833.
40. SCHOFIELD, J. D. and BAIANU, I. C. (1982). *Cereal Chem.*, **59**, 240.
41. BIETZ, J. A. and WALL, J. S. (1972). *Cereal Chem.*, **49**, 416.
42. ORTH, R. A. and BUSHUK, W. (1973). *Cereal Chem.*, **50**, 191.
43. BIETZ, J. A., SHEPHERD, K. W. and WALL, J. S. (1975). *Cereal Chem.*, **52**, 513.
44. PAYNE, P. I., CORFIELD, K. G., HOLT, L. M. and BLACKMAN, J. A. (1981). *J. Sci. Food Agric.*, **32**, 51.
45. LAWRENCE, G. J. and SHEPHERD, K. W. (1980). *Aust. J. Biol. Sci.*, **33**, 221.
46. BIETZ, J. A. and ROTHFUS, J. A. (1971). *Cereal Chem.*, **48**, 677.
47. BIETZ, J. A. and WALL, J. S. (1980). *Cereal Chem.*, **57**, 415.
48. BECKWITH, A. C., NIELSEN, H. C., WALL, J. S. and HUEBNER, F. R. (1966). *Cereal Chem.*, **43**, 14.
49. BIETZ, J. A. and WALL, J. S. (1973). *Cereal Chem.*, **50**, 537.
50. ORTH, R. A. and O'BRIEN, L. (1976). *J. Aust. Inst. Agric. Sci.*, **42**, 122.
51. KAHN, K. and BUSHUK, W. (1978). *Baker's Digest*, **52**, 14.
52. Home-Grown Cereals Authority, *A Marketing Guide for Milling Wheat*, June 1981.
53. FINNEY, K. F. and BARMORE, M. A. (1948). *Cereal Chem.*, **25**, 291.
54. CHAMBERLAIN, N., COLLINS, T. H. and ELTON, G. A. H. (1962). *Baker's Digest*, **36**, 52.
55. WEBB, T., HEAPS, P. W. and COPPOCK, J. B. M. (1971). *J. Food Technol.*, **6**, 47.
56. WALTL, K. (1972). *Getreide Mehl Brot*, **26**, 142.
57. KRULL, L. H. and WALL, J. S. (1969). *Baker's Digest*, **43**, 30.
58. BUNGENBURG DE JONG, H. L. and KLAAR, W. J. (1952). *Kolloidzeitschrift*, **128**, 164.
59. BARMORE, M. (1948). *Baker's Digest*, **22**, 10.
60. WRIGLEY, C. W. (1972). *Cereal Sci. Today*, **17**, 370.
61. PAYNE, P. I., CORFIELD, K. G. and BLACKMAN, J. (1979). *Theor. Appl. Genet.*, **55**, 153.
62. AXFORD, D. W. E., MCDERMOTT, E. E. and REDMAN, D. G. (1978). *Milling Feed and Fertiliser*, **161**, 18.
63. LEE, J. W. and MACRITCHIE, F. (1971). *Cereal Chem.*, **48**, 620.
64. MACRITCHIE, F. (1973). *J. Sci. Food Agric.*, **24**, 1325.
65. EWART, J. A. D. (1980). *J. Sci. Food Agric.*, **31**, 1323.
66. BOOTH, M. R. and MELVIN, M. A. (1979). *J. Sci. Food Agric.*, **30**, 1057.
67. PRESTON, K. R. and TIPPLES, K. H. (1980). *Cereal Chem.*, **57**, 314.
68. JONES, I. K. and CARNEGIE, P. R. (1969). *J. Sci. Food Agric.*, **20**, 54.
69. HIRD, F. J. R., CROKER, I. W. D. and JONES, W. L. (1968). *J. Sci. Food Agric.*, **19**, 602.
70. JONES, I. K. and CARNEGIE, P. R. (1969). *J. Sci. Food Agric.*, **20**, 60.
71. ARCHER, M. J. (1972). *J. Sci. Food Agric.*, **23**, 485.
72. COVENTRY, D. R., CARNEGIE, P. R. and JONES, I. K. (1972). *J. Sci. Food Agric.*, **23**, 587.

73. MAIR, G. and GROSCH, W. (1979). *J. Sci. Food Agric.*, **30**, 914.
74. KOBREHEL, K. and MATIGNON, B. (1980). *Cereal Chem.*, **57**, 73.
75. MACRITCHIE, F. (1979). *J. Food Technol.*, **14**, 595.
76. KHOO, U., CHRISTIANSEN, D.D. and INGLETT, G. E. (1975). *Baker's Digest*, **49**, 24.
77. CUMMING, D. B. and TUNG, M. A. (1975). *Can. Inst. Food Sci. Technol. J.*, **8**, 67.
78. ANGOLD, R. E. (1979). in *Food Microscopy*, Editor J. G. Vaughan, Academic Press, London, New York and San Francisco, p. 75.
79. MECHAM, D. K., SOKOL, H. A. and PENCE, J. W. (1962). *Cereal Chem.*, **39**, 81.
80. TSEN, C. C. (1967). *Cereal Chem.*, **44**, 308.
81. TANAKA, K. and BUSHUK, W. (1973). *Cereal Chem.*, **50**, 590.
82. MAMARIL, F. P. and POMERANZ, Y. (1966). *J. Sci. Food Agric.*, **17**, 269.
83. MEREDITH, O. B. and WREN, J. J. (1969). *J. Sci. Food Agric.*, **20**, 235.
84. TANAKA, K. and BUSHUK, W. (1973). *Cereal Chem.*, **50**, 597.
85. MECHAM, D. K. (1959). *Cereal Chem.*, **36**, 134.
86. MECHAM, D. K. (1980). *Food Technol. in Australia*, **32**, 540 and 582.
87. MECHAM, D. K., COLE, E. G. and SOKOL, H. A. (1963). *Cereal Chem.*, **40**, 1.
88. SOKOL, H. A., MECHAM, D. K. and PENCE, J. W. (1960). *Cereal Chem.*, **37**, 739.
89. TANAKA, K. and BUSHUK, W. (1973). *Cereal Chem.*, **50**, 605.
90. MAURITZEN, C. A. M. and STEWART, P. (1963). *Nature, London*, **197**, 48.
91. REDMAN, D. G. and EWART, J. A. D. (1967). *J. Sci. Food Agric.*, **18**, 15.
92. JONES, I. K. and CARNEGIE, P. R. (1971). *J. Sci. Food Agric.*, **22**, 358.
93. EWART, J. A. D. (1972). *J. Sci. Food Agric.*, **23**, 687.
94. GREENWOOD, C. T. and EWART, J. A. D. (1975). *Cereal Chem.*, **52**, 146r.
95. EWART, J. A. D. (1977). *J. Sci. Food Agric.*, **28**, 191.
96. SIDHU, J. S., NORDIN, P. and HOSENEY, R. C. (1980). *Cereal Chem.*, **57**, 159.
97. POMERANZ, Y. (1980). *Baker's Digest*, **54**(1), 20 and (2), 12.
98. POMERANZ, Y. (1980). *Cereal Foods World*, **25**, 656.
99. MACRITCHIE, F. (1980). *Baker's Digest*, **54**(5), 10.
100. MACRITCHIE, F. (1977). *J. Sci. Food Agric.*, **28**, 53.
101. ANTOGNELLI, C. (1980). *J. Food Technol.*, **15**, 125.
102. MATWEEF, M. (1966). *Bull. Anciens Éleves Écoles Franc. Meun.*, No. 213, 133.
103. MATSUO, R. R., BRADLEY, J. W. and IRVINE, G. N. (1972). *Cereal Chem.*, **49**, 707.
104. DEXTER, J. E. and MATSUO, R. R. (1977). *Can. J. Plant. Sci.*, **57**, 717.
105. MATSUO, R. R. and IRVINE, G. N. (1970). *Cereal Chem.*, **47**, 173.
106. WALSH, D. E. and GILLES, K. A. (1971). *Cereal Chem.*, **48**, 544.
107. WASIK, R. J. and BUSHUK, W. (1975). *Cereal Chem.*, **52**, 322.
108. WASIK, R. J. (1978). *Can. Inst. Food Sci. Technol. J.*, **11**, 129.
109. DEXTER, J. E. and MATSUO, R. R. (1980). *J. Agric. Food Chem.*, **28**, 899.
110. DEXTER, J. E. and MATSUO, R. R. (1978). *Cereal Chem.*, **55**, 44.
111. D'EGIDIO, M. G., FORTINI, S., GALTERIE, G., MARIANI, B. M., SGRULETTA, D. and VOLPI, M. (1979). *Qual. Plant.*, **14**, 333.
112. WASIK, R. J. and BUSHUK, W. (1975). *Cereal Chem.*, **52**, 328.

113. DAMIDAUX, R., AUTRAN, J–C., GRIGNAC, P. and FEILLET, P. (1978). *C. R. Acad. Sci. Paris Ser. D.*, **287**, 701.
114. KOSMOLAK, F. G., DEXTER, J. E., MATSUO, R. R., LEISLE, D. and MARCHYLO, B. A. (1980). *Can. J. Plant Sci.*, **60**, 427.
115. DEXTER, J. E. and MATSUO, R. R. (1977). *Cereal Chem.*, **54**, 882.
116. DEXTER, J. E. and MATSUO, R. R. (1979). *Cereal Chem.*, **56**, 15.
117. SCHOFIELD, J. D., BOTTOMLEY, R. C., KEARNS, H. F., TIMMS, M. F. and DAY, K. W. (1981). 66th Annual AACC Meeting, Denver, Col., *Cereal Foods World*, **26**, Abs. 169, p. 506.
118. MATSUO, R. R., DEXTER, J. E. and DRONZEK, B. L. (1978). *Cereal Chem.*, **55**, 744.
119. HMSO (1980). *Business Monitor PQ211*, 4th Quarter.
120. HODGE, D. G., BOLD, E. R. and WADE, P. (1967). *F.M.B.R.A., Chorleywood, Report*, No. 8.
121. HALTON, P. and GREER, E. N. (1950). *J. Sci. Food Agric.*, **1**, 358.
122. WADE, P. (1970). *Food Trade Rev.*, **40**, 34.
123. WADE, P. (1972). *J. Sci. Food Agric.*, **23**, 333.
124. THACKER, D. and MILLER, A. R. (1980). *F.M.B.R.A., Chorleywood, Bulletin*, No. 1, 19.
125. PRITCHARD, P. E., EMERY, A. H. and STEVENS, D. J. (1975). *F.M.B.R.A., Chorleywood, Report*, No. 66.
126. SARKKI, M.-L. (1979). *J. Am. Oil Chem. Soc.*, **56**, 443.
127. KNIGHT, J. W. (1965). *Wheat Starch and Gluten*, Leonard Hill, London, p. 3.
128. JACKSON, A. (1976). in *Starch Production Technology*, Editor J. A. Radley, Applied Science Publishers, London, p. 155.
129. FELLERS, D. A., JOHNSTON, P. H., SMITH, S., MOSSMAN, A. P. and SHEPHERD, A. D. (1969). *Food Technol.*, **23**, 162.
130. JOHNSTON, P. H. and FELLERS, D. A. (1971). *J. Food Sci.*, **36**, 649.
131. EYNON, L. and LANE, J. H. (1928). *Starch: Its Chemistry, Technology and Uses*, W. Heffner & Sons Ltd, Cambridge, p.145.
132. WALON, R. G. P. (1980). C.P.C. International, US Patent 4217414.
133. DAHLBERG, B. I. (1978). *Stärke*, **30**, 8.
134. KERKONNEN, H. K., LAINE, K. M. ALANEN, M. A. and RENNES, H. V. (1973). British Patent 1397370.
135. VERBERNE, P. and ZWITSERLOOT, W. R. M. (1978). *Stärke*, **30**, 337.
136. VERBERNE, P., ZWITSERLOOT, W. R. M. and NAUTA, R. R. (1976). British Patent 1596742.
137. PHILLIPS, K. L. and SALLANS, H. R. (1966). *Cereal Sci. Today*, **11**, 61.
138. RODGERS, N. E. and GIDLOW, R. G. (1972). British Patent 1357669.
139. DURST, J. R. and WINTERS, W. C. (1974). US Patent 3788861.
140. RAO, G. V., HENRY, W. E. and HAMMOND, D. L. (1976). British Patent 1516038.
141. RAO, G. V., HENRY, W. E. and HAMMOND, D. L. (1976). US Patent 3979375.
142. RAO, G. V. and SHOUP, F. K. (1980). US Patent 4269776.
143. FINLEY, J. W. and HAUTALA, E. (1973). in *Proc. Nat. Conf. on Wheat Utilisation Research—Colorado*, US Agricultural Research Service, ARSW-19, p. 97.

144. FINLEY, J. W. and HAUTALA, E. (1974). US Patent 3828017.
145. FINLEY, J. W. (1976). *J. Food Sci.*, **41**, 882.
146. CLEMENTS, R. L. (1979). *Cereal Chem.*, **56**, 1.
147. BOOTH, M. R. and TIMMS, M. F. (1978). *Assessment of Dried Gluten Quality*, Paper A4-8, Sixth International Cereal and Bread Congress, Winnipeg.
148. FINNEY, K. F. (1943). *Cereal Chem.*, **20**, 381.
149. VOISEY, P. W., MILLER, H. and KLOEK, M. (1966). *Cereal Chem.*, **43**, 408.
150. STENVERT, N. L., MOSS, R. and MURRAY, L. (1981). *Baker's Digest*, **55**, 6.
151. PENCE, J. W., MOHAMMAD, A. and MECHAM, D. K. (1953). *Cereal Chem.*, **30**, 115.
152. JEANJEAN, M. F., DAMIDAUX, R. and FEILLET, P. (1980). *Cereal Chem.*, **57**, 325.
153. BOOTH, M. R., BOTTOMLEY, R. C., ELLIS, J. R. S., MALLOCH, G., SCHOFIELD, J. D. and TIMMS, M. F. (1980). *Ann. Technol. Agric.*, **29**, 399.
154. Top-Scor Products. (1967). British Patent 1083368.
155. KLEINSCHMIDT, A. W. (1979). US Patent 4179525.
156. ABOU-GUENDIA, M. M. (1977). US Patent 4035519.
157. MURRAY, D. W. and SINGER, N. S. (1979). British Patent 1556815.
158. JOHANNSON, H. P. (1978). US Patent 4076845.
159. MITTLEIDER, J. F. (1978). *An Analysis of the Economic Feasibility of Establishing Wheat Gluten Processing Plants in Dakota*, North Dakota Agricultural Experimental Station, Fargo, N.D.
160. HALL, J. B. (1977). *Cereal Foods World*, **22**, 144.
161. BAKER, R. C., DORFER, J. M. and VAHEDRA, D. V. (1972). *Poultry Sci.*, **51**, 1220.
162. KALIN, F. (1979). *J. Am. Oil Chem. Soc.*, **56**, 477.
163. SEIGEL, D. G., CHURCH, K. E. and SCHMIDT, G. R. (1979). *J. Food Sci.*, **44**, 1276.
164. TANAKA, K. (1976). US Patent 4000331.
165. HOSENEY, R. C., FINNEY, K. F., SHOGREN, M. D. and POMERANZ, Y. (1969). *Cereal Chem.*, **46**, 117.
166. PENCE, J. W., ELDER, A. H. and MECHAM, D. K. (1951). *Cereal Chem.*, **28**, 94.
167. PENCE, J. W., WEINSTEIN, N. E. and MECHAM, D. K. (1954). *Cereal Chem.*, **31**, 303.
168. BUSHUK, W., RODRIGUES-BORES, F. J. and DUBETZ, S. (1978). *Can. J. Plant Sci.*, **58**, 923.
169. BLOKSMA, A. H. (1972). *Cereal Chem.*, **49**, 104.
170. FARRAND, E. A. (1964). *Cereal Chem.*, **41**, 98.
171. COLLINS, T. H. (1970). *Baking Industries Journal*, Sept., 52.
172. DUBOIS, D. K. (1980). *Am. Inst. Baking Tech. Bull.*, **2**, Issue 11.
173. HANFORD, J. (1967). *Cereal Chem.*, **44**, 499.
174. REDMAN, D. G. (1977). *F.M.B.R.A., Chorleywood, Bulletin*, No. 5, 177.
175. BUHLE, K. and TIETZE, U. (1969). *Getreide Mehl Brot*, **19**, 57.
176. REDMAN, D. G. (1971). *J. Sci. Food Agric.*, **22**, 75.
177. FRAZIER, P. J. (1979). *Baker's Digest*, **53**, 8.
178. FRAZIER, P. J., BRIMBLECOMBE, F. A., DANIELS, N. W. R. and RUSSELL EGGITT, P. W. (1979). *Getreide Mehl Brot*, **33**, 268.

179. MELVILLE, J. and SHATTOCK, H. T. (1938). *Cereal Chem.*, **15**, 201.
180. CATHCART, W. H. and EDELMANN, E. C. (1944). *Cereal Chem.*, **21**, 575.
181. SANSTEDT, R. M. and HITES, B. D. (1945). *Cereal Chem.*, **22**, 161.
182. MALTHA, P. (1953). *Getreide Mehl Brot*, **3**, 65.
183. TSEN, C. C. (1965). *Cereal Chem.*, **42**, 86.
184. DAHLE, L. K. and MURTHY, P. R. (1970). *Cereal Chem.*, **47**, 296.
185. KUNINORI, T. and MATSUMOTO, H. (1963). *Cereal Chem.*, **40**, 647.
186. GRANT, D. R. (1974). *Cereal Chem.*, **51**, 684.
187. GRANT, D. R. and SOOD, V. K. (1980). *Cereal Chem.*, **57**, 46.
188. ELKASSABANY, M., HOSENEY, R. C. and SEIB, P. A. (1980). *Cereal Chem.*, **57**, 85.
189. GRYNBERG, A., NICHOLAS, J. and DRAPRON, R. (1977). *C. R. Acad. Sci. Paris*, **284**, 235.
190. GRYNBERG, A., NICOLAS, J. and DRAPRON, R. (1978). *Biochimie*, **60**, 547.
191. KOWLESSAR, O. D. (1967). *Gastroenterology*, **52**, 893.
192. DE ROOIJ, F. W. M, VAN DUYN, W., LEMS-VAN KAN, P. H. and HEKKENS, W. Th. J. M. (1977). in *Perspectives in Coeliac Disease*, Editors B. McNicholl, C. F. McCarthy and P. F. Cottrell, MTP Press, Lancaster, p. 17.
193. JOS, J., CHARBONNIER, L., MOUGENOT, J. F., MOSSE, J. and REY, J. (1977). in *Perspectives in Coeliac Disease*, Editors B. McNicholl, C. F. McCarthy and P. F. Cottrell, MTP Press, Lancaster, p. 75.
194. KASARDA, D. D., QUALSET, C. O., MECHAM, D. K., GOODENBERGER, D. M. and STROBER, W. (1977). in *Perspectives in Coeliac Disease*, Editors B. McNicholl, C. F. McCarthy and P. F. Cottrell, MTP Press, Lancaster, p. 55
195. DOHAN, F. C. (1966). *Acta Psychiat. Scand.*, **42**, 125.
196. KLEE, W. A., ZIOUDROU, C. and STREATY, R. A. (1979). in *Endorphins in Mental Health Research*, Editors E. Usdin and W. Bunney, MacMillan, New Jersey, p. 209.
197. BUONOCORE, V., PETRUCCI, T. and SILANO, V. (1977). *Phytochem.*, **16**, 811.
198. SIMMONDS, D. H., BARLOW, K. K. and WRIGLEY, C. W. (1973). *Cereal Chem.*, **50**, 553.
199. KEYS, J. E. (1972). Ph.D. Thesis, University of Maryland, USA.
200. SAUNDERS, R. M. (1975). *Cereal Foods World*, **20**, 282.
201. PULS, W. and KEMP. N. (1973). *Diabetologia*, **9**, 97.
202. WRIGLEY, C. W., ROBINSON, P. J. and WILLIAMS, W. T. (1981). *J. Sci. Food Agric.*, **32**, 433.
203. SHEWRY, P. R., FAULKS, A. J., PRATT, H. M. and MIFLIN, B. J. (1978). *J. Sci. Food Agric.*, **29**, 847.
204. DU CROS, D. L. and WRIGLEY, C. W. (1979). *J. Sci. Food Agric.*, **30**, 785.
205. AUTRAN, J. –C., BUSHUK, W., WRIGLEY, C. W. and ZILLMAN, R. R. (1979). *Cereal Foods World*, **24**, 471.
206. REDMAN, D. G., FERGUSON, S. and BURBRIDGE, K. (1980). *F.M.B.R.A., Chorleywood, Bulletin*, No. 2, p. 63.

Chapter 2

SOY PROTEINS

A. M. Pearson
Department of Food Science and Human Nutrition, Michigan State University, USA

SUMMARY

Typical soybeans consist of roughly 42% protein, 35% carbohydrate, 20% oil and 5% ash. Most of the protein is stored in the cotyledons as protein bodies, while the oil is located in the spherosomes interspersed between the protein bodies. The major proteins are globulins and have molecular weights of over 100 000. After extraction of the oil, about 85% of the protein in the defatted meal can be removed by water extraction at pH 6·5, while adding alkali will remove another 5–10%. The globulins have isoelectric points at a pH range of 4·2–4·6, so that acid is added to insolubilize the protein fraction, while the biologically active compounds, including enzymes, protease inhibitors, hemagglutinins, goitrogens, estrogens, allergens, oligosaccharides and other substances, can be removed by washing.

The soy proteins have the highest nutritional value of all plant proteins for human food, being particularly high in lysine. The protein efficiency ratio of properly processed soy flour and grits generally falls in a range of 2·2–2·3 compared to 2·5 for casein. Soy proteins in combination with other plant and/or animal proteins are complementary. However, a relatively high phytic acid content may bind some minerals, especially zinc and iron, so that more information on the effects of soy proteins on the bioavailability of minerals is needed. On the other hand, soy proteins have been reported to reduce serum cholesterol levels.

The processes of oil extraction and desolventizing–toasting play important roles in production of high quality soybean meal. The major products available for human foods include soy flour and grits, which are used to produce soy concentrates and isolates. Other modifications in processing result in production of spun soy proteins and textured soy products. Soy proteins are widely used in a variety of human foods, including cereal grain foods, dairy-type foods, snack foods, soups, beverages, and meat, poultry and fish products. Their use as meat extenders in the school lunch program and in ground beef for the Armed Forces of the United States will undoubtedly greatly increase the future usage of soy proteins for human food.

INTRODUCTION

The domestic soybean (*Glycine max* (L.) Merrill) is a legume that originated in eastern Asia.[1] Genetically it is thought to have been derived from *Glycine ussuriensis*, which grows wild throughout many parts of eastern Asia.[2] The soybean has been used extensively for human food for well over 2000 years.[2] 'Shih', a bean relish or paste, is mentioned in records dating back to 33 B.C. to 23 A.D. and the use of both dry and fresh soybean sprouts is believed to go back many centuries.[1] Soybeans have been stored as a staple crop and have been used in a variety of foods.

The soybean was brought to the attention of Europeans in 1712 by Engelbert Kaempfer, a German botanist, who discussed in detail the various food products made from it by the Japanese.[1] Soybeans were not mentioned in the United States until 1804,[1] and initially were used primarily as a forage and pasture crop. Harvesting for seed was not common as recently as 50 years ago, with only one-quarter of the total acreage grown in the early 1930s being harvested as beans. Processing of the seed did not begin until 1922, when the total harvested soybean crop amounted to only 4 million bushels.[3] By 1950, 15 744 000 acres of soybeans were planted with an average yield of 21·7 bushels per acre, which had increased to 71 586 000 acres with an average yield of 32·2 bushels per acre in 1979.[4] Thus, there has not only been an increase in the total acreage of soybeans but also an increase in yield, both of which have contributed to the availability of soybeans for processing.

The first processing of soybeans in the United States was carried out to reclaim the oil, with the soybean meal being a by-product of the operation.[3] The meal was initially fed to dairy cows or used for fertilizer. However, the work of Hayward *et al.*[5,6] demonstrated it to be a high quality protein supplement for poultry, provided it was first subjected to treatment with moist heat and properly supplemented. Since that time soybean meal has become widely used for feeding not only to poultry and cattle but also for swine. Since the end of the Second World War, when a world-wide protein shortage was first recognized, emphasis has centered on the development of new technologies to make soy-proteins available and acceptable as human food.

The present chapter briefly reviews the structure and composition of the soybean and the extraction of the oil. Discussion then focuses upon soy flour or grits and new technologies for producing soy concentrates, isolates, spun fibers and extruded products. Compositional changes, the nutritive value and the uses of the various products are reviewed briefly.

SOYBEAN PRODUCTION

Soybeans are widely adapted and have proven to be an excellent crop that provides both oil and protein. Table 1 presents production and yield data for the world and for the major producing countries for selected years from 1939 to 1979. During this period of time, world production

TABLE 1
SOYBEAN PRODUCTION (IN THOUSAND TONNES) AND YIELDS FOR SELECTED COUNTRIES AND THE WORLD FROM 1939 TO 1979[7]

Year	*World*	*United States*	*China*	*Brazil*	*Other countries*	*World yield (kg/ha)*
1939	13 167	2 453	9495	—	1219	—
1949	14 169	6 374	6672	25	1097	1050
1959	25 845	14 511	9525	147	1662	1212
1964	28 188	19 103	6940	305	1840	1177
1969	40 516	30 839	6200	1 057	2420	1425
1974	54 365	33 062	9500	7 876	3927	1387
1979	94 375	61 715	8300	15 200	9160	1857

has increased 7-fold and United States production 25-fold. At the same time the yield per hectare has increased 1·7 times. The most dramatic increase in production has come from Brazil, which only produced 2% of the world soybean crop in 1969 but 10 years later accounted for over 16%.[7]

It has been estimated that some 95% of the soybean crop is processed into oil and meal.[8] About 21% of the world's supply of fats and oils comes from soybean oil, and about 61% of the world's production of high protein meals is obtained from soybean meal.[9] Thus, both soybean oil and meal dominate the markets for oils and for high protein meals, their nearest competitors being tallow and greases (11% of the total) and cottonseed meal (11%), respectively. The importance of soybean meal as a high protein supplement is further emphasized when one recognizes that the combined soybean production of the United States and Brazil accounts for 49% of the potential production of all high protein meals in the world.

Most soybean meal is presently utilized for production of poultry, swine, cattle and other domestic animals. Its use has been widely credited with being responsible for the rapid growth of the livestock and poultry industries of the developed nations in the period since 1940.[10] Production of soy flour, which is a basic constituent in edible soy food products, has increased from 120 million lb per year in 1960 to almost 3 billion lb per year at present.[11] Although there has been a rapid increase in the use of soy proteins for human foods, it is estimated that only 3–5% of the soybean meal produced is further processed into flour and/or other soy protein foods for man.[12]

STRUCTURE OF THE SOYBEAN

Although soybeans differ in size, color and shape, depending upon the variety, commercial soybeans of the United States are generally spherical and yellow.[13] Figure 1 shows some soybeans that are typical of those used for producing oil, meal and protein food products in the United States. Their variability in size and dimensions becomes evident on close examination of the photograph, which contains a scale indicator to facilitate size comparisons.

The soybean is composed of three major structural components, i.e. the hull, the cotyledon and the hypocotyl, which comprise about 8, 90, and 2% of the soybean, respectively.[13] Figure 2 presents a drawing of a cross-section of the soybean showing the seed coat or hull and a portion of the cotyledon. The hull is composed of an outer layer of pallisade cells, a layer of hourglass cells, a small compressed layer of parenchyma cells, a thin layer of aleurone cells and finally a compressed layer of endosperm cells. The cotyledon is covered by a layer of epidermal cells and is

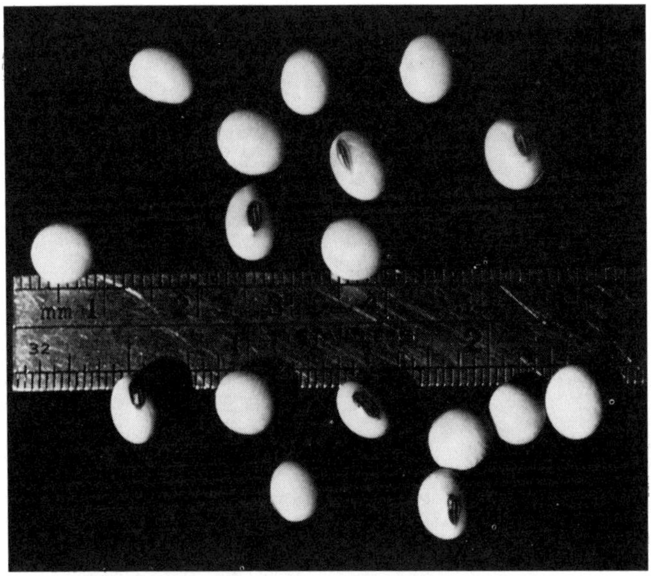

FIG. 1. Photograph of typical soybeans used for production of soy protein products in United States.[13]

composed of elongated pallisade-like cells, which contain both the oil and protein.

Most of the protein is stored in the cotyledon as protein bodies, which are about 2–20 μm in diameter (Fig. 2). The protein bodies survive the grinding of the seed and have been isolated from defatted soybean flour.[8] The oil is located in a smaller structure, which is called the sphereosome (lipid droplet), and is about 0·2–0·5 μm in diameter. The lipid-containing spherosomes are found interspersed between the protein bodies. The structure of part of the soybean is shown in a transmission electron micrograph reproduced in Fig. 3. The protein bodies, the spherosomes and the cell wall can be clearly discerned.

The protein bodies can be classified as storage proteins in contrast to smaller amounts of proteins that comprise an integral part of the cell structure.[15] The storage proteins, which are the major proteins in mature seeds, are few in number and have high molecular weights, i.e. greater than 100 000. In contrast, the structural and functional proteins are far more numerous and have lower molecular weights.[15] They are also usually high in cystine content, and thus may function during inter- and intramolecular bonding.

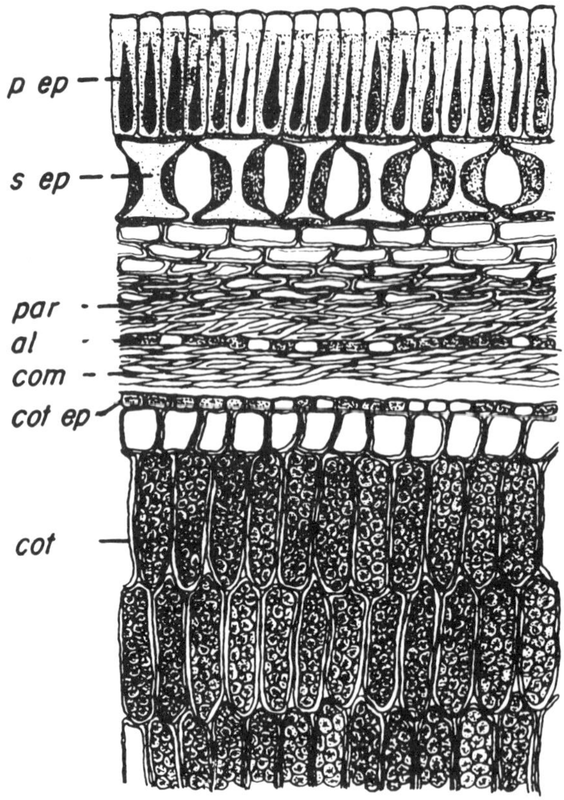

FIG. 2. Drawing of a cross-section of the soybean showing the seed coat or hull and a portion of the cotyledon. Structures in the hull: p ep = palisade cells; s ep = hourglass cells; par = parenchyma; al = aleurone cells; com = compressed cells of the endosperm. Structures in the cotyledon: cot ep = the epidermis of the cotyledon; cot = aleurone cells.[8]

CHEMICAL COMPOSITION OF THE SOYBEAN

The composition of typical soybeans is given in Table 2, and consists of 42% protein, 20% oil, 35% carbohydrate and 5% ash. The carbohydrates can be further subdivided into the soluble and insoluble fractions. The soluble carbohydrates include hexoses, sucrose, raffinose, stachyose and verbascose, while the insoluble fraction contains hemicellulose, cellulose, lignin, pectin and other complex carbohydrates.[13] The 5·5% of

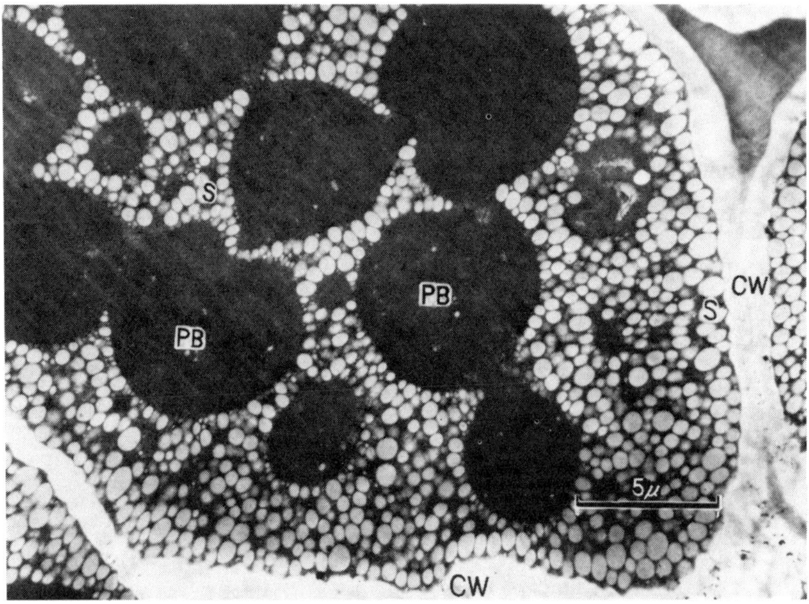

FIG. 3. Transmission electron micrograph of section of mature soybean cotyledon. PB = protein bodies, S = spherosomes; CW = cell wall.[14]

TABLE 2

COMPOSITION (%) OF WHOLE SOYBEAN AND VARIOUS PARTS OF THE SEED[a]

	Protein	*Fat*	*Carbohydrate*	*Ash*
Whole bean	42	20	35	5·5
Cotyledon	43	23	29	5·0
Hull	8·8	1	86	4·3
Hypocotyl	41	11	43	4·4

[a] Adapted from data presented by Wolf and Cowan[8] and Waggle and Kolar.[13]

crude fiber comes from the latter fractions, consisting primarily of lignin, cellulose and hemicellulose.

Mineral and amino acid analyses show that soybeans contain appreciable amounts of a wide spectrum of minerals and amino acids. However, this will be discussed in greater detail later under the nutritional value of various soy protein products, where its significance is of greater importance.

Soy Proteins

Solubility

Soybean proteins are sensitive to physical and chemical treatments that have little or no effect on other food components.[8] Application of moist heat and altering the pH are examples of treatments that can change the physical properties of soy proteins, such as their solubility, viscosity and molecular weight. Such alterations have played an important role in development of soy proteins for human foods and will be discussed in greater detail later.

The most important soy proteins are classified as globulins. Proteins of this class are insoluble near their isoelectric points but dissolve readily on addition of salts, such as sodium or calcium chloride. If the pH, however, is above or below the isoelectric point, globulins will dissolve in aqueous solutions in the absence of salt. Thus, soy proteins have minimum solubility between pH 3·75 and 5·25, whereas they have maximum solubility at pH 1·5–2·5 on the acid side of their isoelectric points and above pH 6·3 on the alkaline side of their isoelectric points. This is clearly illustrated in Fig. 4, which shows the extractability of soy proteins from defatted soybean meal as a function of pH.[16]

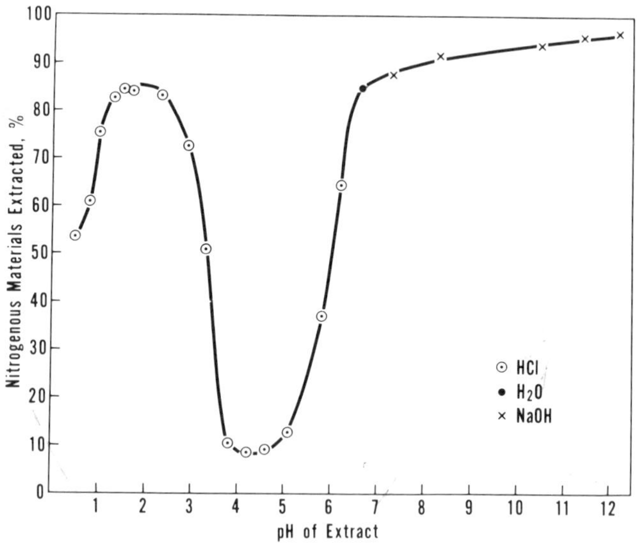

FIG. 4. The extractability of proteins from defatted soybean meal as a function of pH.[8]

About 85% of the protein (nitrogeneous material) in defatted soybean meal is extractable in water at pH 6·5.[8] Adding alkali increases protein extractability by another 5–10%. Addition of acid to lower the pH to 4·2–4·6 results in minimum solubility, as this pH range covers the isoelectric point for the globulins. The insolubility of the globulins at their isoelectric points is used to prepare soy protein isolates, which will be discussed later. Further addition of acid will cause resolubilization of the soybean proteins as the pH drops below the isoelectric points of the globulins.

Characterization

The water-extractable soybean proteins contain a range of different molecular weight components, as shown by their typical pattern of sedimentation on ultracentrifugation (Fig. 5). These proteins have been characterized by a number of workers[17–19] and are now known as the 2 S, 7 S, 11 S and 15 S fractions based on their sedimentation coefficients ($S_{20,w}$).

The proportion of each fraction is given in Table 3, which also shows their major constitutents. Further purification and fractionation will probably show additional components, but the major constituents have

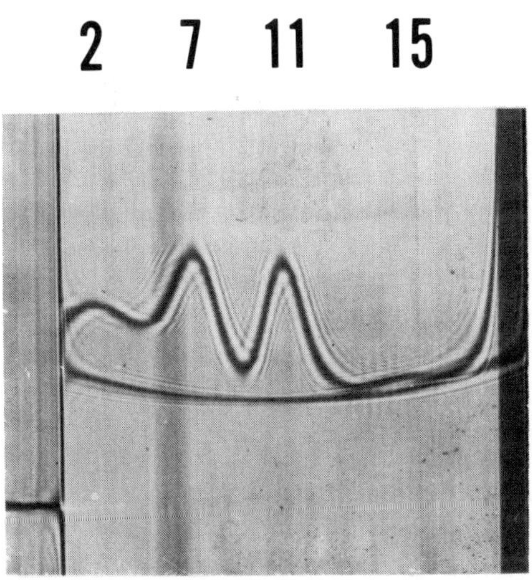

FIG. 5. Ultracentrifuge pattern for water-extractable soybean proteins. Solvent is pH 7·6, 0·5 ionic strength buffer containing 0·01 M mercaptoethanol. Numbers across top correspond to sedimentation coefficients in Svedberg units.[8]

TABLE 3
COMPONENTS IN THE ULTRACENTRIFUGE FRACTION OF WATER-EXTRACTABLE SOYBEAN PROTEINS SHOWING THEIR MOLECULAR WEIGHTS AND PERCENTAGES[8]

Sedimentation value (S_{w20})	*Percent of total*	*Known components*	*Molecular weights*
2S	22	Trypsin inhibitors	8000–21 500
		cytochrome c	12 000
7S	37	Hemagglutinins	110 000
		lipoxygenases	102 000
		β-amylases	61 700
		7S globulin	180 000–210 000
11S	31	11S globulin	350 000
15S	11	—	600 000

undoubtedly been isolated. The 2 S fraction comprises about 22% of the total protein and consists of several trypsin inhibitors, cytochrome c and some unidentified proteins.[8] Their molecular weights range from 8000 to 21 500.

The 7 S fraction accounts for about 37% of the total protein (Table 3) and contains four different types of proteins. These include at least four hemagglutinins, two or more lipoxygenases, β-amylase and a protein, 7 S globulin.[8] The lipoxygenases are of special significance in foods since they catalyze the oxidation of the polyunsaturated fatty acids and may in this way contribute to development of undesirable flavors. The 7 S globulin contains a glycoprotein[20] and comprises over one-half of the 7 S fraction.[21]

The 11 S fraction makes up about 31% of the total protein (Table 3) and so far only a single protein, the 11 S globulin, has been isolated from it. The 11 S globulin has been isolated and purified[18] and shown to have a molecular weight of 350 000. The 15 S fraction makes up 11% of the total soy proteins, but has not been isolated or characterized.[8] Ill-defined peaks are often observed sedimenting just ahead of the main 15 S fraction and have been designated as the >15 S fraction and may be aggregates of this fraction. The molecular weight of the 15 S fraction appears to be over 500 000 on the basis of its sedimentation rate.

Gel filtration of the water-extractable soy proteins results in seven different fractions in contrast to the four obtained by ultracentrifugation.[8] The first fraction appears to contribute to turbidity and has been shown to contain ribonucleic acid.[22] Most of the 11 S fraction is found in the third fraction obtained on gel filtration, which also contains some of the 7 S fraction. The remaining proteins from the 7 S fraction on ultracentrifugation were eluted in gel filtrations fractions 4 to 6, while the 2 S proteins occurred in fractions 7 and 8. The number and complexity of the soybean proteins are clearly illustrated by their gel filtration patterns. Nevertheless, the amount and properties of the 7 S and 11 S globulins make them of the greatest importance from the standpoint of producing edible soy proteins.

Transformations of the 7 S and 11 S Globulins

Both the 7 S and 11 S globulins form disulfide-linked polymers in water extracts prepared from defatted soybean proteins.[8] Further polymerization occurs upon precipitation during isolation of the various soybean proteins. Interactions between the disulfides result in polymer formation, imparting greater viscosity and more turbidity to solutions of the proteins. Both the 7 S and 11 S globulins also are composed of subunits, but behave quite differently. On dialysis of the 11 S protein against 0·01 ionic strength buffer at pH 7·6, its quaternary structure breaks down,[23] while the 7 S protein exists as a dimer under the same conditions.[24] A dimeric structure of two identical monomers made up of six subunits each has been proposed for the 11 S globulin.[24] Disruption of the quaternary structure of the 11 S globulin can also be caused by either high or low pH,[25] by high concentrations of urea,[23, 24] by mixtures of phenol and acetic acid with mercaptoethanol and urea[26] and by temperatures over 80 °C.[26]

There are nine amino terminal residues per mole of 7 S globulin and presumably nine subunits in its quaternary structure.[8] In acid solutions at low salt concentration, the 7 S protein is converted into two slower sedimenting species, which have sedimentation coefficients of 2 S and 5 S. At low pH, salts inhibit the conversion of 7 S globulin into the 2 S and 5 S forms. However, dialysis of the 7 S protein at pH 7·6 and at an ionic strength of 0·5 reverses the sedimentation pattern to produce the 2 S and 5 S proteins. At high pH (0·01 N sodium hydroxide), the 7 S protein changes to form a 0·4 S fraction. These unique changes in structure are exploited in spinning soy protein isolates into fibers and will be discussed later.

Both the 7 S and 11 S globulins are very sensitive to changes in their ionic environment when dissolved in an aqueous system. These effects have been reviewed by Wolf and Cowan,[8] who pointed out that the 7 S protein is especially susceptible to association–dissociation changes as the ionic environment is altered. At pH 7·6 and 0·5 ionic strength, the 7 S globulin has a molecular weight of 180 000–210 000 existing in the monomeric form. Changing the ionic strength to 0·1, however, causes it to sediment as a 9 S component with a molecular weight of 370 000, which is apparently due to dimerization. The 11 S globulin is also converted to a faster sedimenting form on changing the ionic strength from 0·5 to 0·1, but there is a low degree of association. Both the 7 S and 11 S proteins are capable of association–dissociation, but in most food systems the salt concentrations are maintained at low ionic strengths in order to favor association of the two proteins.

Studies have shown that the 7 S and 11 S globulins have different properties, at least in some foods.[8] Tofu (bean curd) made from crude 11 S globulin has a firmer texture than that made from the 7 S crude preparation. This suggests that foods prepared from different soy proteins may have quite different functional properties, which need to be studied further.

Biologically Active Components

There are a number of biologically active components present in the soybean, including enzymes, protease inhibitors, hemagglutinins, goitrogens, estrogens, phytates and other substances, including those compounds that are responsible for flatus. The significance of these compounds in the whole soybean and in various soy protein products has been reviewed in considerable detail.[27–30] Due to the voluminous amount of literature on this subject, the material on each class of biologically active compounds will only be summarized briefly.

Enzymes

Soybeans have been shown to contain a wide spectrum of enzymes, including α- and β-amylases, lipases, lipoperoxidase, lipoxygenase, urease and proteinases.[27] The lipases, lipoperoxidases and lipoxygenases catalyze the oxidation of lipids and may result in flavor changes in improperly processed soybean oil or soy protein products.[27]

Several proteinases have been extracted and isolated from raw soybeans and soy flour.[27] Weil *et al.*[31] obtained six proteinases by water extraction of the whey protein fraction of defatted soy flour, which exhibited papain-like activity. Ofelt *et al.*[32] concluded that the soybean proteinases are not responsible for the softening effect of soy flour on bread doughs by showing that neither oxidizing nor reducing agents had any effect on the proteinase activity of soy flours. Furthermore, the proteinase activity is lower in commercial defatted soy flour than in wheat flour.[27]

Although soybeans have some urease activity, many plant seeds contain greater amounts of urease.[33] The urease activity is largely localized in the hypocotyl, which has nearly twice as much urease activity as is found in the cotyledons.[27] On the other hand, the hulls have a very low level of urease activity.

Protease Inhibitors

Osborne and Mendel[34] first observed that rats grew poorly when fed unheated soybeans. On heating, however, the growth inhibition was largely eliminated. It was later shown that soybeans contain several different factors, which inhibit the action of the digestive enzymes, trypsin and chymotrypsin, thereby decreasing protein digestion.[30] Most of the inhibitory activity is due to the soybean trypsin inhibitor fraction A_2 (SBTI-A_2)[35] SBTI-A_2 was first crystallized and characterized by Kunitz.[36]

Figure 6 demonstrates the mechanism by which the soybean proteases, exert their influence on the utilization of soy proteins.[30] Evidence indicates that soybean trypsin inhibitors accelerate secretion of pancreatic enzymes from the pancreas by means of a feedback mechanism

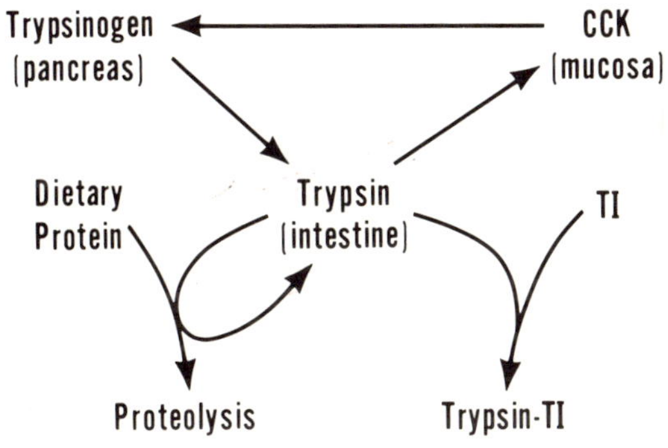

FIG. 6. Regulation of trypsin secretion showing how cholecystokinin (CCK) and trypsin inhibitor (TI) act.[30]

through cholecystokinin. This results in hypertrophy of the pancreas, which is a reversible biological response that does not permanently damage the organ or its function.[30]

According to Rackis[27] raw soybean meal causes growth inhibition, depresses metabolizable energy in the diet, reduces fat absorption, causes enlargement of the pancreas and stimulates hypersecretion of pancreatic enzymes in a number of species. Toasting or heating of the meal improves the nutritional value for a large number of species, including man. The effects of heat and moisture in improving the protein efficiency ratio (PER) of raw soybean meal is clearly shown in Fig. 7. It dramati-

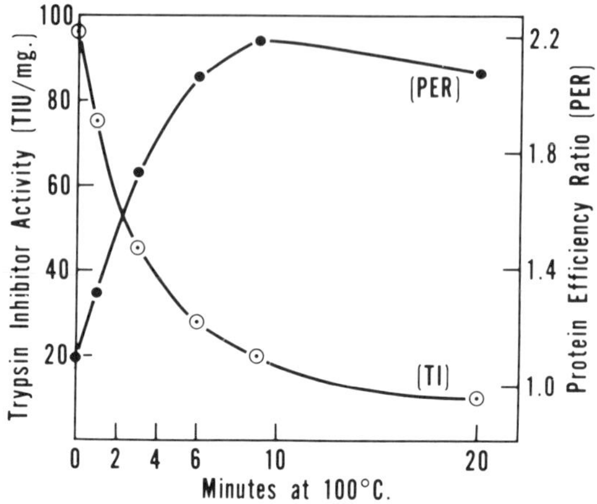

FIG. 7. Effect of heat on trypsin inhibitor activity (TI) and protein efficiency ratio (PER) of soybean meal.[30]

cally demonstrates the effects of heating on destruction of the soybean trypsin inhibitors. Furthermore, it shows that improvement in the PER is greater if heating is carried out at a higher moisture content.

The levels of soybean trypsin inhibitors in a number of different soy protein products are given in Table 4. Although there is some variation in the amount of trypsin inhibitor activity in the different products, it is obvious that all processed products have much lower activity (at least 75% less) than raw soy flour. This clearly illustrates the effectiveness of heat processing in reducing trypsin inhibitor levels. Nevertheless, some products are lower in trypsin inhibitor activity than others. Churella *et al.*[38] have shown that a number of soy-based infant formulas have some residual trypsin inhibitor activity, but do not cause any hypertrophy of the pancreas on feeding them to rats.

Rackis[28] has recently studied the influence of trypsin inhibitory levels on pancreas weights of rats. He found pancreas weights were normal after feeding for 4 weeks on a diet containing raw soy flour in which only 54–68% of the trypsin inhibitor was destroyed. It was calculated that the

TABLE 4
SOYBEAN TRYPSIN INHIBITOR ACTIVITY IN A NUMBER OF SOY PRODUCTS[30]

Item	*TIU/mg*[a]
Raw soy flour	99·0
Toasted soy flour	6–15
Soy protein concentrates	12–26·5
Soy protein isolates	8·5–20·9
Spun soy fibers	12·3
Chicken, meat analog	6·9
Ham, meat analog	10·2
Beef, meat analog	6·5
Textured soy flour	9·8

[a]TIU = trypsin inhibitor units as determined by the method of Kakade *et al.*[37]

average tolerance level (absence of pancreatic hypertrophy on feeding to rats) for soybean trypsin inhibitor was 385 mg/100 g of diet. Furthermore, long-term feeding studies with rats fed commercial preparations of edible soy flour, concentrates or isolates have shown no evidence of pancreatic enlargement. Thus, commercially processed soy proteins used for human foods appear to have no adverse effects on either experimental animals or human beings. This supports the concept that properly processed soy products can be used to supply high quality protein for humans. Nevertheless, more data are needed on the effects of long-term ingestion of soy proteins containing residual levels of trypsin inhibitors on pancreatic function and on histology of the pancreas in man.[28]

Hemagglutinins

Hemagglutinins are named because they cause agglutination or clumping of the red blood cells. Liener[39] has reported that some 800 out of 2663 plant species examined have hemagglutinating activity and are either toxic to animals or inhibit their growth. Soybeans are known to contain several hemagglutinins with a 7-fold variation in the level of activity among 108 different soybean varieties or strains.[40]

Soy hemagglutinins are glycoproteins with a molecular weight of approximately 110 000.[30] They contain about 5% carbohydrate, which consists mainly of mannose and *N*-acetyl-D-glucosamine.

Soybean hemagglutinins were once believed to be responsible for about one-half of the rat growth inhibition observed on feeding raw soybean meal.[41] However, soy hemagglutinins have since been shown to be readily inactivated by heat or pepsin.[42] Birk and Gertler[43] concluded that hemagglutinins contribute but little to the growth impairment associated with feeding of raw soybean meal to chicks and rats. More recent work[44] has conclusively demonstrated that soybean hemagglutinins make only a minor contribution to the poor nutritive value of soy proteins. Thus, soybean hemagglutinins do not appear to be a problem in properly processed soy flour or other soy protein products used for human food.

Goitrogens

Goiter has been observed in rats consuming large quantities of soybean meal.[45] Although addition of iodine effectively eliminated thyroid enlargement, it soon became apparent that soybeans exhibit goitrogenic activity above and beyond that which could be due to their low iodine content.[27] Konijn *et al.*[46] have isolated a goitrogenic agent by fractionating soy flour. It is composed of two or three amino acid residues linked to a sugar moiety. The nature of the goitrogenic substance and its mode of action is not known, although it may prevent intestinal reabsorption of thyroxine, the iodine-containing hormone secreted by the thyroid gland.[30] This goitrogenic substance does, however, cause an increase in iodine uptake by the thyroid and also results in an increase in fecal iodine losses.[47]

The hypertrophy of the thyroid gland appears to be due to an attempt to secrete more thyroxine. Thus, addition of small amounts of iodine to the diet will overcome the goitrogenic effect of soy proteins. Block *et al.*[48] have claimed that the thyroid enlargement associated with high soybean diets is completely reversible on supplying adequate iodine or upon removal of the animals from soybean-based diets.

Estrogens

Soybeans contain a number of isoflavone glucosides including genistin, daidzin and glycitein-7-O-β-glucoside, all of which are known to have estrogenic activity.[30] Knuckles *et al.*[49] have also reported soybeans to contain 120 μg of coumestrol per 100 g, but two-thirds of this is extracted along with the oil. Drane *et al.*[50] concluded that genistin and daidzin are the major contributors to the estrogenic activity of soybeans. They found relatively low levels of estrogenic activity, with values generally falling in the range of 8–37 mg equivalents of diethylstilbestrol (DES) units per gram of soy protein. These values are much lower than those occurring in many other plant proteins. Soybean concentrates and isolates prepared by alcohol extraction exhibit little or no estrogenic activity.[51] Thus, food products made from soy proteins have relatively little estrogenic activity.

Phytates

Phytic acid, which is the hexaphosphate of myoinositol, is an important phosphate compound in plant seeds. Unfortunately many of the phytate salts are poorly digested by animals.[52] Defatted soybean meal contains about 1·5% phytic acid, which is associated mainly with the germ of the intact seed. At acid pH, phytic acid reacts with proteins, and upon neutralization forms insoluble complexes.[52] Thus, processing of soybeans can produce protein–phytic acid–mineral complexes that may reduce the bioavailability of minerals.[53]

Jaffe has reviewed the literature on procedures for eliminating phytate from soy protein products.[52] However, most of the methods are rather difficult to incorporate into current soy protein manufacturing procedures. Soaking of soybeans reduces the phytic acid content through enzymatic hydrolysis by the action of the enzyme phytase, which is an indigenous constituent. In making bread, part of the phytate is hydrolyzed by the enzymatic activity of the yeast. Furthermore, there is evidence that some phytate can be hydrolyzed in the intestinal tract by both animals and man, making the phosphate available, probably by the action of an alkaline phosphatase.

Phytate is tightly bound to soy protein isolates, the binding apparently occurring during heat processing.[52] There is, however, no direct evidence that soy proteins *per se* otherwise influence the bioavailability of minerals.[53] Zinc is the mineral that is most likely to be adversely affected by phytate, although iron, and perhaps copper bioavailability may also be decreased during the processing of soybeans.[52,53] More basic research is needed to ascertain the effects of soy proteins on the bioavailability of minerals for man, and in the development of economically feasible processing procedures that optimize the recovery of the functional soy proteins yet reduce the amount of phytates or their effects of mineral bioavailability.[53] Indiscriminate fortification of soy proteins with minerals is not advisable, however, since it may upset mineral balances and could lead to other mineral deficiencies or excesses.[52]

Oligosaccharides

Flatulence or gas production in the digestive tract is a common problem associated with consumption of soybeans and other leguminous foods.

Rackis[27,29] and Anderson *et al.*[30] have reviewed the problem and suggested processing procedures for soy proteins that may reduce or eliminate it. Elimination of flatulence is a high priority problem for increasing the use of soybeans and other legumes by man, since it is often accompanied by abdominal pain, nausea, cramps, diarrhea, increased intestinal peristalis, borborygmus and social discomfort.[29]

It has been reported that intestinal microflora interact with certain carbohydrates to produce gas, which is primarily carbon dioxide and hydrogen with lesser amounts of methane.[29] An increase in the anaerobic microflora of the intestine is apparently stimulated by foods that are flatus producers.[54,55] The oligosaccharides—verbacose, stachyose and raffinose—are the major contributors to soybean flatulence.[29] These compounds escape digestion and are fermented by the intestinal microflora to produce excess amounts of intestinal gases.

Rackis[29] stated that soy protein processing technology is now being utilized to produce flatus-free products. Hot water treatment, aqueous alcohol extraction or isoelectric protein precipitation insolubilizes the protein fraction and removes the oligosaccharides, thus soy protein concentrates and isolates are essentially free of the flatus-producing constituents. Tempeh and tofu also appear to be essentially free of flatus activity. Soaking, fermentation, enzymatic hydrolysis and germination can be utilized to reduce or eliminate the fermentable carbohydrates responsible for flatulence.[29] It is known that the high molecular weight soybean polysaccharides found mainly in the hull do not normally cause flatulence. However, it is possible that germination may hydrolyze them to form flatus-producing constituents.[29] This could account for the sometimes inconsistent effects of germination on flatulence, which has often been attributed to failure to remove the oligosaccharides.

Unfortunately defatted soy flour, which is the most economical source of soy protein, normally would still contain the flatus-producing oligosaccharides. Though the isolates and concentrates are essentially free of the oligosaccharides responsible for flatulence, the greater amount of processing required for their manufacture increases their cost.

Other Substances

Soybeans may also contain a number of other biologically active compounds, including saponins, possibly sterols and triterpene alochols, and an antithyrotoxic factor, all of which have been discussed by Rackis.[27] Two other substances that have biological activity, i.e. soybean allergenic compounds and lysinoalanine, which is not a constituent of raw soybeans but is often found in processed soy protein products, are discussed briefly below.

Allergenic compounds. Anderson *et al.*[30] have discussed the allergenic substances found in soybeans. They pointed out that problems relating to development of allergies to soybeans may be expected to arise as their use as a human food ingredient increases. Although allergens may have a serious effect on susceptible individuals, they are innocuous when eaten by most people, regardless of the amount consumed.

Glaser and Johnstone[56] pointed out that the incidence of allergenic diseases in children fed soy milk from birth to 6 months was much lower during the first 6 years of life than for siblings and for an unrelated group given cow's milk. Later work, however, has shown only small differences between the incidence of allergenic diseases in children given soy versus cow's milk.[57,58] Soy protein intolerances in children have been observed by a number of investigators,[59–61] which led them to conclude that soy proteins are not hypoallergenic.

Anderson *et al.*[30] after reviewing the literature concluded that less of the population is allergic to soy milk than to cow's milk or to milk made from other legumes. It may be that allergenicity is related to the frequency and amount of exposure to different foods. If this is true, an increase in the incidence of allergies from soy protein foods would be expected as consumption of soy-based food increases. However, it is unlikely to exceed that found for human and cow's milk.

Lysinoalanine. Struthers *et al.*[62] have reviewed the formation, significance and control of lysinoalanine production in soy protein products. Lysinoalanine is formed by abstraction of hydrogen sulfide from cysteine, by dehydration of serine or by hydration of cystine under alkaline conditions to produce dehydroalanine, which then reacts by way of its double bond with the ε-amino group of lysine. This is a well known phenomenon during alkaline treatment of proteins. The quantity of lysinoalanine formed depends on the temperature, the time of exposure to the alkaline pH, the type and concentration of protein, in some instances on the concentration of alkali, and probably on the type of cation in the alkaline solution.[63] Addition of either mercaptoethanol or cysteine to an alkaline protein solution greatly reduces lysinoalanine formation. Some lysinoalanine is probably formed in any protein solution that is subjected to alkaline treatment or in protein products that are subjected to extreme heat.[63] Proper control of temperature and pH during processing can eliminate or reduce formation of lysinoalanine.

Lysinoalanine has not been shown to represent a toxicological hazard to any species except the rat.[63] In the rat, the bound form has no toxicological effect but the free form is an active toxicant. The presence of lysinoalanine in any food in appreciable amounts, either free or bound, indicates destruction of cystine, cysteine and lysine so that the biological availability of the protein may be greatly decreased.[63] The reduced nutritional value of proteins containing relatively high levels of lysinoalanine should be recognized and considered if they are used as major sources of dietary protein. Struthers[63] concluded that soy proteins now on the market do not appear to present any toxicological hazard to man at their current levels of usage.

NUTRITIONAL VALUE OF SOY PROTEINS

The main reason for consumption of soy protein by man is to meet protein requirements, but one cannot ignore the importance of vitamins, minerals, and even energy, since these protein sources are used to replace all or part of other protein foods that must help meet all nutrient needs for humans. Thus, the contribution of soy proteins to each of these classes of nutrients will be briefly reviewed.

Protein

The problems involved in evaluation of protein quality have been reviewed extensively elsewhere.[64–75] Soy proteins for human food have been shown to be of the highest nutritional value of all the well known plant proteins. They are particularly high in their content of lysine (Table 5). Lysine is usually the limiting amino acid in cereal-based diets,[76] so would complement other plant proteins. Properly processed soy flour and grits generally have a PER (protein efficiency ratio) of 2·2–2·3 compared with 2·5 for a casein standard.[77]

In summarizing the nutritional value of soy protein for human food, Williams[77] stated that the limiting amino acid is methionine, while the next limiting one is threonine. Table 5 presents the amino acid composition of soy flour, concentrates and isolates, confirming the low levels for these two amino acids. Williams indicated that supplementation of soy isolates with 1·5% of dl-methionine increases the PER to 2·00–2·45, making it almost equivalent to casein. Not all investigators, however, agree that it is necessary to add methionine to soy diets for children and adults, provided the protein level is adequate.[69b,72,78]

TABLE 5

TYPICAL ESSENTIAL AMINO ACID COMPOSITION OF SOME DIFFERENT SOY PROTEIN PRODUCTS IN g/100 g PROTEIN (MOISTURE-FREE BASIS)[13]

Amino acid	*Soy flour*	*Soy protein concentrate*	*Soy protein isolate*
Isoleucine	4·7	4·8	4·9
Leucine	7·9	7·8	7·8
Lysine	6·3	6·3	6·4
Methionine	1·4	1·4	1·3
Cystine	1·6	1·6	1·3
Phenylalanine	5·3	5·2	5·4
Tyrosine	3·8	3·9	4·3
Threonine	3·9	4·2	3·6
Tryptophan	1·3	1·5	1·4
Valine	5·1	4·9	4·7

Fomon and Ziegler[79] have presented indirect evidence suggesting that supplementation of soy milk diets with methionine improves their utilization by infants. Although Torun[78] stated that evidence supporting the need for supplementing soya proteins for children with methionine is not clear, he concluded that it would be advisable. Earlier Torun stated that it is not practical to use soya as the sole source of protein for preschool children.[80] This view is also supported by Jansen,[71] who pointed out the importance of protein quality in human diets, particularly for young children. Harper[81] has indicated the danger of making drastic changes in dietary intake without fully studying the possible consequences, especially in the diets of young children.

The use of soy proteins in combination with other proteins to meet the nutritional needs of humans is well documented.[65, 66, 82] It has been shown that soy protein can be used to supplement and/or extend meat and fish proteins and to supplement single or mixed vegetable-based protein diets.[66, 67, 83] Soy protein fortified foods have been shown to be useful in relieving malnutrition among specific segments of the population, such as infants and children.[84–87] Thus, mixtures of soy proteins with other food proteins can make a significant contribution to the protein quality of human diets.[67]

Vitamins

Kies[69b] suggested that attention should be focused on the contribution of soy protein foods to the vitamin content of the diet. Liener[76] reviewed the vitamin composition of different soy proteins and suggested that they may play an important role in maintenance of human health. Soy flour, for example, may supply from one-third to one-half of the thiamin, riboflavin and niacin needs of humans if used to furnish one-half of the dietary protein requirement.[76] Heat processing may destroy part of the thiamin as this vitamin is heat labile.

The most notable vitamin deficiency in soy products is the low level of vitamin B_{12},[76] which is virtually absent in all vegetable proteins. According to Liener nutritional studies suggest that raw soybeans are not only deficient in vitamin B_{12} but also contain a heat labile anti-metabolite that increases the requirement of vitamin B_{12}. Since most soybean products consumed by humans are heated, the antimetabolite would not normally be a dietary problem. However, dietary supplementation of soy diets with vitamin B_{12} is necessary.[76]

Minerals

Table 6 presents the mineral composition of several different soy protein products used for human food. According to Waggle and Kolar[13] these

TABLE 6
TYPICAL MINERAL COMPOSITION OF VARIOUS SOYBEAN PRODUCTS (MOISTURE-FREE BASIS)[13]

Element	*Defatted soy flour*	*Soy protein concentrate*	*Soy protein isolate*
Arsenic	0·1 ppm	—	0·2 ppm
Cadmium	0·25 ppm	—	<0·2 ppm
Calcium	0·22%	0·22%	0·18%
Chlorine	0·132%	0·11%	0·13%
Chromium	0·9 ppm	<1·5 ppm	<1·0 ppm
Cobalt	0·5 ppm	—	<1·0 ppm
Copper	23 ppm	16 ppm	12 ppm
Fluorine	1·4 ppm	—	<10 ppm
Iodine	0·01 ppm	0·17 ppm	<10 ppm
Iron	110 ppm	100 ppm	160 ppm
Lead	0·2 ppm	—	<0·2 ppm
Magnesium	0·31%	0·25%	380 ppm
Manganese	28 ppm	30 ppm	17 ppm
Mercury	0·05 ppm	—	<0·5 ppm
Molybdenum	2·6 ppm	4·5 ppm	<3·0 ppm
Phosphorus	0·68%	0·70%	0·76%
Potassium	2·37%	2·1%	960 ppm
Selenium	0·6 ppm	—	0·36 ppm
Sodium	254 ppm	50 ppm	1·1%
Sulfur	0·25%	0·42%	—
Zinc	61 ppm	46 ppm	40 ppm

products contain minerals, such as calcium, iron, copper, phosphorus and zinc, in nutritionally significant amounts. Liener[76] pointed out that the amount and bioavailability of calcium in soybean milk is comparable to that in cow's milk. The calcium content of soybean curd is about 4-fold higher than that of soybean milk because calcium salts are used for precipitating the protein. In whole soybeans, however, much of the calcium is tied up in the form of phytates.

Although soybean proteins contain a high level of phosphorus, about one-half is in combination with phytic acid, the availability of which is not fully known in man.[76] Nevertheless, phytic acid is probably a more serious problem in the bioavailability of zinc, iron and copper than of phosphorus.[52, 53, 88] This, however, has been discussed earlier herein. Kies[69b] has pointed out the paucity of information on the nutritive value of soy proteins for humans, but indicated that we need more knowledge on the bioavailability of the minerals present in different soy protein products.

Energy

Although too little energy is not a common problem in the developed world, the use of whole fat soy products may offer distinct advantages in energy-deficient diets.[66] Table 7 shows the composition of whole soybeans, full fat soy flour and several soy protein products. The data clearly show that whole soybeans and full fat flour may contribute appreciably to energy intake, through both their fat and carbohydrate contents. It also shows that defatted soy flour and soy protein concentrates contain appreciable levels of carbohydrate, even though the fat has been

TABLE 7

TYPICAL COMPOSITION OF SOYBEANS AND VARIOUS SOYBEAN PRODUCTS (%)[13]

	Protein[a]	*Oil*	*Carbohydrates*[b]	*Ash*	*Crude fiber*
Soybeans, whole	42	20	35	5·0	5·5
Full fat soy flour	40	20	35	5·0	5·5
Defatted soy flour	54	1·0	38	6·0	3·5
Soy protein concentrate	70	1·0	24	5·0	3·5
Soy protein isolate	92	0·5	2·5	4·5	0·3

[a]Protein = N × 6·25.
[b]Includes crude fiber.

removed. On the other hand, in countries where too much energy appears to be a problem for humans, the low fat and carbohydrate content of soy protein isolate may make it the product of choice.

The typical carbohydrate composition of soy flour and grits is presented in Table 8. The data show that over half of the carbohydrates are in the insoluble fraction and probably contribute little to energy intake. Furthermore, some 8·6% of the remaining carbohydrates are oligosaccharides and are attacked by bacteria in the digestive tract,[29] thus they probably make little contribution to energy intake. Even though the carbohydrates comprise about 35–38% of soy flour and concentrates, the unavailable portion of this fraction probably results in much less available energy than the composition would indicate. Thus, the fat content or whole soybeans or whole fat soy flour would be the most important energy source in the various soybean foods available to man.

TABLE 8
TYPICAL CARBOHYDRATE COMPOSITION OF SOY FLOUR/SOY GRITS (MOISTURE-FREE BASIS)[13]

	%
Soluble carbohydrates	17
Hexose	Trace
Sucrose	5·7
Raffinose	4·1
Stachyose	4·5
Verbascose	Trace
Insoluble carbohydrates[a]	21
Hemicellulose	
Cellulose	
Lignin	
Pectin	
Other complex carbohydrates	

[a] Value by difference. The amount of each insoluble component is highly variable depending on the method of analysis so individual insoluble carbohydrates are not shown.

Soy Protein and Atherosclerosis

Recent research has demonstrated that replacement of casein with soya as the protein source in cholesterol-free, semi-purified diets for rabbits markedly reduces the level of plasma cholesterol and development of atherosclerosis.[89, 90] The work has since been extended to humans and some reduction in blood serum cholesterol levels has been demonstrated. It has been suggested[90] that partial replacement of animal proteins by soy proteins may reduce the incidence of atherosclerosis in man and may provide a beneficial dietary change for the human population. Although the results of these studies are promising, more information is needed to prove or disprove the anti-atherogenic effects of soy proteins.

PROCESSING OF THE SOYBEAN

Extraction of Oil

Excellent reviews on extraction of the oil from soybeans have been published by Eldridge[91] and Serrato.[92] Before the oil can be extracted, however, the soybeans must be cleaned, cracked and flaked. According to Serrato commercial soybeans may contain up to 5% trash, which must be removed to leave a maximum of 0·5% in the soybeans before processing.[92]

In order to facilitate extraction, the clean beans must be cracked, dehulled and conditioned. Serrato has described the entire operation, which begins by cracking the beans into small pieces averaging about 6–10 US mesh.[92] In actual practice, about 20% are larger than the no. 6 mesh and about 3% are smaller than no. 20 mesh. Dehulling produces about 3–5% of hulls having a maximum residual oil content of 2·5%. The meats next go to a conditioner where the temperature is brought to 65–70 °C, and the moisture content is adjusted to $11 \pm 0{\cdot}5\%$. Serrato concluded that the flaking operation, which rolls the bean meats into thin flakes, is the most important step in the proper extraction of the oil. The flakes should reach the extractor at the correct temperature so as to maintain the extractor temperature at 55 °C.

The flakes must enter the extractor through a vapor seal which prevents the hexane solvent escaping. The extractor consists of a closed vessel designed to wash and extract the oil from the flakes by percolating the solvent through them. Most modern extractors utilize countercurrent extraction since it is a more efficient process. The solution of oil and solvent, which is called miscella, is pumped from the extractor into the evaporator. The hexane is evaporated from the miscella to give an oil with a negative flash-point value of 120 °C. The hexane is re-used while the oil is degummed, refined and bleached.[93]

The remaining hexane is removed from the solvent-extracted flakes in heated vessels. The flakes must be heated while the moisture content is about 20–22% to inactivate urease and to improve the nutritional quality of the resulting soybean meal.[92] This will destroy the trypsin inhibitors as well as improve the flavor and aroma of the meal.[30] The level of urease in the meal is commonly determined to see that the amount of heating is adequate to destroy the trypsin inhibitors.[92] The final heated flakes must have a moisture content of $12 \pm 0{\cdot}5\%$. Properly processed soybean meal will have a final protein content ($N \times 6{\cdot}25$) of 48–50% and 12% moisture. Meals having a lower protein content have not been properly dehulled.[92] The flakes are ground to produce a meal of which 95% will pass a US no. 10 mesh screen with a maximum of 3% passing a no. 80 mesh screen.

Processing of Soybean Meal

Figure 8 presents a flow chart showing the entire process of preparing solvent-extracted soybean meal from raw soybeans, including extraction of the oil into the miscella (oil–hexane mixture). The critical step in controlling the quality of the soybean meal as an animal feed or as an ingredient for human food is the desolventizing–toasting operation.[94] The optimization of the desolventizing–toasting process has been investigated, including factors such as moisture content, temperature and timing, in order to produce high nutritional quality soybean meal.

The effect of temperature and time of the final temperature of soybean meal is presented in Fig. 9. It is notable that the temperature reached over 212°F (100°C) at the end of sparge and then went back down rapidly, reflecting the effects of steam rather than the true meal temperature. Results demonstrated that meal quality was improved by increasing the heating time, the steam jacket temperature and the moisture content. It was also found that the moisture level in the toasting operation was directly affected by the level of hexane in the feed material sent to the toaster. The importance of the toasting temperature and time was verified by measuring urease inactivation and inactivation of the trypsin inhibitors. Mustakas *et al.*[94] demonstrated that residual enzymes, protease inhibitors, protein solubility and meal color can be used as criteria of protein quality and the effectiveness of the toasting process for producing soybean meals.

Edible Soy Flour and Grit Production

Soy flour production has increased from 120 million lb per year in 1960 to over 3 billion lb per year in 1980.[11] The major difference between

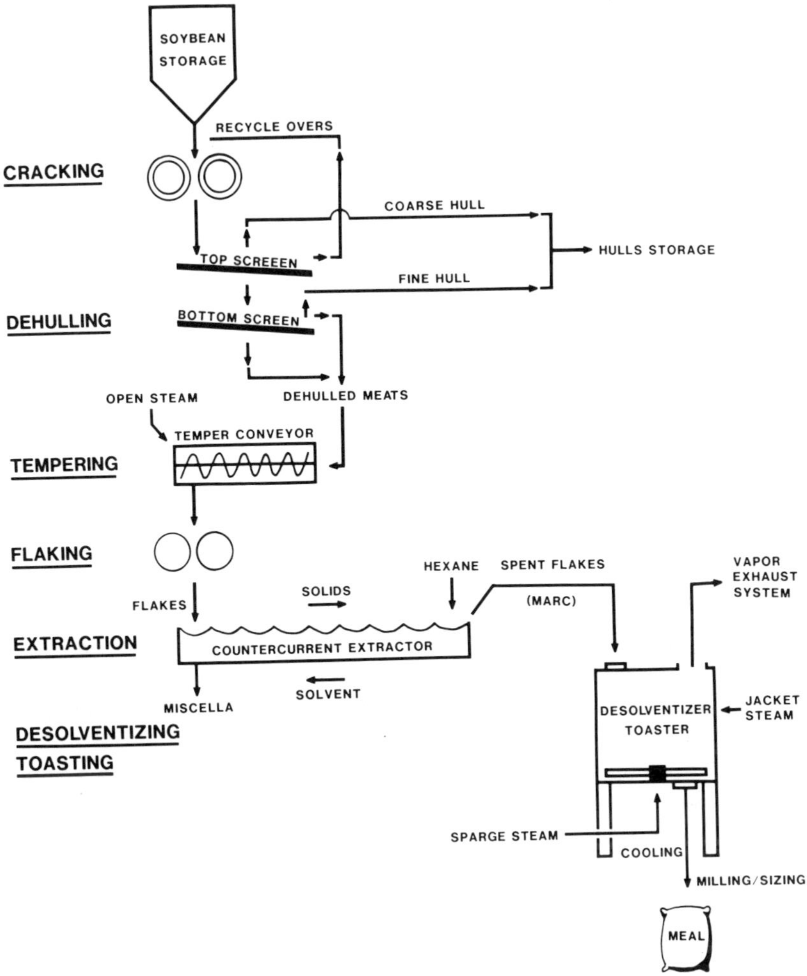

FIG. 8. Flow chart showing the process of preparing solvent-extracted soybean meal from raw soybeans.[94]

production of animal feeds and the production of edible soy flour or further processed products is in the desolventizing process, which uses milder heating conditions for producing desolventized soy flakes. Protein denaturation is minimized with minimum retention times at low temperature and low moisture content. This produces relatively undenatured, light colored soy flakes.[11] The flakes are heated under carefully

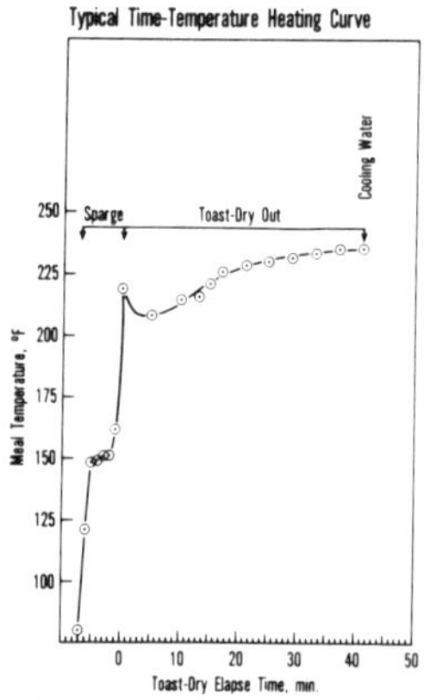

FIG. 9. Effect of temperature and time on the final temperature of soybean meal. Note the effects of sparge on temperature.[94]

controlled conditions to remove any residual solvent and to produce either low, medium or high protein dispersibility index (PDI) flakes still having a light color.

Milligan *et al.* described the processes for desolventizing and heating of the defatted flakes.[11] Most commonly the flakes, which are wet with solvent, are desolventized by passing a superheated vapor stream of the same solvent through them. This evaporates the bulk of remaining solvent from the flakes using turbulent vapor heat transfer at the boiling point of the solvent, and is known as the flash desolventizing system (FDS). The FDS will produce high quality soy flakes with a minimum of protein denaturation and light color. The flakes can be cooked to any desired PDI by simply altering processing conditions of temperature, moisture and time. After flash desolventizing and cooking, the flakes must be cooled prior to further processing into flour and/or grits.

Milligan *et al.* have described the grinding of the soy flakes to produce soy flour and grits.[11] The cooled, desolventized flakes are fed into a hammer mill and ground, after which the finely ground material is conveyed by a mechanical air whizzer to the storage and/or bagging area. The flour typically contains 25% of material that will pass through a 100 or 200 mesh separator. The only difference between flour and grits is in the fineness of grinding.

Soy Concentrates and Isolates

Soy protein concentrates and isolates differ in their protein content, the concentrates containing a minimum of 70% protein (N × 6·25) and the isolates at least 90% protein on a dry basis.[95, 96] The methods used in separating and concentrating the soy protein also differ for concentrates and isolates as can be seen by comparing Figs. 10 and 11, although both procedures generally use defatted soy flakes or flour as the starting material.

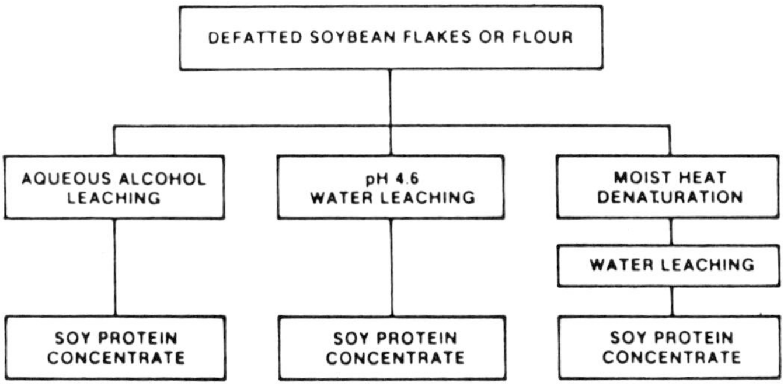

FIG. 10. Procedure for preparing soy concentrates from soy flakes.[96]

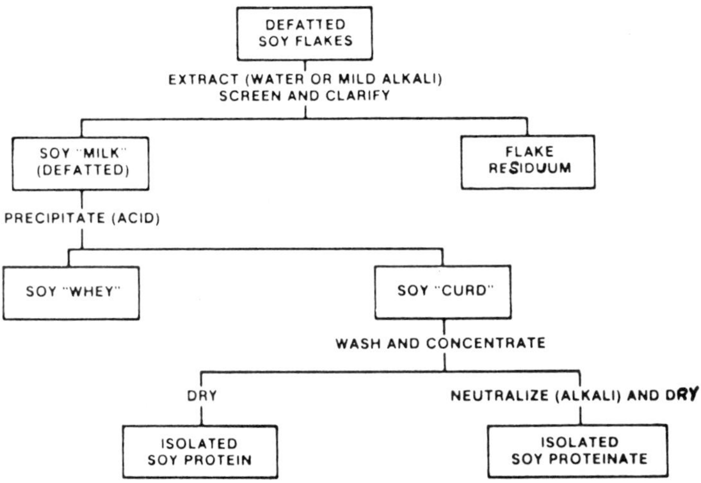

FIG. 11. Procedure for preparation of soy protein isolates. Reproduced from Ref. 3, p. 294.

Soy Protein Concentrates

Ohren[96] outlines three different procedures for preparing soy concentrates. One process uses a 60–80% alcohol extraction; and another involves extraction of the solubles with water acidified to a pH of 4·5 using food grade acid. The third method utilizes water extraction of soy flakes that have been heated or toasted to denature the protein before the extraction process.

The primary objectives in each of these processes are immobilization of the major protein fractions while simultaneously extracting the soluble carbohydrates, the remaining nitrogenous compounds and the other minor constituents from the soy flakes. All of these procedures for preparing soy concentrates are shown in Fig. 10. Details of commercial processes are proprietary secrets and are slightly different for various processors.

The method of extraction of the protein from the flakes can greatly influence the color of the finished concentrates.[96] Generally, concentrates produced from the water extraction procedure are darker in color since the flakes are toasted to denature the protein before extracting the solubles.

Following removal of the solubles, the protein residue is dried in conventional dryers. The wet product from the acid extraction procedure can be dried at its isoelectric point or it can be further modified by neutralization and/or additional heating before drying.[96] Soy concentrates can also be prepared from full fat dehulled beans, but the product is not produced commercially.

Soy protein concentrates have improved flavor characteristics, since the soluble carbohydrates that are responsible not only for the beany flavor, but also for flatulence, have been removed. The oligosaccharides (stachyose, raffinose and verbascose), which have already been discussed as being chiefly responsible for flatulence, are largely removed during the desugaring process.[77] Soy protein concentrates may also be used for further processing, often providing the raw material for thermoplastic extrusion.

Soy Protein Isolates

In preparing soy isolates, the first step involves aqueous extraction of protein from the defatted flakes at neutral or elevated pH (Fig. 11). This separates the protein from the insoluble residue and it is then precipitated from the liquor by acid.[96] The separated curd is washed and concentrated and either dried as soy protein isolate or neutralized with food grade acid to form the isolated soy proteinate and then dried.

Specific functional properties can be achieved by subjecting the protein to various chemical, thermal or enzymatic treatments before drying.[96] These may include modifications to improve whipping ability, gelation, protein binding, solubility, fiber formation, fat and oil emulsification and water absorption, which have advantages in specific food systems. Solubility of the soy protein isolates can be altered by changes in pH, heat, ionic strength, multivalent ions, by interacting with other food ingredients and by use of different solvents.[96] Different combinations of these treatments may be utilized to produce products having specific solubility characteristics.

Chemical analysis of soy isolates reveals that they are essentially pure protein with a minimal ash content and contain only traces of crude fiber.[97] The growth-depressing constituents and the flatulence factors are largely removed during washing of the curd, so that these compounds are not a problem in foods containing soy protein isolates.

Spun Soy Proteins

Boyer was granted the first patent for producing spun soy protein fibers.[98] Figure 12 shows a diagram of the process for producing spun soy protein fibers. The process involves solubilizing the soy protein in

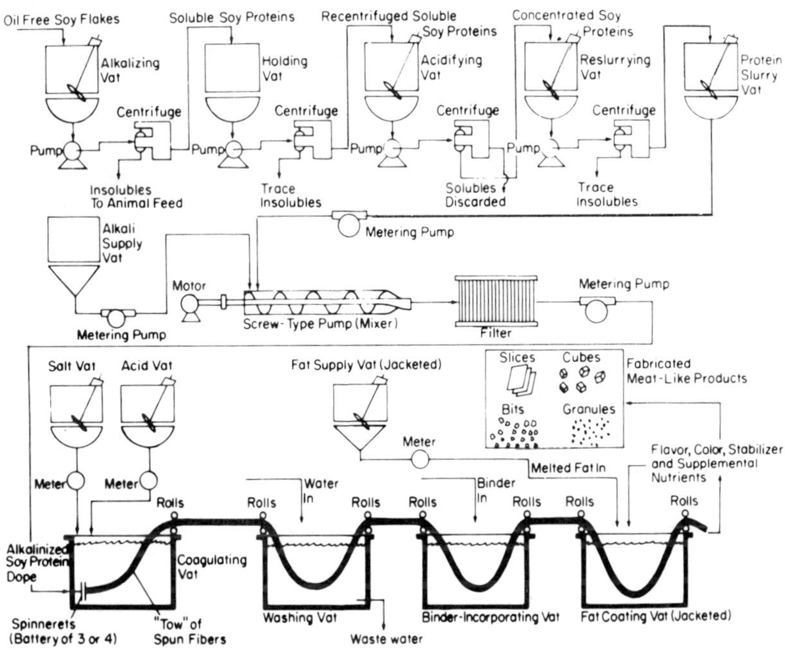

FIG. 12. Diagram showing process for producing spun soy protein fibers.[99]

alkali and forcing the solution through a spinneret similar to the type used in production of synthetic fibers. In this process, the solution of soy protein is continually extruded into an acid bath where fibers are formed. The tow (bundle of fibers) is then placed in a second heated bath, where the fibers are heated and stretched to reduce their diameter to about 75 μm. Typical spun soy protein fibers are shown in Fig. 13. Coloring, flavoring material and a binder (usually egg albumen) are added to produce the desired texture, color and flavor of products produced from spun protein.[97]

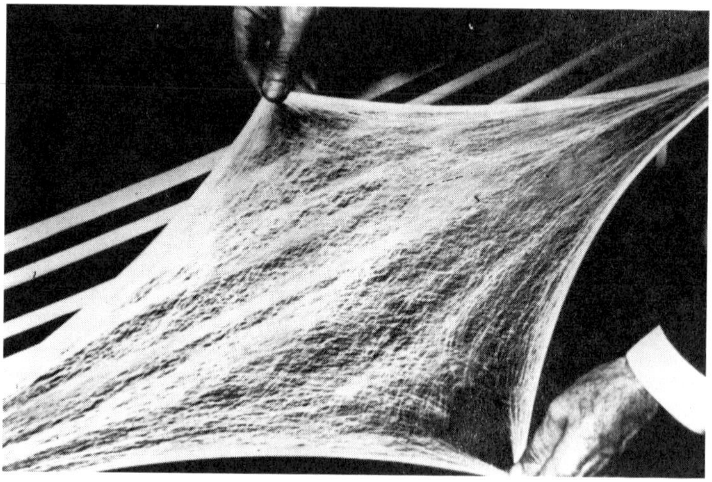

FIG. 13. Photograph showing structure of spun soy protein fibers.[77]

The starting material is generally a soy isolate, although concentrates and even soy flour may sometimes be used. The final products are generally made into meat analogs, such as imitation bacon bits, beef, ham, chicken and seafoods. Some of these products have found a limited market, such as imitation bacon bits, but difficulties in duplicating the natural flavor and the high cost of production have resulted in limitations on its use. Currently, there are only a few producers of spun soy fibers.[100]

Textured Soy Products

Campbell has reviewed the procedure for producing textured soy proteins,[100] which involves a process known as thermoplastic extrusion.

Figure 14 diagramatically shows the process of thermoplastic extrusion as utilized in producing textured soy proteins. In this process, the textured soy proteins are produced by continuous extrusion of a flowable, moist plastic mass of soy proteins through an orifice, forming a texturized product under high pressure and temperature. As the protein passes through the die (orifice), the superheated mixture expands suddenly to produce a porous structure, giving it the name textured soy protein.

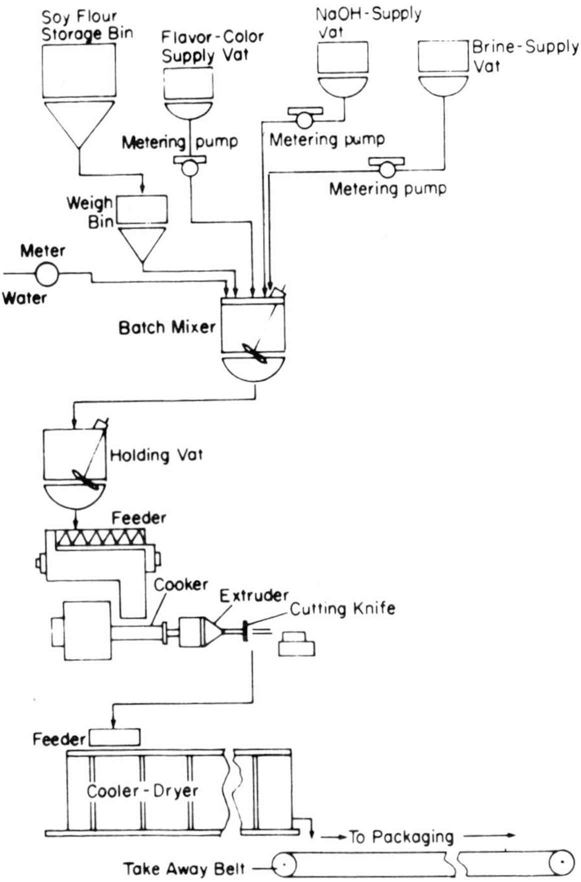

FIG. 14. Diagram showing the process for production of textured soy proteins by thermoplastic extrusion.[99]

To produce texturized soy protein, there must be:

(a) a continuous supply of raw material fed to a cooker–extruder;
(b) preconditioning of the protein feed with water during its conveyance to the extruder;
(c) application of a uniform amount of moisture;
(d) a cooker–extruder that will transform the protein to a dough, give it a colloidal structure and subject it to a high temperature;
(e) the extruder *per se*, which will force the dough into the atmosphere to give it a desired shape and structure;
(f) a drying and cooling system to dry out and cool the finished soy protein.[100]

Figure 15 shows the structure of textured soy protein as seen through the scanning electron microscope.

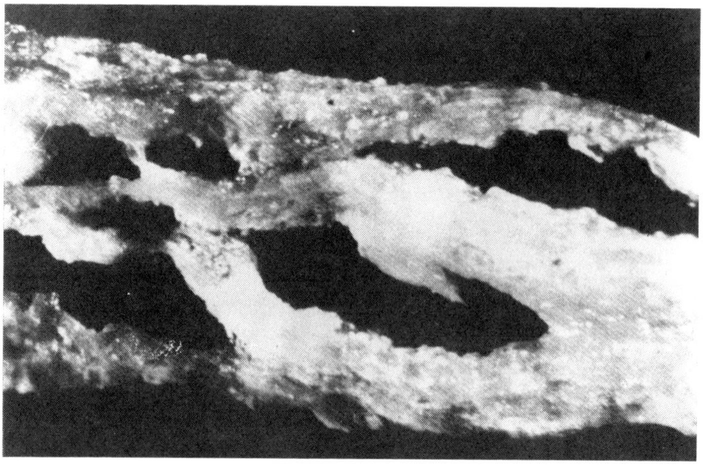

FIG. 15. Scanning electron micrograph of textured soy protein showing its structure. (Courtesy Archer Daniels Midland Company, Decatur, Illinois.)

Soy flour is commonly used to produce textured soy products because the cost is relatively low.[100] Soy concentrates and isolates can also be used. There are many different textured soy protein products on the market, methods of production being varied to produce different physical characteristics in the final products. Most of these products are used as meat extenders, especially for use in ground meat and sausages.

A modification of textured soy proteins has resulted in the development of a textured soy isolate, which has been described by Frederiksen and Heusdens[101] and by Hoer.[102] The process produces soy protein filaments continuously from a heated slurry of soy protein isolate under pressure. The slurry has a solids content of up to 35% by weight. The pH is increased to make the proteinaceous material sufficiently reactive to form elongated protein filaments, which is accomplished by using hydroxides and salts of polyvalent linking agents and phosphate ions. The slurry is placed under pressure and conducted through a heat exchange zone to raise the temperature so it will form elongated protein fibers on passing through an orifice and cooling.

Textured soy protein isolate is used mainly as a meat extender, where it is particularly useful in imparting structure and texture to products lacking textural and structural characteristics. It has been found to be especially useful in mechanically deboned meat and poultry products, which can be used to produce poultry and sausage rolls. The fibrous nature of textured soy isolate can also make it useful in producing meat analogs, but like other meat analogs these products would also need added flavoring.

FOOD USES OF SOY PROTEIN PRODUCTS

Soy proteins are used in many different types of food products. Table 9 shows an estimate of the amount of different soy products that has been or is expected to be utilized for human food in the USA, from 1967 to 2000. Although the data are fragmentary, they do show that there has been a steady increase in the use of all soy protein products for human food. The estimates suggest that there will be a rapid increase in consumption of soy proteins in the United States through the year 2000.

Soy proteins have been used in a variety of foods, including cereal grain foods, dairy-type foods, snack foods, soups, beverages and meat, poultry and fish products. Some of these uses will be briefly reviewed, but coverage cannot be comprehensive.

Cereal Grain Foods

Waggle and Kolar[13] and Dubois and Hoover [103] have reviewed the use of soy proteins in cereal grain products. The largest usage of soy flour has traditionally been in bakery products, which reportedly used some 132 million lb in 1978. It may be used as defatted or full fat flour and as either high fat or lecithinated flour. These flours are widely used in

TABLE 9
ESTIMATED SOY PROTEIN CONSUMPTION BY HUMAN BEINGS IN USA FROM 1967 TO 2000 (IN MILLION KILOGRAMS)[a]

Year	*Flour and grits*	*Concentrate*	*Isolate*	*Textured soy protein*	*TSP in school lunch*
1967	50	8–14	10–16		
1969	55	15	17		
1970		11–14	9–11	11	
1972				23	4
1973				52	5
1974					8
1975				85	51
1978	455			227	
1980	730			490	82
1985				2034	95
1990				4408	122
2000				5539	158

[a] Adapted from Ref. 97 (Copyright © 1976, American Institute of Biological Sciences) using other more recent estimates.

different bakery products being commonly added to bakery items up to a level of 3% by weight. This improves crumb body and resilience, as well as color and toasting characteristics, as a result of its sugar content. Lecithinated soy flour can be used to replace some of the eggs in products such as doughnuts, while full fat soy flour is generally used to enrich sponge and pound cakes.

The greatest usage of soy proteins by the bakery industry is as components of other ingredients, such as milk replacers, according to Dubois and Hoover.[103] Defatted soy flours are commonly used as ingredients in the replacers but concentrates and isolates may also be used. The ingredients used in milk replacers often include sweet dairy whey, defatted soy flour, soy concentrates, soy isolates, corn flour and non-fat dry milk in various combinations. Use of soy proteins in some cereal products tends not only to improve the functional properties but also the nutritional value of the products.

Dairy-type Foods

Soy milk was discussed earlier under nutritional value. It is commonly used as a substitute for either human or cow's milk for babies or young children who have allergies to the former types of milk.[78–80, 87] With some special supplementation it has proved to be a good substitute for cow's milk.

Soy proteins are also widely used in dairy whiteners and artificial creams, either sour or sweet. Waggle and Kolar have discussed these uses in a recent review.[13] Because of their good whipping characteristics soy proteins may be used in frozen whipped toppings. Isolated soy proteins are also used in yogurt, artificial cheese and cheese dips, as well as in some complete breakfast dry mixes to which milk is added before use. Thus, soy protein products are finding many uses in dairy-type foods.

Snack Foods and Soups

Isolated soy proteins and soy concentrates are used in snack foods and cereals.[13] Small quantities of whole soybeans are heated and sold as soy nuts. The use of soy proteins in beverages, soups and a variety of other products has been discussed by a number of workers.[13,104,105] The use of soy proteins in these products is currently limited but the potential for increasing consumption seems to be present.

Meat, Poultry and Fish Products

The relatively high cost of meat, poultry and fish proteins has made the development of extenders and substitutes for these products particularly attractive.[13,97,106] Although meat analogs have been produced in small quantities for special groups who do not eat meat for religious or other reasons, these products have largely failed because of economics, lack of consumer acceptance and their poor flavor.[100] Horan[12] has detailed some of the questions that must be asked relative to the use of spun soy proteins in production of meat analogs.

Even though meat analogs have not been used in appreciable amounts, the use of meat extenders has grown rapidly (Table 9) and is expected to continue to do so. Most of the growth has been due to the acceptance of soy proteins as meat extenders for the school lunch program in the United States.[97] Although not publicly announced, the recent decision to allow use of soy protein extenders in all ground beef used by the Armed Forces of the United States[107] will undoubtedly greatly increase the usage of soy proteins for human food.

Inclusion of soy proteins in ground meat and sausage products has been singularly successful in expanding consumption of soy proteins without any serious effects upon acceptability. Waggle and Kolar[13] and Waggle *et al.*[106] have reviewed the functional and nutritional advantages of adding soy proteins to ground meat and sausages but have indicated the importance of limiting the levels used so as not to impair the acceptability of the extended meat, poultry and fish products. They also discussed the use of soy isolates for injection into intact meat cuts to increase their protein content.

Oriental Soy Proteins

Fukushima[108] recently reviewed the production and use of soy sauce, tofu and other fermented soy products. Although these products have been used for centuries in the Orient, they are minor foods in westernized societies so will not be discussed here. It is possible, however, that the popularity of these foods may increase in the future, thus adding to the demand for soy proteins.

REFERENCES

1. MORSE, W. J. (1950). In: *Soybeans and Soybean Products* (Markley, K. S., Ed.). Vol. I. p. 3. Interscience Publishers, New York.
2. BURNETT, R. S. (1951). In: *Soybeans and Soybean Products* (Markley, K. S., Ed.). Vol. II, p. 949. Interscience Publishers, New York.
3. SMITH, A. K. and CIRCLE, S. J. (1978). In: *Soybeans: Chemistry and Technology* (Smith, A. K. and Circle, S. J., Eds.). Vol. 1, p. 1. Avi Publishing Co., Westport, Connecticut.
4. MCARTHUR, W. C. (1980). *Soybean Production Practices and Costs in the United States.* University of Georgia Agr. Expt. Sta. Res. Rep. 360.
5. HAYWARD, J. W., STEENBOCK, H. and BOHSTEDT, G. (1936). *J. Nutr.*, **11**:219.
6. HAYWARD, J. W., STEENBOCK, H. and BOHSTEDT, G. (1936). *J. Nutr.*, **12**:275.
7. THOMPSON, W. N. (1981). *J. Am. Oil Chemists' Soc.*, **58**:140.
8. WOLF, W. J. and COWAN, J. C. (1975). *Soybeans as Food Source*, revised edition. CRC Press, Cleveland, Ohio.
9. WALTER, A. S. (1976). In: *World Soybean Research* (Hill, L. D., Ed.). p. 657. Interstate Printers and Publishers, Inc., Danville, Illinois.
10. WRIGHT, K. N. (1981). *J. Am. Oil Chemists. Soc.*, **58**:294.
11. MILLIGAN, E. D., AMLIE, J. H., REYES, J., GARCIA, A. and MEYER, B. (1981). *J. Am. Oil Chemists' Soc.*, **58**:331.
12. HORAN, F. E. (1976) In: *World Soybean Research* (Hill, L.D., Ed.). p. 775. Interstate Printers and Publishers, Inc., Danville, Illinois.
13. WAGGLE, D. H. and KOLAR, C. W. (1979). In: *Soy Protein and Human Nutrition* (Wilcke, H. L., Hopkins, D. T. and Waggle, D. H., Eds.). p. 19. Academic Press, New York.
14. SAIO, K. and WATANABE, T. (1986). *J. Food Sci. Technol. Japan*, **15**:290.
15. MARTINEZ, W. H. (1979). In: *Soy Protein and Human Nutrition* (Wilcke, H. L., Hopkins, D. T. and Waggle, D. H., Eds.). p. 53. Academic Press, New York.
16. SMITH, A. K. and CIRCLE, S. J. (1978). In: *Soybeans: Chemistry and*

Technology (Smith, A. K. and Circle, S. J., Eds.). Vol. 1, p. 3. Avi Publishing Co., Westport, Connecticut.
17. NAISMITH, W. E. F. (1955). *Biochim. Biophys. Acta*, **16**:203.
18. WOLF, W. J. and BRIGGS, D. R. (1959). *Arch. Biochem. Biophys.*, **85**:186.
19. WOLF, W. J. (1969). *Baker's Dig.*, **43**(5):30.
20. KOSHIYAMA, I. (1969). *Arch. Biochem. Biophys.*, **130**:370.
21. WOLF, W. J. and SLY, D. A. (1967). *Cereal Chem.*, **44**:653.
22. OBARA, T. and KIMURA, M. (1967). *J. Food Sci.*, **32**:531.
23. WOLF, W. J. and BRIGGS, D. R. (1958). *Arch. Biochem. Biophys.*, 76:377.
24. CATSIMPOOLAS, N. (1969). *FEBS Letters*, 4:259.
25. CATSIMPOOLAS, N., CAMPBELL, T. G. and MEYER, E. W. (1969). *Arch. Biochem. Biophys.*, **131**:577.
26. CATSIMPOOLAS, N., EKENSTAM, C., ROGERS, D. A. and MEYER, E. W. (1968). *Biochim. Biophys. Acta*, **168**:122.
27. RACKIS, J. J. (1978). In: *Soybeans: Chemistry and Technology* (Smith, A. K. and Circle, S. J., Eds.). Vol. 1, p. 158. Avi Publishing Co., Westport, Connecticut.
28. RACKIS, J. J. (1981). *J. Am. Oil Chemists' Soc.*, **58**:495.
29. RACKIS, J. J. (1981). *J. Am. Oil Chemists' Soc.*, **58**:503.
30. ANDERSON, R. L., RACKIS, J. J. and TALLENT, W. H. (1979). In: *Soy Protein and Human Nutrition* (Wilcke, H. L., Hopkins, D. T. and Waggle, D. H., Eds.). p. 209. Academic Press, New York.
31. WEIL, J., PENSKY, A. and GROSSMAN, S. (1966). *Cereal Chem.*, **43**:392.
32. OFELT, C. W., SMITH, A. K. and MILLS, J. M. (1955). *Cereal Chem.*, **32**:53.
33. TAI, K. (1953). *Shiga Agr. College Sci. Rpt. Series*, **1**:55.
34. OSBORNE, T. B. and MENDEL, L. B. (1917). *J. Biol. Chem.*, **32**:369.
35. RACKIS, J. J., SASAME, H. A., MANN, R. K., ANDERSON, R. L. and SMITH, A. K. (1962). *Arch. Biochem. Biophys.*, **98**: 471.
36. KUNITZ, M. (1946). *J. Gen. Physiol.*, **29**:149.
37. KAKADE, M. L., RACKIS, J. J., MCGHEE, J. E. and PUSKI, G. (1974). *Cereal Chem.*, **51**:376.
38. CHURELLA, H. B., YAO, B. C. and THOMPSON, W. B. C. (1976). *J. Agric. Food Chem.*, **24**:393.
39. LIENER, I. E. (1974). *J. Agric. Food Chem.*, **22**:17.
40. KAKADE, M. L., SIMONS, N. R., LIENER, I. E. and LAMBERT, J. W. (1972). *J. Agric. Food Chem.*, **20**:87.
41. LIENER, I. E. (1953). *J. Nutr.*, **49**:527.
42. LIENER, I. E. (1958). *J. Biol. Chem.*, **233**:401.
43. BIRK, Y. and GERTLER, A. (1961). *J. Nutr.*, **75**:379.
44. TURNER, R. H. and LIENER, I. E. (1975). *J. Agric. Food Chem.*, **23**:184.
45. WILGUS, H. S., Jr., GASSNER, F. X., PATTON, A. R. and GUSTAVSON, R. G. (1941). *J. Nutr.*, **22**:43.
46. KONIJN, A. M., GERSHON, B. and GUGGENHEIM, K. (1973). *J. Nutr.*, **103**:378.
47. BECK, R. N. (1958). *Endocrinology*, **62**:587.
48. BLOCK, R. J., MANDL, R. H., HOWARD, H. W., BAUER, C. D. and ANDERSON, D. W. (1961). *Arch. Biochem. Biophys.*, **93**:15.
49. KNUCKLES, B. E., DEFREMERY, D. and KOHLER, G. O. (1976). *J. Agric. Food Chem.*, **24**:1177.

50. DRANE, H. M., PATTERSON, D. S. P., ROBERTS, B. A. and SABA, N. (1980). *Food Cosmet. Toxicol.*, **18**:425.
51. NASH, A. M., ELDRIDGE, A. C. and WOLF, W. J. (1967). *J. Agric. Food Chem.*, **15**:102.
52. JAFFE, G. (1981). *J. Am. Oil Chemists' Soc.*, **58**:493.
53. ERDMAN, J. W., Jr. and FORBES, R. M. (1981). *J. Am. Oil Chemists' Soc.*, **38**:489.
54. RACKIS, J. J., HONIG, D. H., SESSA, D. J. and STEGGARDA, F. R. (1970). *J. Agric. Food Chem.*, **18**:977.
55. ROCKLAND, L. B., GARDINER, B. L. and PIECZARKA, D. (1970). *J. Food Sci.*, **34**:411.
56. GLASER, J. and JOHNSTONE, D. E. (1953). *J. Am. Med. Assoc.*, **153**:620.
57. BROWN, E. B., JOSEPHSON, B. M., LEVINE, H. S. and ROSEN, M. (1969). *Am. J. Dis. Children*, **117**:693.
58. HALPERN, S. R., SELLARS, W. A., JOHNSON, R. B., ANDERSON, D. W., SAPERSTEIN, S. and REISCH, J. S. (1973). *J. Allergy Clin. Immunol.*, **51**:139.
59. KUROUME, T., OGURI, M., MATSUMURA, T., IWASAKI, I., KANBE, Y., YAMADA, T., KAWABE, S. and NEGISHI, K. (1976). *Ann. Allergy*, **37**:41.
60. WHITINGTON, P. F. and GIBSON, R. (1977). *J. Pediatr.*, **59**:730.
61. HALPIN, T. C., BYRNE, W. J. and AMENT, M. E. (1977). *J. Pediatr.*, **91**:404.
62. STRUTHERS, B. J., DAHLGREN, R. R., HOPKINS, D. T. and RAYMOND, M. L. (1979). In: *Soy Protein and Human Nutrition* (Wilcke, H. L., Hopkins, D. T. and Waggle, D. H., Eds.). p. 235. Academic Press, New York.
63. STRUTHERS, B. J. (1981). *J. Am. Oil Chemists' Soc.*, **58**:501.
64. BODWELL, C. E. (1977). In: *Evaluation of Proteins for Humans* (Bodwell, C. E., Ed.). p. 119. Avi Publishing Co., Westport, Connecticut.
65. BRESSANI, R. (1977). In: *Evaluation of Proteins for Humans* (Bodwell, C. E., Ed.). p. 81. Avi Publishing Co., Westport, Connecticut.
66. BRESSANI, R. (1977). In: *Evaluation of Proteins for Humans* (Bodwell, C. E., Ed.). p. 204. Avi Publishing Co., Westport, Connecticut.
67. BRESSANI, R. (1981). *J. Am. Oil Chemists' Soc.*, **58**:392.
68. HACKLER, L. R. (1977). In: *Evaluation of Proteins for Humans* (Bodwell, C. E., Ed.). p. 55. Avi Publishing Co., Westport, Connecticut.
69*a*. KIES, C. V. (1977). In: *Evaluation of Proteins for Humans* (Bodwell, C. E., Ed.). p. 162. Avi Publishing Co., Westport, Connecticut.
69*b*. KIES, C. V. (1979). In: *Soy Protein and Human Nutrition* (Wilcke, H. L., Hopkins, D. T. and Waggle, D. H., Eds.). p. 325. Academic Press, New York.
70. YOUNG, V. R. and SCRIMSHAW, N. S. (1977). In: *Evaluation of Proteins for Humans* (Bodwell, C. E., Ed.). p. 11. Avi Publishing Co., Westport, Connecticut.
71. JANSEN, G. R. (1979). In: *Soy Protein and Human Nutrition* (Wilcke, H. L., Hopkins, D. T. and Waggle, D. H., Eds.). p. 149. Academic Press, New York.
72. SCRIMSHAW, N. S. and YOUNG, V. R. (1979). In: *Soy Protein and Human Nutrition* (Wilcke, H. L., Hopkins, D. T. and Waggle, D. H., Eds.). p. 121. Academic Press, New York.
73. BODWELL, C. E. and MARABLE, N. L. (1981). *J. Am. Oil Chemists' Soc.*, **58**:475.

74. COOK, D. A. (1981). *J. Am. Oil Chemists. Soc.*, **58**:464.
75. LIENER, I. E. (1981). *J. Am. Oil Chemists' Soc.*, **58**:406.
76. LIENER, I. E. (1978). In: *Soybeans: Chemistry and Technology* (Smith, A. K. and Circle, S. J., Eds.). Vol 1, p. 203. Avi Publishing Co., Westport, Connecticut.
77. WILLIAMS, L. D., (1970). In: *Symposium on Vegetable Proteins* (*Protein Symposium 3*). The British Food Manufacturing Industries Research Association, London. p. 23.
78. TORUN, B., VITERI, F. E. and YOUNG, V. R. (1981). *J. Am. Oil Chemists' Soc.*, **58**:400.
79. FOMON, S. J. and ZIEGLER, E. E. (1979). In: *Soy Protein and Human Nutrition* (Wilcke, H. L., Hopkins, D. T. and Waggle, D. H., Eds.). p. 79. Academic Press, New York.
80. TORUN, B. (1979). In: *Soy Protein and Human Nutrition* (Wilcke, H. L., Hopkins, D. T. and Waggle, D. H., Eds.). p. 101. Academic Press, New York.
81. HARPER, A. E. (1977). *J. Nutr. Educ.*, **9**:154.
82. BRESSANI, R. NAVARRETE, D. A., ELIAS, L. G. and BRAHAM, J. E. (1979). In: *Soy Protein and Human Nutrition* (Wilcke, H. L., Hopkins, D. T. and Waggle, D. H., Eds.). p. 313. Academic Press, New York.
83. HOPKINS, D. T. and STEINKE, F. H. (1981). *J. Am. Oil Chemists' Soc.*, **58**:452.
84. CHAVEZ, J. F. (1981). *J. Am. Oil Chemists' Soc.*, **58**:448.
85. SGARBIERI, V. C., GARRUTI, R. and GUZMAN, E. C. (1981). *J. Am. Oil Chemists' Soc.*, **58**:522.
86. AGUILERA, J. M. and LUSAS, E. W. (1981). *J. Am. Oil Chemists' Soc.*, **58**:514.
87. TORUN, B. (1981). *J. Am. Oil Chemists' Soc.*, **58**:460.
88. O'DELL, B. L. (1979). In: *Soy Protein and Human Nutrition* (Wilcke, H. L., Hopkins, D. T. and Waggle, D. H., Eds.). p. 187. Academic Press, New York.
89. CARROLL, K. K., HUFF, M. W. and ROBERTS, D. C. K. (1979). In: *Soy Protein and Human Nutrition* (Wilcke, H. L., Hopkins, D. T. and Waggle, D. H., Eds.). p. 261. Academic Press, New York.
90. CARROLL, K. K. (1981). *J. Am. Oil Chemists' Soc.*, **58**:416.
91. ELDRIDGE, A. C. (1978). In: *Soybeans: Chemistry and Technology* (Smith, A. K. and Circle, S. J., Eds.). Vol. 1, p. 144. Avi Publishing Co., Westport, Connecticut.
92. SERRATO, A. G. (1981). *J. Am. Oil Chemists. Soc.*, **58**:157.
93. WIEDERMAN, L. H. (1981). *J. Am. Oil Chemists. Soc.*, **58**:159.
94. MUSTAKAS, G. C., MOULTON, K. J., BAKER, E. C. and KWOLEK, W. F. (1981). *J. Am. Oil Chemists' Soc.*, **58**:300.
95. MEYER, E. W. and WILLIAMS, L. D. (1976). In: *World Soybean Research* (Hill, L. D., Ed.). p. 904. Interstate Printers and Publishers, Inc., Danville, Illinois.
96. OHREN, J. A. (1981). *J. Am. Oil Chemists' Soc.*, **58**:333.
97. PEARSON, A. M. (1976). *BioScience*, **26**:249.
98. BOYER, R. A. (1954). US Patent 2,682,466.
99. ZIEMBA, J. V. (1969). *Food Eng.*, 41:72.
100. CAMPBELL, M. F. (1981). *J. Am. Oil Chemists' Soc.*, **58**:336.

101. FREDERIKSON, C. W. and HEUSDENS, W. (1972). US Patent 3,662,671.
102. HOER, R. A. (1972). US Patent 3,662,672.
103. DUBOIS, D. K. and HOOVER, W. J. (1981). *J. Am. Oil Chemists' Soc.*, **58**:343.
104. MORALES, J., BOURGES, H. and CAMACHO, J. L. (1981). *J. Am. Oil Chemists' Soc.*, **58**:374.
105. PEREIRA, L. and DE CAMPOS, S. D. S. (1981). *J. Am. Oil Chemists' Soc.*, **58**:355.
106. WAGGLE, D. H., DECKER, C. D. and KOLAR, C. W. (1981). *J. Am. Oil Chemists' Soc.*, **58**:341.
107. GIFFEE, J. W. (1981). Personal communication.
108. FUKISHIMA, D. (1981). *J. Am. Oil Chemists' Soc.*, **58**:346.

Chapter 3

RAPESEED PROTEIN FOR FOOD USE

FRANK W. SOSULSKI
University of Saskatchewan, Saskatoon, Canada

SUMMARY

Rapeseed is a major world source of vegetable oil but the fibrous meals have limited value as animal feeds. The proteins contain an excellent balance of essential amino acids but a satisfactory system for dehulling the seeds has not yet been developed. The recent developments of low-glucosinolate and yellow-seeded cultivars represent a substantial improvement in the nutritive value of the meal for monogastric animals. However, the concentrations of thyroid-active agents in dehulled flours from these cultivars are still too high for safety in human consumption. Even the residual levels of glucosinolates in rapeseed protein concentrates appear to cause some deleterious physiological effects. The economic feasibility of manufacturing protein concentrate or isolate has not been favorable due to low product yields, dark colors and bitter flavors, the presence of sinapine and phytic acid, and the lack of strong functional properties in the products.

INTRODUCTION

Rapeseed is a member of the brassica family, which includes several oilseed species, as well as root and vegetable crops. In Europe, most of the oilseed crop is winter rape (*Brassica napus*) while summer rape (*B. napus*) and turnip rape (*B. campestris*) are grown in Canada. Oriental and brown mustard (*B. juncea*), black mustard (*B. nigra*) and yellow mustard (*B. hirta*) are grown on a small scale as condiments and oilseeds.

Rapeseed from the Far East may be summer turnip rape, brown mustard or mixtures of these species.[1]

Rapeseed ranks fourth in importance as a source of vegetable oil and protein meal, after soybean, cottonseed and sunflower. Annual production is nearly 4 million tonnes of oil and 6 million tonnes of oilseed meal. The contributions of rapeseed and mustard as local sources of oil and protein are substantial because the crop is grown widely in most parts of the world especially in temperature and warm temperate zones.

Brassica seeds are minute, especially the cultivars grown in Canada (Table 1). Among world cultivars, the seed coats represent 12–25% of the seed weight, the colors being black, brown or yellow. Cell layers under the seed coat retain chlorophyll until the seed is nearly mature, the green color being an adverse quality factor in the crude oil. The green cell layers of immature or frozen seeds will also discolor the defatted meal.

The endosperm tissues of rapeseed are largely absent or crushed in the mature seed and the cotyledons of the embryo represent the bulk of the seed. Thin cell walls composed of pectinous material surround each cotyledon cell. Lipid droplets are scattered throughout the cytoplasm, accounting for about 50% of the cotyledon weight. Aleurone grains, measuring 2–10 μm in diameter, are the major protein bodies in the cell.

CHEMICAL COMPOSITION

The moisture content of mature rapeseed should be 6–8% for safe storage. Due to unfavorable weather conditions, a portion of the crop is harvested at over 9% moisture which has adverse effects on oil and meal quality due to bin-heating and microbiological growth. The seed and meal become dark in appearance and protein solubility is reduced substantially.

The oil contents of the *Brassica* species and biotypes and quite variable. The values range from 42 to 50% for winter rapeseed, 37 to 47% for summer rapeseed, 40 to 48% for winter turnip rapeseed and 36 to 46% for summer turnip rapeseed.[1] Brown and oriental mustards are generally lower in oil content than turnip rapeseed while yellow mustards contain less than 30% oil (Table 1).

The hulls of rapeseed are not removed before oil extraction due to losses of oil in the hulls and attached embryo. Therefore, meals exhibit high crude fiber levels, more so in the small-seeded rapeseed than the larger yellow-seeded mustards (Table 1). Protein contents in the meal

TABLE 1

SEED CHARACTERISTICS, PRODUCT YIELD AND COMPOSITIONS OF *Brassica* SPECIES[2]

Brassica species and common name	*Seed characteristics*			*Protein product*	*Yield of product (g/100 g seed)*	*Protein (N × 6·25) (%)*	*Crude fiber (%)*	*Ash %*	*Color of product*	*Isothiocyanates (mg/g meal)*				*Oxazolid-inethione*
	Weight (mg)	*Oil (%)*	*Hull (%)*							*p-OH benzyl*	*Allyl*	*Butenyl*	*Pentenyl*	
B. napus, rape	4·0	45·4	22	Defatted meal	54	40·0	11·5	6·5	Grey	—	—	2·1	0·6	7·5
				Dehulled flour	32	50·0	3·7	7·2	Cream					
				Protein isolate	16	83·0	0·1	2·0	Brown					
B. campestris, turnip rape	2·1	41·5	23	Defatted meal	56	37·6	11·9	7·1	Grey	—	—	2·0	1·5	2·4
				Dehulled flour	33	47·5	4·6	8·0	Cream					
				Protein isolate	17	85·2	0·1	2·1	Brown					
B. juncea, oriental mustard	3·2	37·8	17	Defatted meal	60	47·4	6·5	5·5	Cream	—	10·0	—	—	—
				Dehulled flour	43	54·0	3·7	6·0	Cream					
				Protein isolate	22	93·2	0·1	1·5	Tan					
B. carinata, abyssinian mustard	2·6	34·9	23	Defatted meal	63	45·4	8·8	6·4	Grey	—	4·9	—	—	—
				Dehulled flour	40	53·7	5·0	7·1	Cream					
				Protein isolate	22	88·3	0·2	2·0	Brown					
B. nigra, black mustard	1·7	29·3	—	Defatted meal	68	46·3	7·7	5·7	Cream	—	7·0	—	—	—
				Dehulled flour	low	53·4	4·2	5·9	Cream					
				Protein isolate	24	91·7	0·1	2·0	Tan					
B. hirta. yellow mustard	7·9	26·3	18	Defatted meal	73	47·1	7·7	5·6	Cream	30·1	—	—	—	—
				Dehulled flour	55	56·0	3·4	5·9	Cream					
				Protein isolate	31	92·5	0·3	2·0	Cream					

tend to be low, ranging between 35 and 40% (8·5% moisture basis) in commercial car lots in Canada. The high ash levels of 6–8% also contribute to the low digestible energy in the meal.

Despite other adverse features of the meal, there is considerable interest in developing rapeseed as a food protein because of its excellent balance of essential amino acids. Lysine content of rapeseed protein averages about 6·0% (Table 2) as compared to about 6·5% in comparable soybean samples. In addition, rapeseed proteins have 3·0–4·0% of the sulfur-containing amino acids, methionine and cystine. The proportions of these essential amino acids are closer to FAO requirements[3] for humans than any other vegetable protein available in quantity.

There are a wide range of nitrogen-containing compounds in rapeseed, about 80% being true proteins (Table 3). Bhatty *et al.*[5] isolated two major proteins from rapeseed, a neutral high molecular weight fraction (12 S) and a basic 1·7 S fraction having a molecular weight of 13 800. In rapeseed meal the 12 S glycoprotein contained about 13% carbohydrate consisting of arabinose, galactose, glucose, inositol, glucosamine and mannose.[6] The complexing of sugars with protein would be due, in part, to Maillard condensation reactions during oil extraction and meal desolventization. Inositol would arise from the globoid bodies in the aleurone grains which apparently contain phytic acid.

The carbohydrate content of rapeseed meals is relatively low. Sucrose and stachyose are the major oligosaccharides (Table 3). The highly branched polysaccharides in the cotyledon cell walls yield arabinose, xylose, galactose and galacturonic acid after hydrolysis. Cellulose, hemicellulose and lignin are major constituents of the hulls.

The meals typically contain about 1% of phosphorus, of which 70–80% occurs as phytic acid (myoinositol hexaorthophosphate) or its calcium–magnesium salt, phytin.[4] Phytic acid also forms stable complexes with basic amino acid residues in proteins. This decreases the solubility of rapeseed proteins and affects their electrical properties. Phytin phosphorus is not available for digestion by the monogastric animal and interferes with the absorption of multivalent cations such as iron and zinc.

The *Brassica* species are characterized by the presence of sulfur-containing compounds called glucosinolates or thioglucosides which may represent a high proportion of the meal sulfur (Table 3). The glucosinolates of rapeseed, gluconapin, glucobrassicanapin and progoitrin, are hydrolyzed by the seed enzyme myrosinase (thioglucoside glucohydrolase) to give glucose, sulfate, isothiocyanates and L-5-vinyl-2-

TABLE 2

AMINO ACID COMPOSITION OF DEHULLED FLOURS OF *Brassica* SPECIES REPORTED AS % OF THE FLOUR PROTEINS, DRY BASIS[2]

Amino acid	*B. napus*	*B. campestris*	*B. juncea*	*B. carinata*	*B. nigra*	*B. hirta*
Tryptophan	1·1	1·1	1·0	1·0	1·1	1·0
Lysine	6·0	5·9	5·1	5·4	5·7	4·6
Histidine	2·9	2·6	2·8	2·8	2·7	2·5
Ammonia	3·3	2·8	2·7	3·0	3·0	3·1
Arginine	6·5	5·9	7·8	7·8	7·0	8·3
Aspartic acid	7·2	7·3	7·1	7·0	7·0	7·1
Threonine	4·2	4·4	4·2	4·0	4·1	4·1
Serine	4·4	4·4	4·3	4·3	4·2	4·1
Glutamic acid	21·7	21·7	22·9	22·5	22·2	22·4
Proline	6·2	6·8	6·7	6·8	7·0	6·4
Glycine	5·5	5·5	5·3	5·4	5·4	5·6
Alanine	4·6	4·7	4·2	4·5	4·8	4·7
Half-cystine	2·0	2·2	2·2	2·2	2·2	2·3
Valine	4·8	4·9	4·6	4·6	4·8	4·7
Methionine	1·7	1·8	1·6	1·5	1·7	1·6
Isoleucine	3·8	3·8	3·8	3·7	3·9	3·9
Leucine	7·3	7·3	7·1	7·0	7·1	7·1
Tyrosine	2·6	2·5	2·4	2·3	2·0	2·2
Phenylalanine	4·0	3·9	4·0	4·0	4·0	4·1

TABLE 3
AVERAGE COMPOSITION OF MEALS OF *B. napus* AND *B. campestris* CULTIVARS, % DRY WEIGHT[4]

	B. napus	*B. campestris*
Nitrogen-containing fractions		
True protein (N × 6·25)	34·0	34·0
Peptides, free amino acids	2·0	2·0
Nucleic acids	3–4	3–4
Glucosinolates	1–2	1–2
Hull nitrogen	3·0	3·0
Carbohydrate constituents		
Sucrose	7·4	7·0
Raffinose	0·3	0·3
Stachyose	2·5	2·4
Galactinol	0·1	0·1
Digalactosyl glycerol	0·1	0·1
Arabans, arabinogalacturonic acid	4–5	4–5
Hemicellulose	3	3
Cellulose	4–5	4–5
Starch	<1	<1
Phytic acid	3–5	—
Minerals		
Calcium	0·65	—
Phosphorus	1·1	—
Sodium	0·01	—
Potassium	1·3	—
Magnesium	0·05	—
Zinc	0·57	—
Copper	0·07	—
Sulfur	0·7–1·8	—

oxazolidinethione (Figs. 1 and 2). Under certain conditions the intermediates formed by the catalyst can rearrange into nitriles, thiocyanates and inorganic sulfur compounds. The isothiocyanates and oxazolidinethione, in particular, are implicated in a number of physiological disorders in animals, including thyroid enlargement. The nitriles can cause liver and kidney disorders such as liver hemorrhagic syndrome.

Rapeseed species contain progoitrin, although the level of oxazolidinethione in *B. campestris* is much less than occurs in most cultivars of *B. napus* (Table 1). Some of the goitrogenicity of rapeseed meals may be associated with the recently discovered indolylisothiocyanate which yields thiocyanate *in vivo*.[7]

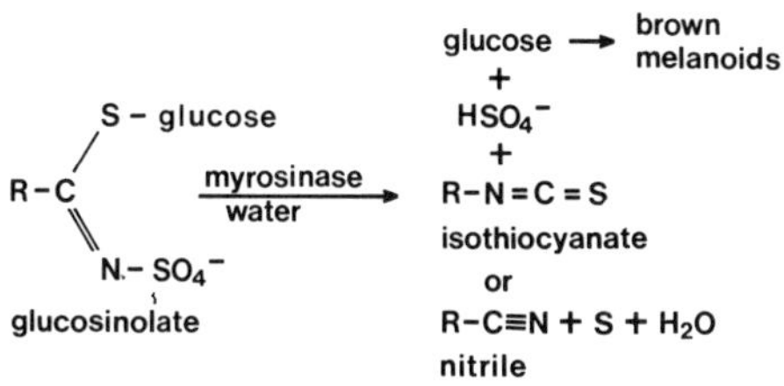

FIG. 1. General structure of glucosinolates and the products formed by enzymatic hydrolysis.

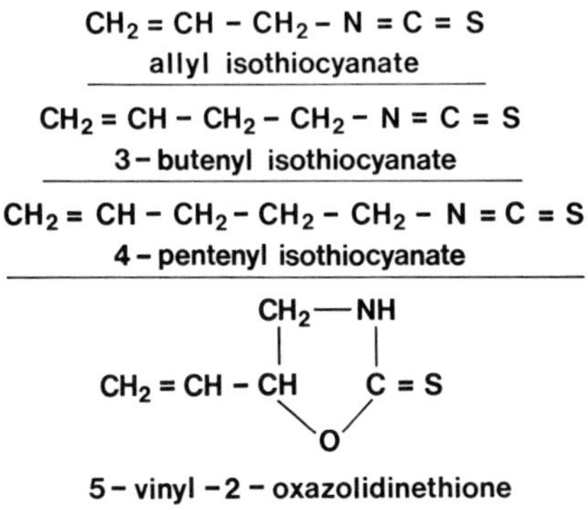

FIG. 2. Structures of the principal isothiocyanates and oxazolidinethione in rapeseed and mustard meals.

During rapeseed crushing it is important to avoid glucosinolate hydrolysis because the aglucones are readily soluble in the extracted oil, causing flavor problems and poisoning the catalysts used for hydrogenation. It is common practice to inactivate the myrosinase by heating the seed at 80–85 °C for 15 min. In the subsequent oil extraction, the intact glucosinates remain with the meal and have only limited physiological effects on monogastric animals unless myrosinase is reintroduced into the animal diet from another source.

The mustard species do not contain oxazolidinethione (Table 1) and, in the preparation of condiments, the myrosinase is activated to produce the hot volatile oils of mustard. Recently allyl isothiocyanate from black mustard was identified as a carcinogen in male rats.[8] This may reopen the question of the safety of glucosinolates in human nutrition.

Breeding for Improved Seed Chemistry

Canadian plant breeders have pioneered the improvement of the nutritive value of rapeseed cultivars. Prior to 1970, the lipids in commercial cultivars contained high proportions of erucic acid among the fatty acids in the triglycerides and other seed lipids (Table 4). Erucic acid in the diet has been associated with excessive deposition of fat in the muscles of young animals. The conversion to cultivars which store almost no erucic acid in the seed oils was largely completed in Canada in 1974 and in Europe by 1978.

TABLE 4

PROGRESS IN THE IMPROVEMENT OF GRAIN YIELD, PROTEIN AND OIL CHARACTERISTICS OF CANADIAN RAPESEED CULTIVARS[9]

Year released	*Cultivar*	*Relative yield* (%)	*Oil composition*		*Meal composition*		
			Content (%)	*Erucic acid* (%)	*Protein* (%)	*Glucosinolates* ($mg\,g^{-1}$)	*Fiber* (%)
B. napus							
1943	Argentine	100	—	40	—	12	—
1966	Target	110	42·9	40	44·8	12	—
1968	Oro	107	40·8	1	42·9	12	—
1973	Midas	116	42·8	1	42·4	12	—
1974	Tower	109	41·4	1	46·6	< 2	—
1977	Regent	116	41·9	1	46·7	< 2	—
B. campestris							
1964	Echo	—	40·0	27	43·2	7	13
1973	Torch	—	39·0	1	42·9	7	13
1978	Candle	—	41·2	1	43·6	1	9·5

Rapeseed meal is discounted as a protein supplement for animal feeds because of the presence of glucosinolates, low levels of digestible protein and energy, and high crude fiber and hull contents. The recent conversion to low glucosinolates in Canada has removed one of the major constraints in rapeseed meal utilization by livestock and poultry (Table 4). Currently, plant breeders are achieving a significant reduction in crude fiber content of the meal by incorporating genes for yellow seed coats into black or brown cultivars. Each advance in seed quality has been made with new cultivars which were less satisfactory in yield, oil or protein content. However, these defects have been overcome by further plant breeding and selection. The various improvements have greatly enhanced the utilization of rapeseed meal in swine and, to a lesser extent, poultry rations.

Recently, plant breeders in Australia[45] have discovered genetic sources of low erucic acid in mustard (*B. juncea*). This species is relatively high yielding and the oil content could be improved by selection to the level of rapeseed cultivars (Table 1). Therefore, consideration is being given to developing yellow-seeded, low fiber genotypes of mustard as an oilseed crop. The glucosinolates in *B. juncea* could be removed from the meal or flour by steam distillation, a process which is not effective in the removal of oxazolidinethione.

RAPESEED FLOURS

Most rapeseed cultivars have black or brown seed coats which contribute dark colors and coarse texture to the meals. The high proportion of hulls in the seed results in defatted meals which are composed of 25–35% hulls. Removal of hulls is an essential step in the production of a food grade product in order to reduce fiber levels, improve flour appearance and upgrade the protein levels to over 50% (Table 1). Thus, there would be a substantial yield of hulls for disposal either industrially or as ruminant feed. It is noteworthy that the dehulled flours would still contain 3·5–4·5% crude fiber and 7–8% ash (Table 1).

In early experiments, air classification of commercial rapeseed meals gave a 20–24% yield of high protein fraction.[10] Due to the low recovery of protein and dark brown color in the flour, the product was considered unsuitable for supplementation of foods.

Jones *et al.*[11] investigated mechanical processes for dehulling rapeseed before oil extraction. A Bauer Disc Mill consisting of stationary and rotary splined discs was effective in cracking the seeds with minimum pulverization of cotyledons. The hulls-plus-fines fraction obtained by air classification, representing 20–30% of the seed, was fractionated on gravity table, sieves or air flotation units to provide hulls, fines and an inseparable mixed fraction. Oil contents of the hull fractions varied between 18·7 and 21·0% while the mixed fractions contained 22·0% or more of lipid. Due to the losses of oil in the hull-rich fractions, the crushing industry has not adopted the process on any significant scale. Front end dehulling of the type described has been incorporated into schemes for production of rapeseed protein concentrate by several investigators.[12–14]

Sosulski and Zadernowski[15] patented a process for fractionating the defatted meal into hull and flour fractions using liquid cyclones. The defatted meal before desolventization was finely ground in wet mills to pass a 150 mesh (Tyler) screen. The finely ground marc was then suspended in hexane or other non-polar solvent and pumped through a liquid cyclone or suitable centrifuge to yield relatively pure flour and hull fractions. Due to the hazards of milling the hexane-soaked marc, an alternative procedure would be to pin-mill the expeller meal before solvent extraction and cyclone fractionation (Fig. 3). The latter pro-

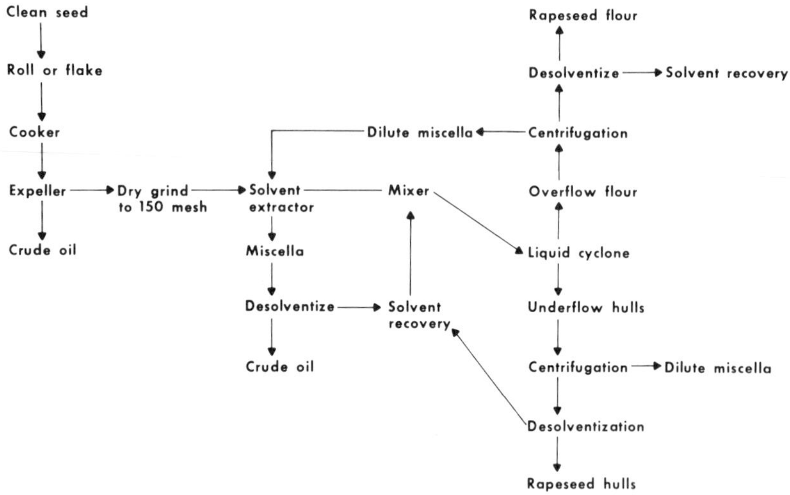

FIG. 3. Prepress plus solvent extraction scheme using dry fine grinding before oil extraction, followed by solvent extraction and liquid cyclone fractionation of the marc.[16]

cedure has other advantages since the cyclone step can remove up to 14% of residual oil in the meal and only a short solvent extraction would be required after the expeller treatment.

The yield and composition of the flour and hull obtained from fractionation of the marc from an all-solvent plant and a prepress plus solvent plant are given in Table 5. During the cyclone operation, meal-to-hexane ratios of 1:5 were optimal for separation of the seed components including the protein and fiber. The flours were free of hulls and the hull fraction contained about 2% flour (Fig. 4). The protein enrichments were about 7·5% and the crude fiber reductions about 6·3%. If lower meal-to-hexane ratios are employed in the cyclone, the separation of crude fiber would be less effective (Table 5).

TABLE 5
YIELD AND COMPOSITION OF LIQUID CYCLONE FRACTIONS OBTAINED AFTER FINE GRINDING OF COMMERCIAL MEALS[16]

Type of oil extraction plant	*Meal-to-hexane ratio*	*Fractionated products*[a]	*Yield ratio*	*Protein (%)*	*Oil (%)*	*Fiber (%)*	*Ash (%)*
All-solvent	—	Meal	100	39·6	3·1	11·8	6·8
	1:5	Flour	65	46·9	1·0	5·7	6·9
		Hulls	34	27·0	0·9	24·1	6·5
Prepress + solvent	—	Meal	100	37·6	3·1	14·7	6·8
	1:5	Flour	66	45·1	1·1	8·2	7·2
		Hulls	34	25·9	0·5	29·3	6·3
	1:3	Flour	67	42·4	1·3	10·3	7·0
		Hulls	33	29·2	0·5	25·0	6·8

[a] The marcs were desolventized before analyses; values reported are on a dry basis.

FIG. 4. Appearance of the hull and flour fractions obtained from liquid cyclone fractionation of pin-milled rapeseed meal.

The relatively efficient fractionation of the meal components resulted from their differential densities. The weight ratio of hulls to flour was about 1·6:1 on the basis of hull densities of 0·55 g cm^{-3} and flour density of 0·34 g cm^{-3}. While the liquid cyclone separation was based on differential densities, air classification efficiency was determined to a greater extent by particle size and shape (see Chapter 5).

After desolventization, the dry products were difficult to handle because of their fine particle size. Sosulski and Zadernowski[16] pelleted the flour and hulls, and reground the material to conduct animal feeding trials on each product. In poultry experiments, the metabolizable energy of rapeseed flour samples averages 2200 kcal kg^{-1} as compared with 1860 for rapeseed meal and 2250 for soybean meal.

The developers did not consider the flour as suitable for food use due to the presence of glucosinolates and phenolic constituents. Extraction of the flour to remove these undesirable constituents would reduce the yield of protein product substantially from the present flour yield of 36 %, on a seed basis, and would probably not be an economic process.

Utilization of Hulls

Rapeseed hulls contain a high proportion of the seed nitrogen, due in part to the presence of portions of embryo and endosperm. Finlayson[17] separated the hulls from cotyledonous material by extensive fractionation and determined the true nitrogen content to be 1·6–1·8 % of the hulls. Extraction of the hulls with 0·1 M sulfuric acid or sodium hydroxide failed to release more than 1 % of the hull nitrogen. This result suggested that much of the hulls would resist digestion by monogastric animals. The availability of the hull energy and nitrogen to the ruminant has not been determined on pure hull fractions.

Industrial uses for hulls obtained by front-end dehulling before oil extraction have not been demonstrated, due to problems associated with the oil and cotyledon components. Bender[18] has established two potential uses for defatted hulls based on the phenolic and polyphenolic compounds which represent approximately 20 % of the hulls. The hulls have been successfully incorporated into phenolic resin glues used in plywood production. They can also be used as a component of phenolic moulding compounds for Bakelite resins.

PROTEIN CONCENTRATES

Because of the need for removal of glucosinolates and phenolic compounds, much attention has been given to extraction procedures that result in protein concentrates. Early procedures were devised for aqueous extraction of glucosinolates from rapeseed meal,[19] ground rapeseed[10] and, by diffusion, from intact rapeseed[20] without removal of hulls. Aqueous extraction of commercial meal resulted in a dark coarse material which was unsuitable as a food. When applied to low-temperature-treated meals, the losses of soluble meal proteins were high.[21] Losses of oil and protein, which occurred in the ground seed procedure, were controlled effectively by diffusion extraction (Table 6).

TABLE 6

YIELD AND COMPOSITION OF PRODUCTS AND LOSSES IN THE EXTRACTS FOLLOWING AQUEOUS OR ETHANOLIC EXTRACTIONS OF RAPESEED MEAL OR SEED[21]

Aqueous or ethanolic extraction process	*Protein product*[a]			*Aglucones in product ($mg\,g^{-1}$ product)*			*Losses in extract (% of total seed)*		
	Yield (% seed)	*Protein (%)*	*N solubility (%)*	*BI*	*PI*	*OZT*	*Solids*	*Nitrogen*	*Oil*
Untreated control	54·7	39·1	65·6	4·6	1·4	5·2	—	—	—
Extraction of meal[19]	37·8	33·9	60·8	0·00	0·00	0·00	17·9	1·2	—
Extraction of ground seed[10]	36·8	40·6	64·2	<0·05	<0·05	0·00	25·0	1·2	6·7
Extraction of intact seed[20]	45·3	48·3	39·5	0·01	<0·05	0·00	12·0	0·3	0·1

[a] Products were not dehulled.
[b] BI = butenyl, PI = pentenyl isothiocyanate; OZT = vinyl oxazolidinethione.

However, the latter procedure required high temperatures and long diffusion times for the complete removal of progoitrin from the seed.[2] Two 30-min extractions at 80 °C were needed to detoxify the low-glucosinolate cultivar, Tower,[2] even in the recent adaptation of the diffusion process in which the hulls are removed to yield a protein concentrate.[14]

Recent modifications of the Eapen *et al.*[10] process have involved mild heat treatments, careful myrosinase inactivation, solvent oil extraction and complete hull removal as well as glucosinolate extraction. Similar processes developed in Canada[12] and Sweden[13] involve the sequence of steps illustrated in Fig. 5. The material balance reported by Anjou and Ohlson[13] was:

Fraction	*Yield (%)*
Hulls fraction	17
Fines (pressed)	6
Dry matter from evaporated leaching water	15
Oil	37
Rapeseed protein concentrate (RPC)	25

The original seed was reported to contain 43% oil and 23% protein so losses of oil and protein into the leach water or with hulls were substantial. Also the yield of protein concentrate was low, yet this product would have to bear a major portion of the costs of the aqueous extraction process. The processes have not been adopted by commercial crushers on any significant scale.

A detoxified lipid–protein concentrate has been prepared by water–ethanol extraction.[22] The preparation contained 48% protein and 31% lipid, and only 0·04 mg g^{-1} of aglucones of the glucosinolates. Unfortunately, the recoveries of protein and oil in the concentrate were only 50% and 70% respectively, and recent studies[23] have not improved the yields substantially.

Nutritive Value of RPC

Because of the inherent nutritional problems in feeding rapeseed meal to livestock, numerous nutritional tests have been conducted on the rapeseed protein concentrates. Protein efficiency ratios (PER) obtained by Jones[12] were consistently as good as casein. Anjou and Ohlson[13] have reported PER values of 3·0–3·5 (when adjusted to casein control = 2·5), net protein utilization (NPU) of 87–90 and true digestibility of 95–100.

The role of RPC as a protein supplement in foods was investigated by

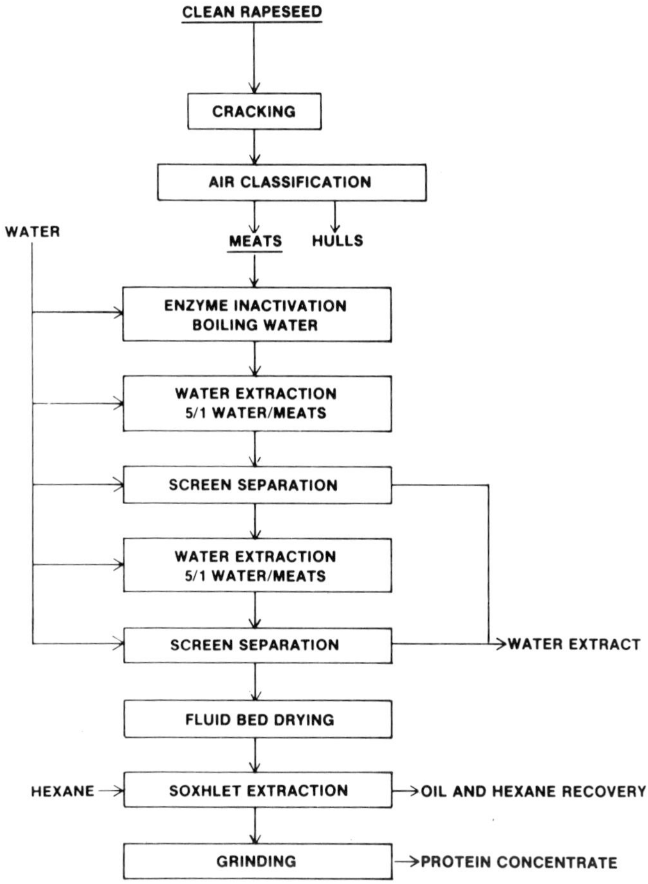

FIG. 5. Procedure for preparing rapeseed protein concentrate (RPC) by aqueous extraction of dehulled meats.[12]

Olson *et al.*[24] in a feeding trial with weanling male rats. The palatability, weight gains and protein nutritive value of Tower RPC was determined as the sole source of protein and in equal protein blends of RPC with wheat flour, field pea flour and defatted ground beef. The protein contents of the protein sources, which ranged from 16·0 to 95·2%, were adjusted to 10% with cerelose and other diet constituents. Based on the amino acid compositions of the proteins, lysine was determined to be the first limiting amino acid in the RPC + wheat blend while the sulfur-containing amino acids were limiting in RPC and the other blends

(Table 7). Palatability and growth performance of rats on the RPC diet was superior to the casein diet and second only to the RPC–beef blend in the experiment. While the average daily gain for RPC–wheat blend was lower than for the other diets, the PER was a substantial improvement over values for wheat flour alone. The protein digestibility of 80% for RPC was much less than was claimed by Anjou and Ohlson[13] but is consistent with other reports.[25] The blend of RPC with the legume, field pea, was similar to RPC alone, but blends with wheat or meat were much higher in protein digestibility. The low digestibility of the RPC is consistent with the high crude fiber (5·3%) and ash (9·7%) contents of this protein-enriched (61·3% protein) product.

Short-term feeding trials are inadequate for assessing the safety of RPC in human foods. Concern about the presence of thyroid-active substances in rapeseed has precluded their acceptance for use in food products until satisfactory toxicity trials have been completed. The principal antithyroid compounds in rapeseed are isothiocyanates, which compete with iodine for entry into the thyroid gland, and oxazolidinethione, which retards the incorporation of iodine into thyroid hormones.

Loew *et al.*[26] reported the results of sub-acute toxicity tests which involved the feeding of RPC at high levels in the diets of two animal species, growing dogs and rats, for 90 days. Two RPC samples containing 0·90 and 0·03 mg of total glucosinolates per gram of RPC, including 0·29 and 0·02 $mg\,g^{-1}$ of oxazolidinethione, were incorporated into semisynthetic diets at 20 and 40% of the total dietary protein. The levels of RPC in the diets ranged from 6 to 15% of the dry matter. A low order of antithyroid activity was detected for both diets containing RPC samples but thyroid iodine uptake, thyroid weight and thyroid histology remained normal. No treatment-associated abnormalities were found in the dogs. However, in rats the serum glutamic–pyruvic transaminase activity was decreased in proportion to the level of RPC. The investigators concluded that the antithyroid effects of RPC were substantially reduced, but not eliminated, in comparison with rapeseed flours.

The occurrence of anorexia and loss of weight during late pregnancy, high fetal mortality,[27] and reduced serum tocopherol and low density lipids[28] has been observed in rats fed essentially glucosinolate-free protein concentrates. McLaughlan *et al.*[29] have found the toxic effects of RPC in pregnant rats to be associated with zinc deficiency. Since RPC preparations contain significant amounts of phytate (Table 8), it was deduced that this metal-binding constituent played the major role in reducing the bioavailability of dietary zinc. Although RPC contains

TABLE 7

RELATIONSHIPS BETWEEN AMINO ACID COMPOSITION AND THE FEED INTAKE, WEIGHT GAIN, PROTEIN DIGESTIBILITY AND PROTEIN UTILIZATION OF RAPESEED (RPC)-BASED DIETS BY RATS[24]

Protein source in diet	*Amino acid composition of diet (% of rat requirements)*				*Average daily feed intake*	*Average daily gain*	*Feed:gain ratio*	*Protein digestibility*	*PER adjusted to casein*
	Lys	*Met + Cys*	*Thr*	*Try*	(*g*)	(*g*)	($g\,g^{-1}$)	(%)	2·50
Casein	88	<u>63</u>[a]	75	171	11·35	3·33	3·42	93	2·50
RPC	66	<u>55</u>[a]	73	65	12·07	3·39	3·47	80	2·58
RPC + wheat	<u>44</u>	<u>50</u>[a]	59	65	11·15	2·45	4·56	86	1·76
RPC + field pea	76	<u>47</u>[a]	71	65	11·79	3·04	3·89	79	2·26
RPC + beef	75	<u>52</u>[a]	70	67	12·87	4·12	3·13	87	2·70

[a] First limiting amino acid(s) underlined.

TABLE 8
EFFECTS OF PHYTIN IN RAPESEED AND MUSTARD PROTEIN CONCENTRATES ON SERUM ZINC LEVELS IN RATS[12]

Dietary protein source	*Composition of dietary protein source*		*Serum zinc ($\mu g\ ml^{-1}$)*
	Phytin (%)	*Zinc ($\mu g\ g^{-1}$)*	
Casein	—	60	1·5
Bronowski	5·27	96	0·9
Target	6·68	113	1·0
Tower	7·32	121	0·7
Yellow mustard	6·41	113	0·8
Span	7·18	108	0·6
Karlshamn	7·50	132	0·6

substantial amounts of zinc, the various RPC preparations in Table 8 reduced serum zinc levels in proportion to phytin rather than to glucosinolate concentrations. The addition of zinc to the drinking water completely relieved the symptoms of anorexia in pregnant rats.

Rapeseed meals and flours contain other constituents which have deleterious effects on animal performance. Hobson-Frohock *et al.*[30] associated the occurrence of fishy odors in eggs from brown-shell layers to the presence of trimethylamine. The concentration of this compound was proportional to the levels of rapeseed meal in the rations. The source of trimethylamine was later identified as sinapine, which represents about 1·2–2·3% of rapeseed meal.[31]

While the glucosinolates are responsible for the mustard-like flavor and odor of rapeseed, the bitter flavors and poor palatability of the meal are ascribed to the high concentrations of sinapine. Sinapine is the choline ester of sinapic acid (Fig. 6). Other phenolic acids, especially the more reactive *o*-dihydroxyphenolic acids, are present in rapeseed products. They readily oxidize to *o*-quinones under alkaline extraction conditions in the preparation of protein isolates.[32] The phenolic acid–protein complex develops the brown color which is characteristic of rapeseed protein isolate, even when adjusted to pH 4·5 (Table 1).

PROTEIN ISOLATES

Efforts have been made to separate rapeseed proteins from undesirable constituents of the meal or flour by traditional procedures employed for

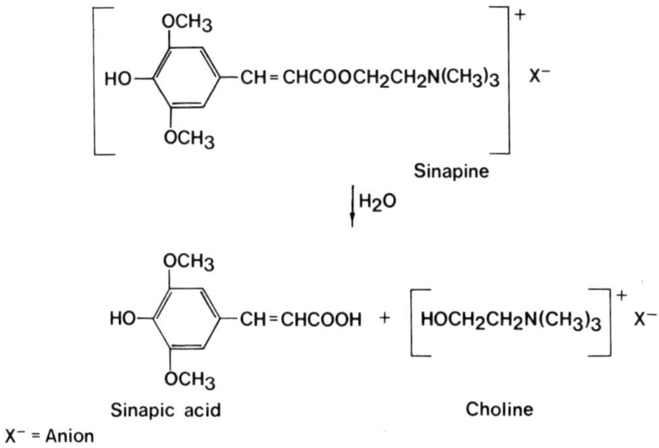

FIG. 6. Sinapine and its hydrolytic cleavage products.[33]

the corresponding isolation of soybean protein. Protein isolates are prepared by dissolution of the protein in alkali, removal of insoluble residue by centrifugation, and recovery of the soluble proteins by acid precipitation. Sosulski and Bakal[34] found that, in six cultivars of *B. napus* and *B. campestris*, only 80–85% of the seed nitrogen could be extracted with dilute alkali, and 25–30% of the nitrogen remained in the whey after acid precipitation at pH 4·5. Therefore only one-half of the flour protein was recovered in the isolate and the product was relatively low in protein content (Table 1). The color of the material was dark brown and the flavor bitter, probably due to phenol–protein interactions. The ash contents of the isolates (Table 1) were low, due to the elimination of phytic acid by extraction at pH 11 as prescribed by Gillberg and Törnell.[35] The low recovery of protein in the isolates from rapeseed flour has been a common experience.[36–38]

In contrast to the above results, El Nockrashy *et al.*[39] extracted 94% of the meal nitrogen using a meal-to-alkaline solvent ratio of 1:25. A two-stage precipitation procedure at pH 6·0 and then pH 3·6 yielded 76·7% and 24·3% respectively of the extract nitrogen. The isolates contained 92·9 and 98·6% respectively of protein after washing and drying the curds. These results have yet to be confirmed by other investigators.

Three-stage extraction procedures were employed by Kodagoda *et al.*[37] and Keshavarz *et al.*[40] The latter investigators extracted the meal at pH 8, 2 and 10, obtaining protein fractions containing 48·9–92·2% of protein, which represented 12·0, 19·0 and 31·3% respectively of the original meal protein. The major protein fraction was brown in appearance.

Complexing agents have been used to improve the recoveries of the alkali-soluble protein during the acid precipitation step. One approach was to co-precipitate rapeseed and acid cheese whey but nitrogen recoveries in the curd varied from 47 to 54 % of the total nitrogen in both products.[41]

Gillberg and Törnell[35] investigated the solubility pattern of rapeseed protein and phytic acid over the pH range 1–13. Phytic acid was most insoluble at pH 7 and 11, so phytate-free protein solutions could be prepared at the latter pH. Only 35 % of the extracted nitrogen was precipitated at pH 4·9 but the yield was increased to 75 % by adding sodium phytate. Attempts were made to find more acceptable precipitation acids. Hexametaphosphate, carboxymethyl cellulose, polygalacturonic acid and alginate were found to give essentially quantitative recovery of dissolved protein during acid precipitation.[42]

A protein concentrate or isolate plant would have a serious waste effluent problem.[1] The first stage of purification involves centrifugation, sedimentation or filtration to remove fines and colloidal protein and carbohydrate material. The large volumes of filtrate contain glucosinolates and their hydrolytic products in addition to large quantities of carbohydrates, proteins and minerals. Normally, the costs of concentrating the effluent can be partially recovered from the sale of the concentrate as a highly digestible feed. However, rapeseed effluent could not be disposed of as a source of nutrients and alternate methods of disposal have not been elaborated in the literature.

FUNCTIONAL PROPERTIES AND UTILIZATION

The end uses of protein ingredients are largely dependent on functional properties related to solubility in water or salt solutions, water and fat binding, oil emulsification, viscosity and gelation. Bland flavors and light colors are important sensory characteristics required for most applications. Because cereals constitute the major food staples in the diet, high levels of lysine are prerequisites for most protein ingredients.

In functionality tests, the rapeseed flours showed excellent water and fat absorption, oil emulsification, whipping and viscosity characteristics compared with soybean flour (Table 9). The rapeseed protein concentrates were low in nitrogen solubility, oil emulsification, whippability

TABLE 9

NITROGEN SOLUBILITIES AND FUNCTIONAL PROPERTIES OF SOYBEAN AND RAPESEED PRODUCTS[43]

Protein product	*% Nitrogen solubility at pH 6*	*pH of dispersion*	*Water absorption (%)*	*Fat absorption (%)*	*Oil emulsification (%)*	*Volume increase on whipping (%)*	*Viscosity of heated 10% slurries*	*Color of slurry*
Soybean								
Flour	10·7	6·6	223	130	12	66·9	133–3330	Light brown
Concentrate	6·4	4·9	331	202	10	133·5	6240–41 600	Light tan
Isolate	—	—	—	—	—	321·4	> 83 300	Off-white
Rapeseed								
Tower meal	33·6	5·9	294	235	34	362·0	16 600–20 825	Greenish brown
Tower flour	44·8	6·0	265	298	46	333·2	5000–16 600	Light green–brown
Midas flour	46·0	5·3	223	362	20	346·2	6640–31 540	Light green–brown
Torch flour	44·6	5·4	215	310	29	358·6	11 620–33 200	Light brown
Tower RPC[20]	6·3	6·8	341	413	17	212·3	333–300	Brown
Tower RPC[12]	4·0	6·4	398	389	10	112·5	333–1500	Light brown
Tower isolate	12·7	6·8	310	380	38	365·0	6240	Brown

and viscosity, and showed little improvement in color over the flours. The rapeseed protein isolate represented no improvement in functional properties over the less concentrated products, and failed to exhibit the gelation characteristic of the soybean isolate, Promine D, which was used as the control. Kodagoda *et al.*[37] found that rapeseed protein isolates and concentrates have lower specific foam volumes, but greater stability than egg white protein.

Rapeseed protein concentrates, and to a lesser degree isolates, had adverse effects on bread loaf volume when incorporated into the formulation at the 5% replacement level.[37] Emulsifiers improved the loaf volume and structure of the blends but crumb and crust colors were darker than in the controls.

Frankfurters or wieners supplemented with rapeseed meal, flour and protein concentrates at the 20% level showed less shrinkage during the smokehouse treatment and were substantially firmer than the meat control prior to cooking.[44] The cooking yields of rapeseed-supplemented wieners were also high but the cooked wieners became softer than the controls. A panel of five judges ranked the cooked rapeseed wieners as inferior to soybean-supplemented wieners in organoleptic properties (Table 10). Even the rapeseed protein concentrates would require further processing to improve their color, flavor and texture.

In summary, rapeseed flours show desirable functional properties but have no prospects as food ingredients because of adverse organoleptic

TABLE 10

ORGANOLEPTIC QUALITY OF COOKED WIENERS[44]

Protein additive	*Color score*[1]	*Flavor score*[2]	*Texture score*[2]
None (control)	$5{\cdot}6_a$[3]	$7{\cdot}4_a$	$6{\cdot}6_a$
Soybeam			
Flour	$5{\cdot}4_a$	$6{\cdot}2_b$	$5{\cdot}8_{ab}$
Concentrate	$5{\cdot}2_a$	$5{\cdot}8_b$	$6{\cdot}6_a$
Rapeseed			
Meal	$1{\cdot}0_c$	$1{\cdot}0_e$	$2{\cdot}0_d$
Flour	$2{\cdot}3_{bc}$	$2{\cdot}3_d$	$3{\cdot}5_{cd}$
Concentrate[20]	$3{\cdot}4_b$	$4{\cdot}0_c$	$4{\cdot}6_{bc}$
Concentrate[12]	$3{\cdot}4_b$	$4{\cdot}8_c$	4·6

[1]6-point scale (6 = like most; 1 = dislike most).
[2]9-point scale (9 = like most; 1 = dislike most).
[3]Two means with no letter in common differ significantly at the 95% level.

properties as well as the presence of physiologically active glucosinolates. The potential for protein concentrates is diminished by the need for extensive extraction, low product yields and poor market value for by-products. Due to concerns about their safety and acceptability, rapeseed protein concentrates may be limited to applications such as in pet foods and milk replacers for calves. The production of the protein isolate is not economically feasible, due to low yields, toxic by-products and lack of enhanced functional properties. The prospects for future utilization of rapeseed proteins in foods would be greatly encouraged by the breeding of new cultivars devoid of glucosinolates and sinapine, and containing more protein and less crude fiber and phytic acid. An alternative would be to develop mustard, especially *B. juncea*, as an oilseed crop since the large seed provides a higher protein yield in the flour, and processing treatments to remove glucosinolates are less expensive.

REFERENCES

1. OHLSON, R. and SEPP, R. (1975). In *Food Protein Sources*, Ed. Pirie, N. W. Cambridge University Press, Cambridge, p. 65.
2. SOSULSKI, F. (1978). *Proc. 5th. Int. Rape. Conf. Malmo, Sweden*, June 12–16, **2**, 365.
3. FAO/WHO (1973). *Energy and Protein Requirements*. Tech. Report Series No. 522. World Health Organization, Rome.
4. FINLAYSON, A. J. (1977). In *Rapeseed Oil, Meal and By-product Utilization*, Ed. Bell, J. M. Publ. 45, Rape. Assoc. of Canada, Winnipeg, p. 124.
5. BHATTY, R. S., MACKENZIE, S. L. and FINLAYSON, A. J. (1968). *Can. J. Biochem.*, **46**, 1191.
6. GILL, T. A. and TUNG, M. (1978). *Cereal Chem.*, **55**, 180.
7. MCGREGOR, D. I. (1978). *Can. J. Plant Sci.*, **58**, 795.
8. ANON. (1981). *Food Chem. News*, June 29, p. 59.
9. DOWNEY, R. K. and KLASSEN, A. J. (1977). In *Rapeseed Oil, Meal and By-product Utilization*, Ed. Bell, J. M. Publ. 45, Rape. Assoc. of Canada, Winnipeg, p. 4.
10. EAPEN, K. E., TAPE, N. W. and SIMS, R. P. A. (1969). *J. Am. Oil Chem. Soc.*, **46**, 52.
11. JONES, J. D., CHAN, C. and BENDER, F. (1976). Paper No. 1545, Annual Meeting, Can. Inst. Food Sci. Technol. May 30–June 2, Ottawa, Canada.
12. JONES, J. D. (1977). In *Rapeseed Oil, Meal and By-product Utilization*, Ed. Bell, J. M. Publ. 45, Rape. Assoc. of Canada, Winnipeg, p. 112.
13. ANJOU, K. and Ohlson, R. (1978). *Rapeseed Protein Concentrate*. Karlshamns Research Laboratory, Sweden.
14. KOZLOWSKA, H., BOGACZYNSKI, K., ZADERNOWSKI, R., LOSSOW, B. and SZEBIOTKO, K. E. (1978). GDR Patent 127,952.

15. SOSULSKI, F. W. and ZADERNOWSKI, R. (1980). Can. Patent 1,089,849, Nov. 18, 1980.
16. SOSULSKI, F. and ZADERNOWSKI, R. (1981). *J. Am. Oil Chem. Soc.*, **58**, 96.
17. FINLAYSON, A. J. (1974). *Can. J. Anim. Sci.*, **54**, 495.
18. BENDER, F. (1980). *Research on Canola Seed, Oil, Meal and Meal Fractions.* 6th Prog. Report, Canola Council of Canada. Publ. **57**, p. 257.
19. BALLESTER, D., RODRIGO, R., NAKOUZI, J., CHICHESTER, C. O., YANEZ, E. and MONCKEBERG, F. (1970). *J. Sci. Food Agric.*, **21**, 143.
20. SOSULSKI, F. W., SOLIMAN, F. S. and BHATTY, R. S. (1972). *Can. Inst. Food Sci. Technol. J.*, **5**, 101.
21. KOZLOWSKA, H., SOSULSKI, F. W. and YOUNGS, C. G. (1972). *Can. Inst. Food Sci. Technol. J.*, **5**, 149.
22. EKLUND, A., ÅGREN, G. and LANGLER, T. (1971). *J. Sci. Food Agric.*, **22**, 650.
23. ROGULSKI, W. and WITCZAK, F. (1977). *Zesz. Probl. Nauk Roln.*, **15**, 222.
24. OLSON, J. P., SOSULSKI, F. W. and CHRISTENSEN, D. A. (1978). *Can. Inst. Food Sci. Technol. J.*, **11**, 93.
25. JOSEFSSON, E. (1972). In *Rapeseed*, Ed. Applequist, L. A. and Ohlson, R. Elsevier Publ. Co., New York, p. 354.
26. LOEW, F. M., DOIGE, C. E., MANNS, J. G., SEARCY, G. P., BELL, J. M. and JONES, J. D. (1976). *Toxicol. Appl. Pharmacol.*, **35**, 257.
27. EKLUND, A. (1973). *Nutr. Rep. Int.*, **7**, 647.
28. EKLUND, A. and ÅGREN, G. (1978). *Proc. 5th Int. Rape. Conf. Malmo, Sweden*, June 12–16, **2**, 144.
29. MCLAUGHLAN, J. M., JONES, J. D., SHAH, B. G. and BEARE-ROGERS, J. L. (1975). *Nutr. Rep. Int.*, **11**, 327.
30. HOBSON-FROHOCK, A., LAND, D. G., GRIFFITHS, N. M. and CURTIS, R. F. (1973). *Nature*, **243**, 304.
31. MUELLER, M. M., RYL, E. B., FENTON, T. and CLANDININ, D. R. (1978). *Can. J. Anim. Sci.*, **58**, 579.
32. SOSULSKI, F. (1979). *J. Am. Oil Chem. Soc.*, **56**, 711.
33. AUSTIN, F. L. and WOLFF, I. A. (1968). *J. Agric. Food Chem.*, **16**, 132.
34. SOSULSKI, F. W. and BAKAL, A. (1969). *Can. Inst. Food Sci. Technol. J.*, **2**, 1.
35. GILLBERG, L. and TÖRNELL, B. (1976). *J. Food Sci.*, **41**, 1063.
36. GIRAULT, A. (1973). *J. Sci. Food Agric.*, 1973, **24**, 509.
37. KODAGODA, L. P., NAKAI, S. and POWRIE, W. D. (1973). *Can. Inst. Food Sci. Technol. J.*, **6**, 266.
38. RUTKOWSKI, A. and KOZLOWSKA, H. (1981). *Preparaty Zywnościowez z Bialka Roślinnego.* Wydawnictwa Naukowo-Techniczne, Warszawa.
39. EL NOCKRASHY, A. S., MUKHERJEE, K. D. and MANGOLD, H. K. (1977). *J. Agric. Food Chem.*, **25**, 193.
40. KESHAVARZ, E., CHEUNG, R. K. M., LUI, R. C. M. and NAKAI, S. (1977). *Can. Inst. Food Sci. Technol. J.*, **10**, 73.
41. THOMPSON, L. U. (1977). *Can. Inst. Food Sci. Technol. J.*, **10**, 43.
42. GILLBERG, L. and TÖRNELL, B. (1976). *J. Food Sci.*, **41**, 1070.
43. SOSULSKI, F., HUMBERT, E. S., BIU, K. and JONES, J. D. (1976). *J. Food Sci.*, **41**, 1349.
44. SOSULSKI, F., HUMBERT, E. S., LIN, M. J. Y. and CARD, J. W. (1977). *Can. Inst. Food Sci. Technol. J.*, **10**, 9.
45. KIRK, J. T. O. and OVAM, R. N. (1981). *J. Aust. Inst. Agric. Sci.*, **74**, 51.

Chapter 4

LUPIN SEED PROTEINS

Paolo Cerletti
Department of General Biochemistry, University of Milan, Italy

SUMMARY

The protein content of some species of lupin seed is the highest among legumes. Albumins and globulins are present. The molecular properties of these proteins, as they appear in various species, are discussed. The origin of the many different subunits of the legumins and vicilins is attributed to fragmentation of one original parent polypeptide: this explains in part the micro-heterogeneity observed at subunit level. The effects of proteolytic enzymes associated with the globulins and the immunochemical behaviour of proteins from different species are also described.

Relevant nutritional aspects are (a) deficiencies in methionine and cysteine, and the related joint regulation by the sulphur content of fertilizers of the levels of sulphur-poor vicilins versus legumins; (b) reduced in vitro *digestibility, which is improved by denaturation; (c)* in vivo *assays indicating how to best exploit the nutritional potential of lupin for food and feed.*

The functional properties are also examined, and these indicate interesting possibilities for the use of lupin proteins.

INTRODUCTION

Lupin is one of the legume seeds for which substantial information on constituent proteins has been obtained in recent times. Studies on its proteins span 100 years starting with the pioneering work of Ritthausen,[1]

Osborne and Campbell[2] and Osborne and Harris,[3] who showed that two types of globulin were present. In the mid-1950s physicochemical studies produced new data on the purification of the globulins, their physical size and the interactions they undergo in solution[4,5] but little work followed immediately. In the last decade a renewed interest in lupin as a potential source of protein for food and feeds has prompted systematic investigations on the molecular characteristics of the proteins, in parallel with their nutritional and functional properties and their distribution in the various lupin species.

Lupin deserves considerable attention as a protein source because of the high yield of protein in the seed, which compares favourably with soybeans and in some species is highest among legumes. Further, its geographical distribution covers temperate areas where the soybean does not grow. Considerable advantages over other legume seeds are the low levels of toxic and antinutritional factors: alkaloids in wild varieties may reach 2·5% of dry seed weight,[6] but in sweet varieties they decrease to less than 0·01%,[7] which makes the flour or its components acceptable as such for animal or even human use. Haemagglutinins, trypsin inhibitors, isoflavones and favism factors are stated to be absent, saponins are present (1·1–1·7%), and in some species erucic acid may represent up to 2% of the component fatty acids.[6] Flatus-inducing factors (in practice mainly oligosaccharides) are similar to those in soybeans for *Lupinus albus* and *L. angustifolius*, but are considerably higher in *L. luteus and L. mutabilis.*[8]

Alkaloids can be removed by several procedures but the choice depends on economic considerations which relate to the properties of the lupin as an oil producer. Indeed it has become clear that, since protein productivity of the plant is always high, production of an adequate amount of oil can be an important factor in promoting the use of lupin as a protein crop. It has been calculated[9] that species which produce less than 15% (dry weight) oil do not economically justify oil separation. Sweet varieties may serve for producing fibreless flour for human and animal feeding. Bitter varieties, with more than 0·2% alkaloid, require grinding, followed by separation of an alkaloid-free fatty protein concentrate for human use. This process, however, results in the facile oxidation of polyunsaturated fatty acids in the oil: its use is therefore subject to some limitations. On the other hand, bitter varieties with oil contents higher than 16–18% justify alkaloid and oil extraction, with production of defatted and debittered flours for human and animal use. Prospects in the field are promising since the average seed oil productivity of 8–10% is raised in *L. mutabilis* above 20%[10] and this opens challenging goals for selective breeding.

In the following text, the molecular characteristics of lupin proteins are examined together with related nutritional aspects and functional properties, with a view to possible technological applications.

PROTEIN YIELD AND EXTRACTION

The amount of protein in the seed, its distribution in different seed parts, and protein productivity per hectare under the conditions used in the field trials reported, are given in Table 1.

TABLE 1
PROTEIN CONTENT AND DISTRIBUTION IN LUPIN SEED

Species and cultivar	*Kernel (% of seed)*	*Protein N × 6·25 (%)*		*Protein yield (kg ha⁻¹)*	*Ref.*
		Seed	*Kernel*		
L. albus, W. $B._2$	81·4	34·3	41·2	1585	11
L. luteus, Weiko III	73·9	40·0	51·0	1238	11
L. angustifolius					
Uniwhite		29·6	38·1		11, 12
Uniharvest		36·2			12
Unicrop	76·4	28·0	37·0	858	11
L. cosentini	71·0	32·4	42·0		11
L. mutabilis					
AZ 733	84·7	45·8		535	10
AZ 1012d	89·1	40·1		380	10

Lupin is one of the richest sources of protein, if referred to seed weight. Protein is mostly in the kernel. The seed coat contains only about 2–3% of the total.[12] Protein is probably deposited in protein bodies, as for other legume seeds, but a systematic study has not been done for lupin. When a homogenate of *L. angustifolius* seeds was sedimented on a sucrose gradient, some of the major protein constituents, namely conglutin γ and conglutin β, but not conglutin α, were present in a layer containing typical protein bodies[13] and corresponding in density with that containing protein bodies in *Phaseolus aureus*.[14]

Proteins are in general extracted from decorticated ground seeds defatted by repeated treatment with pentane[4, 15] or hexane.[13] The particle size of the meal has a marked effect on the quantity of protein recovered: as it falls below 30 mesh the proportion of nitrogen solubilised changes sharply from less than 40 % to more than 80 %, both in alkaline and in acid media. Also, the amount of water available is important: as it increases, the protein extracted asymptotes to a line representing the total amount of protein available for extraction.[16]

Stirring defatted meal in water at the natural pH (about 5·0–5·5) of the suspension for 12–24 h at 4 °C solubilises about 10 % of total proteins, i.e. the albumins.[13, 15, 17] The rest is recovered by shaking for 24 h at 4 °C with 1·0–1·7 M sodium chloride.[13, 15] Proteins are also solubilised by adjusting the dispersion to pH 9 without using high salt concentrations, as was done by Sgarbieri and Galeazzi.[18] However, under conditions of low salt and high pH, extensive disaggregation of the proteins occurs and there is no certainty that they will subsequently reform to their native state.

Proteins have also been extracted from the defatted meal directly with sodium chloride in the conditions mentioned above. They can then be fractionated into a fraction soluble in water, the albumins, and one soluble in salt, the globulins.[4]

INDIVIDUAL PROTEIN FRACTIONS

Albumins

This group of proteins has been investigated far less extensively than the globulins, and its components have not been individually characterised. This is because of their diversity, and because together they form only a minor part of the total seed protein.

Albumins represent 12·8 % of seed proteins in *L. albus*[15, 19] and only 5–10 % in *L. angustifolius*.[17] The amount of water-soluble nitrogen is in some cases considerably higher, and may vary depending on species. Part of this is low molecular weight material which is dialysed off against distilled water and may represent as much as two-thirds of the total.[17] As will be detailed later, albumins display proteolytic activity and therefore part of the non-protein nitrogen may originate by hydrolysis during dialysis.

The bands separated from the albumins on electrophoresis in cellulose acetate differ from the other seed proteins; at least five are evident in *L. albus*[8] and a similar number in *L. angustifolius*.[17] SDS electrophoresis on polyacrylamide gel (SDS–PAGE) separates about 20 components in *L. albus*[19] and in *L. angustifolius*,[17] with molecular weights ranging from 117 kD to 6 kD,[19] which differ from those given by the globulins. Probably not all molecular species are detected under the conditions mentioned. Albumins include a number of molecules which represent the functional protein of the seed. Many of them are enzymes which metabolise the substances deposited in the seed. An example of this is the high proteolytic activity, which will be discussed in a subsequent section.

The amino acid composition of the albumins isolated from *L. albus*[20, 21] and *L. angustifolius*[17] is shown in Table 2 together with that of globulins. The albumins of the two species differ more than do the globulins. Their composition has little in common with that of the globulins except for high aspartic and glutamic acids, or amides, which is, however, a rather common feature in plant proteins. Notable is their higher content of lysine and of sulphur-containing amino acids, especially methionine, which indicates a better nutritional value than the globulins.

Globulins

Proteins extracted at high ionic strength are globulins, comprising the storage protein of the seed, and quantitatively representing the major protein component. The complexity of this group has become clear as the analytical approaches used have been refined. In the original studies, the extract was analysed by ultracentrifugation or electrophoresis, as such or after ammonium sulphate precipitation followed by dialysis. *L. angustifolius* yielded two components: one, with a molecular weight (MW) of 336 kD and sedimentation constant 11·6 S at ionic strength (I)=0·31, pH 7·0, dissociated to a species of 170 kD, 7·2 S, at I=0·1, pH 9·0. The other component of 7·8 S, MW 181 kD, did not change by altering the pH or ionic strength. *L. luteus* produced three components: two of them, of 11·6 S and 7·9 S, which behaved like those already mentioned, and a third one of 2·8 S. This small component was found also in *L. albus* together with a species of 12·2 S which did not dissociate on changing pH and ionic strength, and one of 8·7 S (I=0·31, pH 7·0) which at I=0·1 and pH 9·0 associated to a 12·5 S species.[4, 5]

In subsequent work by Blagrove and Gillespie[13] on *L. angustifolius* cv. Uniwhite, the globulin extract was fractionated by ammonium sulphate precipitation or by chromatography on celite with ammonium sulphate gradient solubilisation, into three major components, conglutins α, β and γ and a 2 S globulin. Conglutin α and β correspond respectively

TABLE 2
AMINO ACID COMPOSITION OF LUPIN FLOUR, ALBUMINS AND GLOBULINS (%)

	Flour		*Albumins*		*Globulins*	
	L. albus[a]	*L. angustifolius*[b]	*L. albus*[a]	*L. angustifolius*[b]	*L. albus*[a]	*L. angustifolius*[b]
Lys	4·36	5·09	5·40	7·41	3·90	3·94
His	2·13	2·94	2·09	2·21	1·76	2·44
Arg	11·06	11·48	12·57	7·23	11·81	14·09
Asp	10·76	10·89	11·95	10·43	10·55	11·16
Thr	3·86	3·79	4·96	5·47	3·08	3·21
Ser	5·22	5·57	5·30	4·99	5·35	5·43
Glu	25·07	22·18	26·61	18·88	27·40	23·92
Pro	4·06	4·36	4·07	5·38	4·38	4·49
Gly	3·96	4·55	2·00	6·19	3·68	4·22
Ala	3·04	3·51	3·62	6·70	2·69	2·87
$\frac{1}{2}$ Cys	2·13	1·10	1·70	1·14	1·48	0·72
Val	3·35	4·08	3·62	5·66	3·16	3·42
Met	0·51	0·45	0·52	1·42	0·27	0·22
Ile	4·26	4·17	3·80	3·74	4·50	4·22
Leu	7·31	7·16	7·62	5·81	7·36	6·97
Tyr	4·57	4·39	1·73	3·44	5·58	4·21
Phe	3·86	4·25	3·15	3·40	3·85	4·46
Trp	0·44	n.d.	0·19	n.d.	0·49	n.d.

[a] From Ref. 20.
[b] Calculated from Ref. 17.

to the 11·6 S and the 7·8 S components reported by previous authors. Congluten α at 4 °C yields soluble and cold-insoluble components, while four subfractions were separated from conglutin β by arbitrary ammonium sulphate cuts. The conglutins, but not their subfractions, are also resolved by cellulose acetate electrophoresis.

In a different approach the globulins were separated by ion exchange chromatography followed by gel filtration.[15] A summary of the procedures applied is given in Fig. 1. The full procedure requires 2 days in addition to the time needed for dialysis and freeze-drying. Twelve fractions are resolved, homogeneous on gel filtration, ion exchange chromatography on DEAE cellulose and cellulose acetate electrophoresis, and having a characteristic molecular weight and amino acid composition. They appear therefore as individual proteins. Globulins 9a and 9b, which are eluted together as fraction 9 from DEAE cellulose and are separated by gel filtration, differ in many of their properties. Separation on gel filtration depends on the widely different molecular weight, whereas the interaction which makes them move together in ion exchange, either before or after purification,[15] is probably electrostatic in nature and is related to their different acidities. Purified 9b and 9a interact in a 5:1 molar ratio, which corresponds to their ratio in the total globulin extract: this suggests that they may be associated also in the seed.

Some major characteristics of the fractions separated are reported in Table 3. A few of them, namely conglutin γ and proteins 4, 6, 8, 9a and 9b, make up 91·9% of the total globulins in *L. albus* and are quantitatively constant in the extracts, whereas the other globulins are only minor components and some of them vary considerably in quantity. All globulins contain covalently bound carbohydrate, the amount and quality of which are typical for each protein.[15] Conglutin γ and globulins 2, 5 and 7 preserve their specific quantitative carbohydrate composition although they may vary in amount in the seed. In 3, 10 and 11, protein and sugar vary independently. The minor fractions are likely to contain more than one protein.

Globulins 4, 5, 6 and 7 are similar in many respects, as are globulins 8 and 9a. This allows us to assign them, in analogy with other legume seeds, to two homogeneous groups, the vicilins and the legumins. Owing to their similarity the proteins of each of these groups are generally not resolved by the separation techniques used by previous workers.

The ionic character of the globulins has an important role in their solubility behaviour and related properties. Acidity increases with the

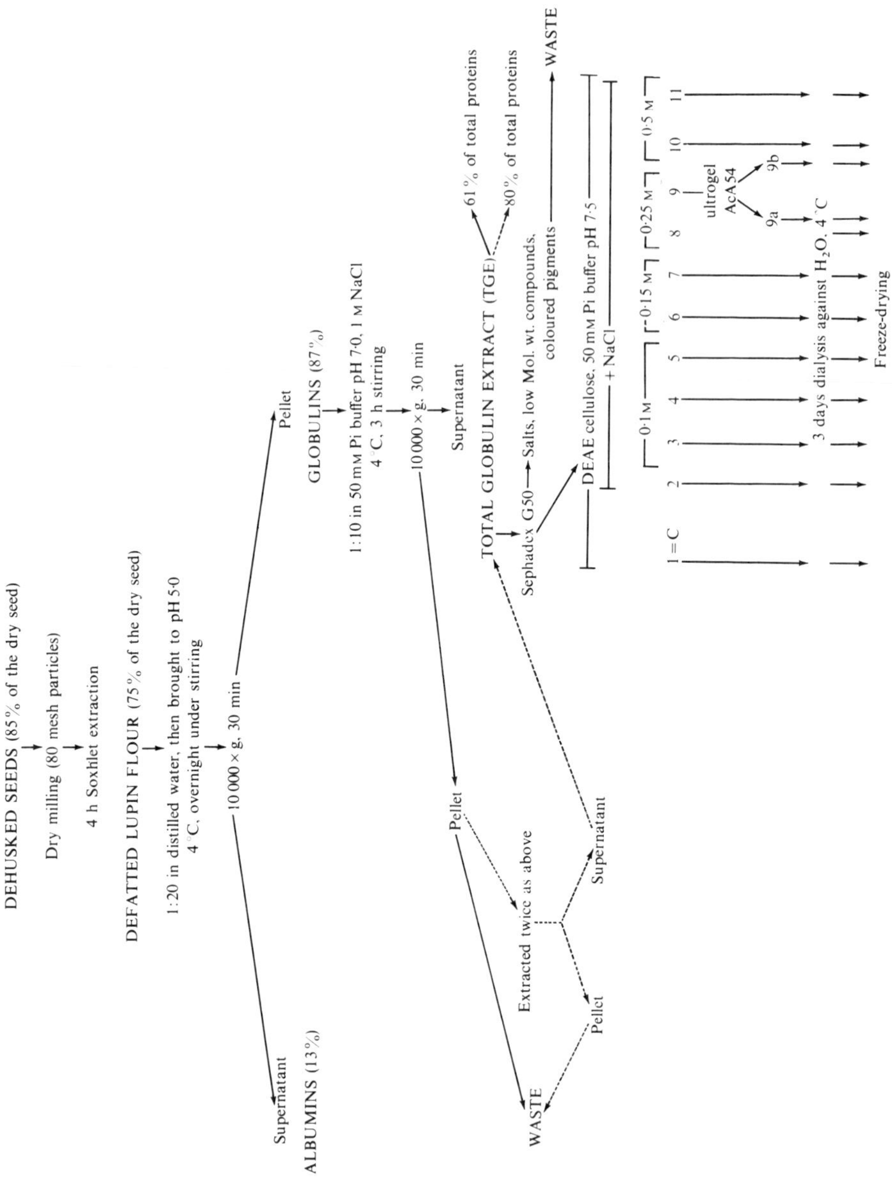

FIG. 1. Extraction and purification of proteins from *L. albus* seeds.

TABLE 3
MOLECULAR WEIGHT AND DISTRIBUTION OF PROTEIN AND NEUTRAL SUGAR IN THE GLOBULINS OF *Lupinus albus*[a]

	Globulin	Electrophoretic component	Mol. Wt. × 10^{-3}	Protein		Neutral sugar			
				% of total globulins	v.c.	% of total	v.c.	mg/100 mg peptide	v.c.
	1[b]	γ	199	6·0	0·13	11·7	0·37	4·3	0·07
	2	—	n.d.	0·5	0·40	8·4	0·50	36·9	0·04
	3	—	123	0·7	0·57	4·9	0·39	15·5	0·50
						25·0	0·02		
Vicilins	4	β	270	10·0	0·15	10·9	0·09	2·4	0·04
Vicilins	5	β	275	2·1	0·19	3·5	0·38	3·7	0·03
						14·4	0·10		
Vicilins	6	β	335	30·2	0·03	28·9	0·06	2·1	0·09
Vicilins	7	β	143	1·9	0·21	2·9	0·06	3·4	0·09
Legumins	8	α	240–470	21·2	0·06	14·5	0·12	1·5	0·07
Legumins	9a	α	240–470	12·0	0·00	4·4	0·11	0·8	0·15
	9b	δ	44	12·5	0·04	4·0	0·05	0·7	0·14
	10	—	226	2·6	0·46	1·8	0·12	1·5	0·53
	11	—	n.d.	0·4	0·25	3·9	0·47	21·5	0·49

v.c. = variability coefficient.
[a]from Refs. 15 and 22. [b]Conglutin γ.

elution order from DEAE cellulose, being lowest in conglutin γ and highest in 9b: as shown in Table 4 it closely reflects the relative amounts of basic and dicarboxylic amino acids in the molecules and the amidation level. All these factors vary coherently to raise acidity from conglutin γ to 9b; the contribution of dicarboxylic amino acids is decisive in view of the high content of glutamic acid in most globulins.

The following sections show that corresponding proteins or protein fractions isolated from different lupin species and/or by different authors have in general analogous molecular properties. Relevant differences appear only in the apparent molecular weight of the native globulins: to some degree at least this may depend on the experimental approach used.

Legumins

Protein structure. Globulins 8 and 9a, the two legumins, correspond to the cold-insoluble and cold-soluble fractions of conglutin α[13] and to the

TABLE 4
CHEMICAL BASIS OF THE IONIC CHARACTER OF GLOBULINS[a]

Globulin	*Asx + Glx*	*Amide nitrogen*	*Amidation (%)*	*Unsubstituted Asp + Glu*	*His + Lys + Arg*
1[b]	158	134	84·8	24	102
4	260	176	67·7	84	109
6	256	166	64·8	90	112
8	244	n.d.	n.d.	n.d.	107
9a	238	133	55·9	105	95
9b	353	202	57·2	151	83

[a] The data refer to *L. albus* and are calculated from values reported in Tables 6, 9 and 10. Unless otherwise specified, they are given as μmol/100 mg protein.
[b] Conglutin γ.

component with sedimentation constant 11·6 S, mentioned already, which reversibly dissociates into a 7·2 S protein.[4] Each legumin yields, on gel filtration, two non interconvertible forms of apparent molecular weights 240 and 470 kD, respectively.[22] A similar situation is confirmed by ultracentrifugation studies but with somewhat lower values of molecular weight: overestimation is possible by gel filtration, especially in glycoproteins. The two molecular forms have same amino acid composition but differ in types of subunit. The protein structure explaining these data will be discussed in that which follows. The molecular weights were determined in *L. albus* by gel filtration experiments; sedimentation equilibria gave lower values for conglutin α from *L. angustifolius*. Some overestimation of molecular size is a known possibility in gel filtration.

Both legumins contain several subunits some of which have interpeptide disulphide bonds. As shown in Figs. 2 and 3 subunits and protomers differ in the two globulins. The two types of conglutin α are similar in subunit pattern,[13] the cold-soluble fraction being more like globulin 9a and the cold-insoluble one resembling globulin 8.

In each legumin no simple stoichiometry appears between the various subunits: it is therefore impossible to accommodate all of them in a single structure. On the other hand the molecules isolated by gel filtration are homogeneous in many respects, viz. sedimentation behaviour, gel filtration, ion exchange chromatography, electrophoresis, amino acid and sugar composition.[15] A situation occurs therefore in which the overall composition of the molecules is substantially similar though they contain different subunits. We therefore propose that in each legumin the determined subunits originate from the breakdown of a single parent

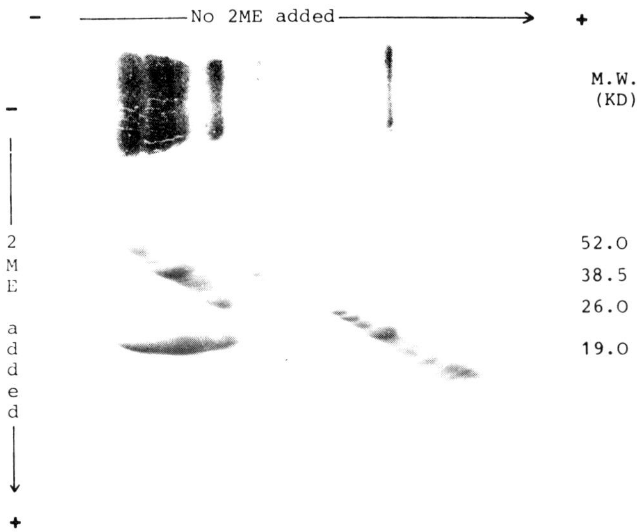

M.W. (KD) and molar ratios of the major subunits and protomers

64·5	55·5	44·0	21·0
13	64	17	6
52·0			
10			
	38.5		
	64		
		26·0	
		20	
19·0	19·0	19·0	21·0
14+	68+	18+	6+

+Calculated values.

FIG. 2. Subunit and protomer composition of globulin 8 on two-dimensional SDS–PAGE, first under non-reducing then under reducing conditions (2ME = 2-mercaptoethanol).

polypeptide as shown in Figs. 4 and 5. The fragments of proteolysis all remain present in the molecule forming a component unit of the oligomeric structure: one large and one small ‘peptide’, linked by an interchain disulphide. On SDS–PAGE they liberate the determined subunits. Under reducing conditions the interpeptide disulphide is split and the known pattern of protomers is formed. Two-dimensional isoelectric focusing–SDS–PAGE indicates that the large peptides are more acidic in

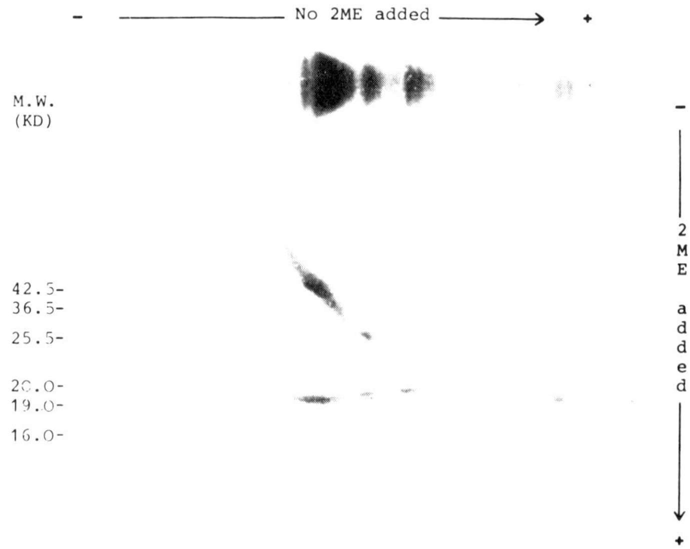

M.W. (KD) and molar ratios of the major subunits and protomers

61·5	55·5	45·0	36·5	21·0	16·0
8	32	15	19	10	16
42·5					
4					
	36·5				
	33				
		25·5			
		17			
19·0	19·0	19·0	19·0	21·0	16·0
15	58	27	19⁺	10⁺	17

⁺Calculated values.

FIG. 3. Subunit and protomer composition of globulin 9a (as Fig. 2).

character than the one of 19·0 kD:[22] this may correspond to a high content of glutamate and aspartate, which together represent 35 % of total amino acids in the protein.[15]

In view of the identity in molecular weight of the two legumins, and of their close similarity in amino acid composition, we propose that they are both derived from the same original polypeptide, split by proteases in different positions.

The proteolysis proposed probably occurs in the seed. Significant

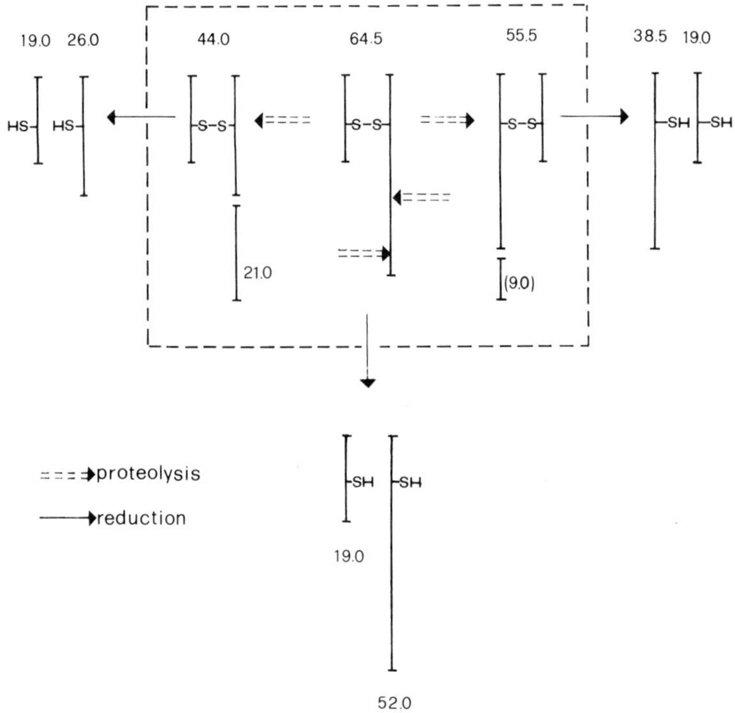

FIG. 4. Subunit and protomer formation by proteolysis in globulin 8. Subunits present in the protein in the dormant seed are those enclosed by the dashed line. Protomers liberated by reduction are indicated outside it. Determined molecular weights in kD are given. Hypothetical small fragments, not seen in SDS–PAGE, are those with molecular weights within brackets.

changes during the experimental manipulations have to be excluded.[38] Proteases associated with the seed globulins, which are described later, might be the agent for these events. Interestingly, the proteolytic attack apparently occurs only on the larger peptide. The position of the bound sugar and the acidic character of the peptide segments may play a role in determining the site of action of the protease. The isolated peptides may differ moderately in different preparations, indicating a more or less advanced breakdown of the original peptide.[56]

If now we come to the quaternary structure, in globulin 8 the small association appears to contain three or four component units of molecular weight 64·5 kD and the large one six or eight depending if one refers to the molecular weight determined by gel filtration or by ultra-

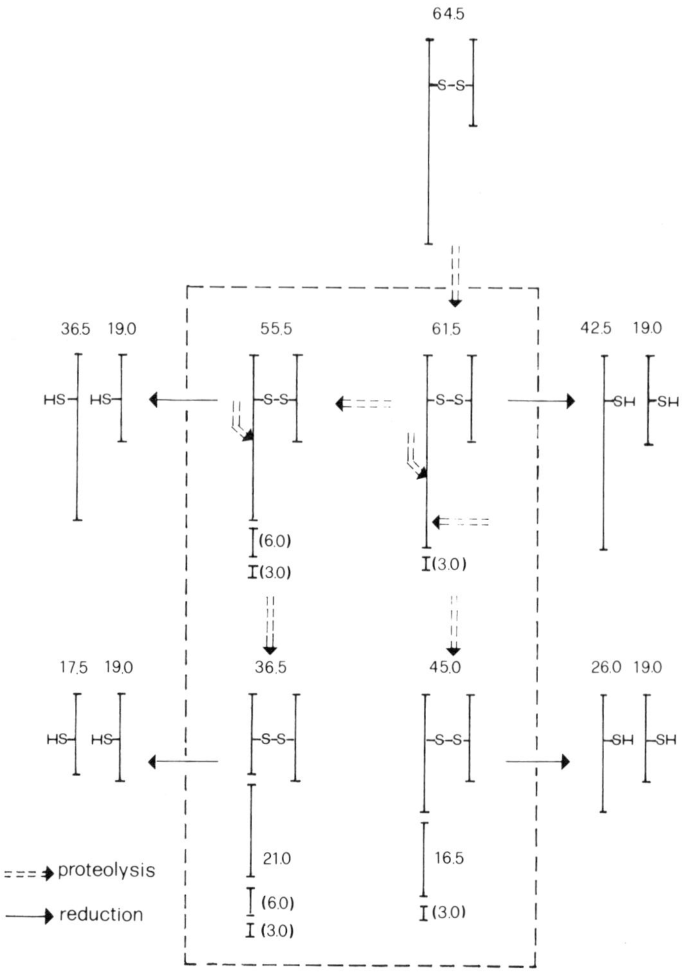

FIG. 5. Subunit and protomer formation by proteolysis in globulin 9a. Subunits present in the protein in the dormant seed are these enclosed by the dashed line. The original parent polypeptide and protomers liberated by reduction are indicated outside it. Determined molecular weights in kD are given. Hypothetical small fragments, not seen in SDS–PAGE, are those with molecular weights within brackets.

centrifugation. The type of peptides present in the subunits seems to influence the yield of either the small or the large association.[56] For globulin 9a the same ratios are observed with a unit of molecular weight 61·5 kD. The molecular weights and molar ratios of subunits and their protomers allow us to calculate the compositions and relative amount of component units, as shown in Table 5. Hydrophobic interactions are important binding forces, as shown by the dissociating effect of detergents and urea. Values of molecular weight determined for the major protomers are very close to those found by other authors for corresponding proteins of other legumes.

Amino acid composition. The amino acid composition of the legumins in *L. albus*[15] is compared in Table 6 with that of the cold-insoluble and cold-soluble fractions of conglutin α of *L. angustifolius* cv. Uniwhite and with the fractions with sedimentation coefficient 11·6 S of *L. angustifolius* and *L. luteus*.[23] All these various proteins appear far more similar to each other than did the albumins (Table 2). The high content of glutamic and aspartic acid has already been noted for albumins and is a rather common feature in seed proteins. It may be involved in forming a coordinated sheet of water around each protein molecule in the water-poor medium; the dehydration which follows protein deposition allows tighter protein packing since the protein-to-solvent interactions lost are substituted by protein–protein interactions. This may also involve some conformational change. As shown in Table 4 the relative amounts and degrees of amidation of dicarboxylic amino acids are mainly responsible for the acidity levels of the various proteins.

The major differences between legumins are in their arginine, threonine, methionine, serine and leucine contents. For other amino acids differences are smaller but nevertheless real: most of them are repeated in the relative amounts of the same components in the cold-insoluble and cold-soluble fractions of conglutin α. This confirms the correlation of legumin 8 with the cold-insoluble fraction and legumin 9a with the cold-soluble one.

The amino acid compositions of the storage proteins of the seed is mostly of interest from a nutritional standpoint: it was thought that it mighı be variable because the storage function of the protein supposedly did not dictate stringent molecular features. This is, however, not correct since the proteins in question are submitted to strict structural requirements in view of their close packing in a water-deficient medium and their subsequent transformations during germination. Indeed, in other

TABLE 5

SUBUNIT COMPOSITION OF SOME GLOBULINS OF *L. albus*[a]

	Mol. wt. (kD) of the native protein	*Mol. wt. (kD) of the protomers (S–S bound)*	*% of total*	*Number of subunits*
Globulin 8	240, 470	(44·5 + 19·0)	14	3 or 4⇆6 or 8
		9·0[b] + (38·5 + 19·0)	68	
		21·0 + (26·0 + 19·0)	18	
Globulin 9a	240, 470	(42·5 + 19·0)	5	3 or 4⇆6 or 8
		6·0[b] + (36·5 + 19·0)	46	
		6·0[b] + 21·0 + (17·5 + 19·0)	25	
		16·5 + (26·0 + 19·0)	24	
Conglutin γ (globulin 1)	199	(28·0 + 16·0)		4
Globulin 9b	44	(10·5 + 10·5)		1
		13·0		2

[a] from Refs. 15 and 22.

[b] Not seen in SDS–PAGE.

TABLE 6
AMINO ACID COMPOSITION OF LUPIN LEGUMINS (%)

	L. albus[a]		*L. angustifolius*			*L. luteus*[c]
	8	*9a*	*Conglutin α*[b]		*11·6 S*[c]	*11·6 S*
			Cold-insoluble	*Cold-soluble*		
Lys	3·98	3·60	3·03	3·08	3·41	3·71
His	1·86	2·06	2·20	2·17	2·66	2·53
Arg	11·86	9·98	12·64	11·78	12·61	10·10
Trp	0·96	1·19	1·06	1·16	1·42	1·41
Asp	10·12	10·39	10·28	10·55	11·32	11·76
Thr	3·85	4·78	3·25	3·60	3·23	3·26
Ser	4·91	4·11	4·50	4·38	4·15	4·01
Glu	24·76	23·56	27·21	25·96	22·32	23·26
Pro	4·38	4·34	5·02	4·83	4·19	4·18
Gly	3·91	4·25	4·32	4·46	3·74	4·02
Ala	2·63	2·84	3·10	3·43	2·95	3·04
½ Cys	1·26	1·14	0·47	0·60	1·58	1·52
Val	3·81	3·88	3·75	3·98	3·77	4·09
Met	0·18	0·54	0·14	0·19	0·44	0·39
Ile	5·81	5·40	4·07	4·36	4·91	5·05
Leu	7·54	8·61	7·09	7·36	7·42	7·68
Tyr	5·34	4·80	3·86	3·87	3·27	3·33
Phe	3·96	3·78	4·00	4·22	4·69	4·62
Amide		2·26	—	—	1·87	2·03
Index for potential bitter taste	1197	1183				
MEAA	82·82	83·97				

[a] Data from Ref. 15.
[b] Calculated from data of Ref. 13.
[c] Calculated from data of Ref. 23.

legumes, e.g. *Pisum*, a rigorous mechanism has been established for biosynthesis, and for conservation of sequence in the primary structure.[24]

Nature of the glycoprotein. Legumins are the globulins with the lowest level of covalently bound carbohydrate (Table 3) both in *L. albus*[15] and in *L. angustifolius*.[25] The component monosaccharides are typical for the group but their ratios differ in the two legumins, as does the amount of sugar per milligram of peptide (Table 7). The relatively high content of *N*-acetyl glucosamine is also typical of legumins (Table 3).

TABLE 7

MONOSACCHARIDE COMPOSITION OF CARBOHYDRATE IN GLOBULINS (%)

Proteins 1 to 9b refer to *L. albus*,[15] conglutins α, β and γ to *L. angustifolius*[25]

Sugar			*Vicilins*				*Legumins*			
	γ	*1*	*4*	*5*	*6*	β	*8*	*9a*	α	*9b*
Fucose	7·0	—	—	—	—	—	—	—	—	—
Xylose	—	3·9	—	—	—	—	trace	trace	—	3·3
Arabinose	—	20·8	—	—	—	—	trace	trace	—	7·5
Galactose	24·4	61·2	—	trace	trace	18·7	2·9	20·1	38·2	31·2
Mannose	49·5	10·7	96·3	72·0	94·2	65·6	79·9	68·9	44·5	41·5
Glucose	—	2·8	—	22·9	—	—	6·3	1·3	—	13·1
N-Acetyl glucos-amine.	19·1	0·6	3·7	5·1	5·8	13·3	10·9	9·7	14·6	3·4

Eaton Mordas and Moore[26] suggest for *L. angustifolius* one single oligosaccharide chain of mol. wt. 4 kD per molecule of conglutin α, with a mannose to galactose ratio of about 4. Also in *L. albus* legumins, mannose far exceeds galactose.[15] The distribution of sugar in the various legumin protomers is shown in Table 8. Carbohydrate is not attached to

TABLE 8

SUGAR DISTRIBUTION[a] IN PROTOMERS OF LEGUMINS FROM *L. albus*[b]

Globulin 8		*Globulin 9a*	
Mol. wt. (kD)	*Sugar*	*Mol. wt. (kD)*	*Sugar*
55·6 } 47·2 }	68·1		—
42·1	20·5	45·6 } 42·1 }	61·6
24·4	7·1	25·3	30·4
21·4	4·1	21·5	6·9
18·8	0	19·0	0

[a] Sugars and molecular weights refer to peptides liberated after treatment with β-mercaptoethanol. For each protomer, sugar is given as percent of total.

[b] From Ref. 56.

the small disulphide-linked peptide. If reference is made to the origin of the protomers from the parent polypeptide indicated in Figs. 4 and 5, remembering that data refer to two different cultivars of *L. albus*, it also appears unlikely that, in the major polypeptide, sugar is bound far from the interchain disulphide(s).

Vicilins

Molecular properties. Globulins 4, 5, 6 and 7, the vicilins, correspond to the component with sedimentation coefficient 7·8 S[5] and to conglutin β. This protein was resolved by arbitrary ammonium sulphate cuts into four subfractions. Subunits in each subfraction appear quite similar but, taking them in order of increasing solubility, there is a progressive increase in the proportion of low, and a decrease in the proportion of the high molecular weight components.[13] On the other hand, as shown in Fig. 6, globulins 4, 5, 6 and 7 display a pattern on SDS–PAGE which is

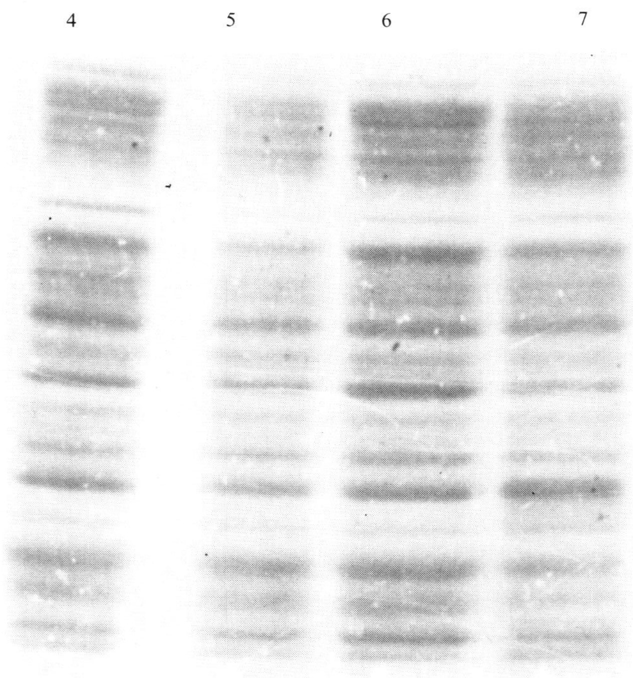

M.W. (kD)	Mole, % 4	5	6	7
64·0	15	15	18	18
43·0	9	5	9	8
34·0	11	11	10	12
28·0	11	9	11	8
21·0	14	14	12	19
16·5	27	25	20	17
13·5	13	21	20	18

FIG. 6. Subunit separation by SDS–PAGE of the vicilins of *L. albus*. The molecular weights and quantitative distributions are also given.

qualitatively identical and shows only minor differences in the quantitative distribution of subunits.

Differences in subunit composition of different species have been described. One major component of conglutin β, a 30 kD subunit, is present in *L. angustifolius* and other Old and New World *Lupinus* species but is absent from the American *L. elegans* and *L. mutabilis*.[27]

In vicilins all subunits contain only one peptide; absence of interchain disulphide bonds is the natural consequence of the lack of cysteine in these molecules. Subunits with corresponding molecular weight in separate proteins probably have similar amino acid compositions in view of the overall amino acid similarity of the vicilins. The relative amounts of the various subunits in each molecule do not indicate a definite structure which might include them all. In globulin 7, the sum of molecular weights of the major subunits is larger than the molecular weight of the protein itself, and it is therefore impossible to accommodate them all into a single molecule. It is likely that the peptides in question originate, as with the legumins, from different fragments of the same polypeptide, so that their associations yield the same overall amino acid composition and molecular weight. However, the high number of peptides resolved, and the molar percentages they display, do not allow as straightforward an interpretation of their associations as was possible for the legumins.

In view of the similarity of subunits of the various vicilins in a given species, splitting apparently occurs in all of them in the same or almost the same positions, in line with the similarities of these molecules.

Some of the major subunits are accompanied in very small amount by a number of peptides of similar molecular weight. This heterogeneity is a further cause of difficulty in evaluating the structure of these globulins.

The basic unit of the molecule, formed by the set of peptides produced in the splitting of the parent polypeptide, associates in different ratios to yield the various oligomeric vicilins. This is probably the basis for the molecular identity of each vicilin, which further relies on minor differences in amino acid composition and in diversity of the bound carbohydrate.

Values of molecular weight shown in Table 3 for vicilins appear somewhat higher than was reported for conglutin β of *L. angustifolius*.[4] There may be some overestimation due to the assay used.

Amino acid composition. The amino acid composition of the vicilins in *L. albus* is compared in Table 9 with that of conglutin β from

TABLE 9

AMINO ACID COMPOSITION OF LUPIN VICILINS (%)

	L. albus[a]				*L. angustifolius*[b]	*L. luteus*[c]
	4	*5*	*6*	*7*	*Conglutin β*	*7·4 S*
Lys	4·10	3·78	3·51	3·08	4·20	3·54
His	1·21	1·79	1·35	1·28	1·75	1·62
Arg	12·80	12·35	13·77	13·35	12·25	11·12
Trp	0·04	0	0	0	0·12	—
Asp	13·43	12·17	11·74	12·86	13·82	13·56
Thr	3·04	2·85	2·63	2·71	2·70	2·77
Ser	5·65	6·61	5·53	4·91	5·66	5·31
Glu	23·34	25·27	24·66	27·31	22·70	21·16
Pro	4·04	4·64	4·08	3·70	4·05	3·89
Gly	3·31	4·09	3·90	3·19	3·60	2·94
Ala	2·88	2·73[c]	2·66	2·75	2·57	2·78
½ Cys	trace	0·03	0	0	0·27	1·51
Val	3·09	2·39	2·95[b]	3·13	3·14	3·73
Met	trace	trace	0	0	0	—
Ile	4·89	3·39	4·09	4·46	5·34	5·38
Leu	7·31	6·49	7·00	6·84	7·55	8·58
Tyr	4·49	5·27	6·69	6·32	5·24	4·86
Phe	4·91	4·41	4·36	3·14	5·04	4·93
Amide	2·99		2·82		—	2·30
Index of potential bitter taste	1132	1090	1159	1135		
MEAA	33·23	25·12	25·91	26·39		

a, *b*, *c*, as in Table 6.

L. angustifolius cv. Uniwhite[13] and of the component with sedimentation constant 7·4 S in *L. luteus*.[23] The overall similarity of these proteins, except for absence of cysteine and somewhat lower values of isoleucine and tyrosine in *L. albus*, is remarkable.

The two major vicilins, proteins 4 and 6, differ significantly only in aspartic acid and tyrosine content. Proteins 5 and 7 which, in terms of subunit composition, molecular weight and behaviour in ion exchange, appear strictly related to 4 and 6 respectively, are also like these as far as amino acids are concerned.

The situation with respect to dicarboxylic amino acids is similar to that of the legumins.

Nature of the glycoprotein. In *L. albus*, covalently bound carbohydrate is more relevant in vicilins than in legumins[15] (Table 3), and in *L. angustifolius*, conglutin β is the globulin with the highest bound sugar content.[25] According to Eaton Mordas and Moore each molecule of this protein contains two oligosaccharide chains of molecular weight 5 kD.[26]

In the vicilins of *L. albus* sugar is present mainly as mannose, with some *N*-acetyl glucosamine, but globulin 5 contains also a considerable level of glucose (Table 7). In *L. angustifolius*, mannose is the major component but galactose and glucosamine are also present.

Conglutin γ, i.e. globulin 1

Molecular structure. Globulin 1 of *L. albus* corresponds to the conglutin γ isolated from various species. Gel filtration assays determined for globulin 1 an apparent molecular weight of 199 kD,[22] which indicates that it is a tetramer formed by equal monomers of 45·0 kD (Table 5). This latter value was obtained by gel filtration and by SDS–PAGE.[28] In *L. angustifolius* cv. Uniwhite conglutin γ is reported as a hexamer of 280 kD, determined by sedimentation equilibrium. An apparent molecular weight of 320 kD was obtained by calibration of a Sepharose CL-6B column with protein standards. The oligomeric protein dissociates at pH 4·8 to the monomer, which shows a molecular weight of 47 kD by sedimentation equilibrium and 43–45 kD on SDS–PAGE.[29] The native protein contains approximately 15% α-helix, 35% β-structure, 18% β-turns and 32% unordered regions.[29]

In both species the monomer contains two disulphide-linked protomers of molecular weight respectively 16·0 and 28·0 kD in *L. albus*[28] and 16·5–17 and 28–30 kD in *L. angustifolius*. In this species the lower molecular weight chain has an apparent isoelectric point near 6·9 whereas the value for the larger chain is near 8·0. There is also a minor band with the same molecular weight as the larger polypeptide, with an apparent isoelectric point near 7·8. If the amount of protein loaded on the gel is greatly increased, further minor bands are observed which appear to be charge isomers of both the small and large chains.[29] Also, in *L. albus* the light protomer is more acidic in character, and both peptides display charge but not molecular weight heterogeneity: the isopeptides of the heavier protomer have isoelectric points which range from 6·2 to 6·8 whereas those of the lighter one are in the range from 5·6 to 5·9.[28]

Amino acid composition. The amino acid composition of globulin 1 (conglutin γ) of *L. albus* is compared in Table 10 with that of the same

TABLE 10

AMINO ACID COMPOSITION OF LUPIN GLOBULINS OTHER THAN LEGUMINS AND VICILINS

	Globulin 1 = conglutin γ				*L. albus*[b]	*L. luteus*[d]	*L. albus*[b]	
	L. albus[b]		*L. angustifolius*[c]		*9b*	*2 S*	*3*	*11*
			Residues per mole Small Subunit[e]	*Large Subunit*[f]				
Lys	6·34	4·72	6	11	1·02	1·01	5·90	7·85
His	5·53	5·60	6	15	1·84	0·62	3·40	2·57
Arg	4·04	3·84	5	11	11·14	8·92	12·41	9·91
Trp	1·11	1·13	1	3	0·03	—	1·67	n.d.
Asp	11·81	11·63	18	25	9·81	9·52	9·25	7·08
Thr	7·43	5·93	7	15	1·36	1·12	4·68	3·60
Ser	8·79	7·63	16	12	5·07	4·66	7·13	6·95
Glu	10·10	9·65	11	25	41·06	37·09	21·15	12·10
Pro	5·98	5·55	5	16	2·00	2·39	5·07	5·12
Gly	4·05	5·06	16	3	2·00	2·08	4·85	6·95
Ala	4·21	4·17	11	4	0·83	1·23	3·23	3·86
$\frac{1}{2}$ Cys	2·62	2·78	3	7	6·55	7·19	0·83	0
Val	6·30	6·30	14	9	1·73	1·48	3·96	2·96

Met	0·75	1·20	2	2	0·50	0·42	0·28	3·99
Ile	4·91	4·11	7	8	3·33	3·55	3·40	5·66
Leu	9·35	10·84	13	27	8·76	10·72	5·96	11·71
Tyr	3·02	4·02	4	11	1·33	0·77	3·63	4·89
Phe	4·82	5·83	9	13	1·10	3·66	3·57	5·79
Amide	2·28		—	—	3·34	3·59		
Index for potential bitter taste	1202				856		1088	1322
MEAA	99·46				35·12		80·77	91·50

[a] All data are given as percent, except for the large and small subunit of conglutin γ, for which residues per mole are given.
[b] Data from Ref. 15.
[c] Calculated from Ref. 13.
[d] Calculated from Ref. 23.
[e] Calculated from Ref. 30.
[f] Calculated as follows: residues per mole conglutin γ (calculated from Ref. 13 and 29)–residues per mole small subunit.[30]

protein from *L. angustifolius*, for which is also listed the number of amino acid residues per mole of the two component polypeptides. The two proteins are similar, with some differences in lysine and threonine and to a smaller extent serine, leucine and glycine. The high level of leucine, threonine, and aromatic amino acids is noteworthy as compared to the other globulins considered so far, together with lower glutamate.

For the smaller subunit of conglutin γ the amino acid sequence was also determined,[30] as shown in Fig. 7.

10 20
Ser-Thr-Ser-Tyr-His-Gly-Ser-Gly-Glu-Ile-Gly-Gly-Ala-Leu-Ile-Thr-Thr-Thr-His-Pro-

30 40
Tyr-Thr-Val-Leu-Ser-His-Ser-Ile-Phe-Glu-Val-Phe-Thr-Gln-Val-Phe-Ala-Asn-Asn-Met-

50 60
Pro-Lys-Gln-Ala-Gln-Val-Lys-Ala-Val-Gly-Pro-Phe-Gly-Leu-SCMCys-Tyr-Asp-Ser-Arg-Lys-

70 80
Ile-Ser-Gly-Gly-Ala-Pro-Ser-Val-Asp-Leu-Ile-Leu-Asp-Lys-Asn-Asp-Ala-Val-Trp-Argt-

90 100
Ile-Ser-Ser-Glu-Asn-Phe-Met-Val-Gln-Ala-Gln-Asp-Gly-Val-Ser-SCMCys-Leu-Gly-Phe-Val-

110 120
Asp-Gly-Gly-Val-His-Ala-Arg-Ala-Gly-Ile-Ala-Leu-Gly-Ala-His-His-Leu-Glu-Glu-Asn-

130 140
Leu-Val-Val-Phe-Asp-Leu-Glu-Arg-Ser-Arg-Val-Gly-Phe-Asn-Ser-Asn-Ser-Leu-Lys-Ser-

150
Tyr-Gly-Lys-Thr-SCMCys-Ser-Asn-Leu-Phe-Asp-Leu-Asn-Asn-Pro

FIG. 7. Complete amino acid sequence of the S-carboxymethylated derivative of the small subunit of conglutin γ from *L. angustifolius*.[30]

Nature of the glycoprotein. Conglutin γ of *L. albus* has the highest bound sugar of the globulins.[15] In *L. angustifolius*, it also has a high carbohydrate content. The size of the oligosaccharide chain in this protein is less clearly established than for the other conglutins: a chain of molecular weight 1·3 kD has been proposed for each subunit.[26] In *L. albus* all the sugar is bound to the large polypeptide.[28]

The composition of the sugars is more varied than for legumins and vicilins: pentoses and hexoses are present, with galactose as the major component. Only traces of glucosamines are found.[15] In *L. angustifolius*, mannose is the prevailing monosaccharide and important amounts of glucosamine are present[25] (Table 7).

Globulin 9b

Globulin 9b corresponds to the 2 S components found in *L. angustifolius*[5, 13] and in *L. luteus*.[5] A 2 S globulin has also been reported for *L. albus*.[5, 31] Protein 9b is the only globulin with a relatively low molecular weight: it liberates two types of subunit, one of 13 kD and one of 21·6 kD, this latter formed by two disulphide-linked protomers of 10·5 kD each. Since the apparent molecular weight of the protein is 44 kD, the heavier and lighter subunits appear in a 1:2 stoichiometric ratio.[28]

Protein 9b is the most acidic of the globulins. As shown in Table 4 this corresponds to an unusually high content of non-amidated glutamic acid and low content of basic amino acids.

The amino acid composition is similar in *L. albus* and *L. luteus* (Table 10): besides the presence of very high levels of dicarboxylic amino acids the high cysteine and absence of methionine are typical.

Bound sugar is scarce; pentoses and hexoses are present but their relative amounts differ from those found in globulin 1 (conglutin γ) (Table 7).

Other Globulins

Globulin fractions 2, 3, 10 and 11 correspond to minor seed proteins. Except for fraction 10, they are likely to contain proteoglycans because of the very high bound carbohydrate. Each fraction is probably made of more than one protein: this has been clearly established for fraction 3.[38] This fraction contains an active protease which is probably involved in the nicking of polypeptide chains of storage proteins. The enzyme represents less than 1/5 the total protein in the fraction.

The amino acid compositions of proteins 3 and 11 are shown in Table 10. They differ from any of the major proteins. In globulin 11, note the high lysine, leucine and methionine, no cysteine and low glutamate.

FACTORS DETERMINING PROTEIN COMPOSITION

Heterogeneity of Proteins

Lupin proteins behave as homogeneous species at the molecular level but are micro-heterogeneous at subunit level. Indeed, in all the major globulins isoelectric focusing (IEF) followed by SDS–PAGE in the two-dimensional techniques, shows that a number of bands which can be separated by IEF have the same molecular weight, thus indicating heterogeneity of charge. Heterogeneity in molecular weight is revealed by gel filtration in 8 M urea followed by SDS–PAGE: in the gel filtration fractions, the latter technique resolves several similar protomers having molecular weights close to that of the fraction itself. These do not derive from the breakdown of a larger unit but probably were present as alternatives in the molecules.[28] The situation is relatively specific: in *Lupinus albus* the subsunits of conglutin γ and the 19·0 kD peptide of both legumins display only charge heterogeneity.[57]

Heterogeneity is an intrinsic characteristic of the proteins, being also observed when those from a single seed are analysed. The multiplicity of bands increases only slightly in proteins isolated from the flour of a genetically selected cultivar. Flours of commercial origin apparently display fewer bands than individual seeds.

This is because the differences between individual seeds are swamped when bands present in all seeds dominate the profile, as in flour preparations.[22, 32]

The heterogeneity described may depend on different mechanisms. It may reflect real differences in the composition of the peptide which have been observed, e.g. for the storage proteins of *Vicia faba*, *Pisum sativum*, etc. However, such differences probably contribute only partially, since conservative behaviour has been observed for seed proteins from *Pisum*. In view of the mechanism previously outlined for the origin of experimentally determined subunits, slightly different splittings of the original parent polypeptide might produce similar, but not identical, subunits. In this hypothesis, both multiplicity of subunits and heterogeneity are post-translational effects related to different splittings of the original polypeptide. The different location of proteolysis might also depend on position, chain length or composition of bound sugar, all resulting from post-translational events. It is interesting that in legumins the 19·0 kD peptide is homogeneous and does not contain bound sugar; the carbohydrate appears to be associated with that part of the major polypeptide which is more involved in subunit heterogeneity. Finally, heterogeneity may also be related to the amidation of dicarboxylic amino acids. It has already been shown that the level of amide influences the charge properties of the proteins: local differences between molecules may affect the behaviour on IEF of separated subunits. Deamidation is also a post-translational event, and charge differences have been observed in newly synthesised polypeptides from seeds of *Pisum sativum*.[24]

Species Variability of Proteins

One aspect related to heterogeneity is the genotype-dependent differentiation between proteins. The protein level in seeds of 53 *L. angustifolius* genotypes varied: 60 % of the samples contained 32–38 % (dry weight) protein, measured as total nitrogen × 6·25, but a smaller proportion (28 %) contained 38–44 %. There was no apparent correlation between protein level and size. Cellulose acetate electrophoresis of the globulins from 99 genotypes showed considerable variation both in the proportions of conglutins α, β, γ and in their relative electrophoretic mobilities. This corresponds to differences in the number, proportions and molecular size of the subunits originating from the total seed globulins and in those isolated from the separated conglutins.[17] Differences are more marked between different species than between cultivars of the same species.[27]

Studies on F_1, F_2 and F_3 families showed that the proportion of conglutin γ in the storage proteins varied inversely with the proportion of the subunit of 30 kD of conglutin β. The proportion of the β subunit in any seed depends largely on its own genotype, and not on the genotype of the maternal plant. A *beta subunit controller* was postulated, which, probably in addition to one or more other loci, has regulating and processing functions. The high level of conglutin γ, which is particularly valuable for its high nutritional quality, is also controlled by several loci, making it more difficult to manipulate in a plant breeding programme.[33]

Environmental Effects on Protein Pattern

A number of lupin species having different proportions of conglutin α, β and γ in their seed were grown either under sulphur deficiency or with adequate sulphur fertilisation, i.e. their pots received sulphur-free nutrient or a solution containing 1 mM sulphate. Protein level in the seeds was only marginally affected by sulphur supply. However, the ratio of total sulphur to total nitrogen in the seed was greatly depressed under sulphur deficiency conditions. This large change in sulphur–nitrogen ratio corresponded to suppression of the synthesis of conglutins α and γ, which contain significant amounts of sulphur-containing amino acids. The level of protein was maintained by increased synthesis of conglutin β, which normally contains no methionine and very little cysteine. These changes in the proportions of the proteins were reflected in the amino acid analyses for the globulin extracts.[34] Apparently, in *L. angustifolius*, the conglutin β which is increasingly synthesised in conditions of sulphur deficiency is largely if not entirely made up of high molecular weight subunits[34, 35] whereas with other species, like *L. elegans*, the ratio between subunits does not change.[34]

Similar effects were observed in field trials. The level of total nitrogen in seeds of *L. angustifolius* cv. Uniharvest was not significantly modified by the rate of sulphur, potassium and phosphorus fertilisers applied. Nevertheless, addition of phosphorus increased the amount of cysteine as also did a high rate (100 kg/ha) of sulphur. Potassium had hardly any effect.[36]

Associated Proteases

In many of the molecular aspects of proteins so far discussed the action of proteolytic enzymes has been assumed to be responsible for the effects observed. Proteases are indeed present in the dormant seed. As shown in Table 11, highest activity in *L. albus* is displayed by the albumins and by globulin fraction 3. A weak effect is observed also with the other globulins and the question is whether it is due to contamination by the mentioned proteases.[37, 38]

The enzymes purified from the albumins and from globulin fraction 3 appear similar in many respects. Both have optimal activity around pH 8 and, with respect to inhibitors, behave as SH proteases. On isoelectric focussing the protease from the albumins displays isoelectric point at pH 4·6 and the one from fraction 3 at pH 4·5. The apparent molecular weight is 63 kD for the enzyme from albumins and 73 kD for the one from fraction 3. On SDS–PAGE both enzymes display same peaks but differ in relative amounts.[38]

It can be excluded that these enzymes have major effects during extraction and isolation of proteins in the conditions used for *L. albus*.[38] It is therefore likely that the proteases act in the seed.

In *L. angustifolius* all three salt-extracted conglutins contain a lower relative proportion of high molecular weight subunits than those extracted directly with 2% SDS. This has been attributed to degradation by enzymes originating in the seed.[17] Indeed this may be the case, since part of the extraction was done at room temperature and proteases are very sensitive to temperature effects.

During germination of lupin, as in other legumes, a strong proteolytic activity takes place. It differs however from the one just discussed. The protomer pattern is abruptly modified in *L. albus* around the eighth germination day. Both for vicilins and for legumins other protomers disappear and components around 17 and 12 kD accumulate.[39] In this

TABLE 11
PROTEOLYTIC ACTIVITY ASSOCIATED WITH THE VARIOUS GLOBULINS OF *L. albus*[a]

Globulin fraction	*Var. LA 7110*[b]	*Var. Multolupa*[a]
Conglutin γ	0·169	0·045
4	0·132	0·106
6	0·028	0·068
8	0·075	0·122
9a	4·27 (fractions 9a, 9b, 3, 5, 7, 10+11 together)	0·033
9b		0·017
3		22·8/109[c]
5		0·240
7		0·035
10+11		0·037
Total globulin extract	0·287	0·151
Lupin albumins	31·1	29·4/146[c]
Trypsin (2 × crystallised, Sigma)	1710	

[a] From Ref. 38.
[b] From Ref. 32.
[c] After purification.
nmoles 0·33 mM benzoyl–arginyl–paranitroanilide (BAPNA) hydrolysed per min per mg protein at 37 °C and pH 7·0 are given.

species and in *L. angustifolius* vicilins, i.e. conglutin β, are hydrolysed first followed by legumins (conglutin α) and, in *L. angustifolius*, by conglutin γ which remains unmodified in *L. albus*. A protease has been isolated from the cotyledons on the fifth day after germination. It is a protein of 27·5 kD, with isoelectric point 9·0 and optimum pH 4·0 using gliadin as substrate.[40]

Immunochemistry

A chemotaxonomic approach to lupin has been carried out by differentiating the protein fractions by immunological methods. It has been found that the genus constitutes a serologically uniform taxonomic unit that can be divided into several sections. Most distant are *L. luteus*, *L. albus*, *L. angustifolius* and *L. perennis*. Quantitative serological tests, however, do not reveal significant differences, treating the Mediterranean and American lupins separately.[41]

The proteins of *L. albus* and *L. mutabilis* were divided into six, and those of *L. luteus* and *L. angustifolius* into four, serologically different fractions. The proteins of Mediterranean lupins appeared to be serologically closely related, whereas the proteins of *L. mutabilis*, despite displaying a certain degree of antigenic identity with those of *L. angustifolius*, had comparatively few antigenic determinants in common with the proteins of the two remaining Mediterranean species.[42]

In cases of species showing differences in immunological pattern, F_1 fertility becomes reduced. Species belonging to unlike sections display great differences in serological patterns. Crossing only produces poorly developed seeds incapable of development. In cases of species showing great differences in the compositions of their proteins, after foreign pollination seeds are not developed at all. Differences in the serological pattern between non-intercrossing species may be of the same order as differences between related genera. For example, the difference between *L. albus* and *L. nanus* is of the same order as that between *L. luteus* and the *Thermopsis* genus.[41]

The antigenic properties are modified during seed ripening, germination and seedling development;[43] this reflects the breakdown of reserve proteins.

NUTRITIONAL ASPECTS

When nutritional aspects are under consideration, both the total amount of protein produced and its nutritional quality, in terms of component amino acids and metabolic availability, are important.

Essential Amino Acids

Nutritionally, vicilins are poor, lacking both sulphur-containing amino acids and tryptophan, and being low in valine, lysine and threonine. This produces a low MEAA index (Table 9). Legumins are better, having only minor deficiencies of lysine, sulphur-containing amino acids and valine: indeed their MEAA index is above 80% (Table 6). Globulin 9b is peculiar, having no tryptophan, low lysine, threonine, methionine, valine and aromatic amino acids, but very high cysteine and glutamic acid. This unbalanced composition results in a low MEAA value (Table 10). Globulin 1 has a well balanced amino acid composition, which results in an MEAA value of 99%. The index for potential bitter taste calculated according to Ney[44] is always below 1300, the critical level for beginning of bitterness, except for globulin 11 which, however, is still in the marginal range.

The total globulin extract is low in methionine and tryptophan but still has an MEAA index of 74.[20] Due to the presence of albumins, which display a more balanced amino acid composition, the amino acid pattern of lupin flour is better than for total globulins, giving a higher MEAA value.[20]

When different varieties are compared for essential amino acids (Table 12) they all appear to behave similarly, within the limits of data from

TABLE 12

TOTAL PROTEIN AND ESSENTIAL AMINO ACID CONTENT[a] OF WHOLE LUPIN SEED[b]

	Lupinus species				*Soybean*
	Albus	*Angustifolius*	*Luteus*	*Mutabilis*	
Total protein	35·4	31·2	44·0	42·9	35·0
Ile	5·12	3·84	4·55	4·68	4·54
Leu	8·53	6·63	8·96	7·42	7·68
Lys	5·56	5·17	6·10	5·95	6·38
Met + Cys	2·78	2·29	3·11	2·67	2·59
Phe + Tyr	8·87	7·11	6·91	7·91	8·08
Thr	4·14	3·54	3·99	4·01	3·86
Trp	0·94	0·91	0·86	0·84	1·28
Val	4·95	4·03	4·25	4·01	4·80
MEAA	74·31	62·21	70·65	68·33	69·47

[a] Proteins are given as percent of whole seed; amino acids as g per 16 g nitrogen.
[b] Data from Ref. 60.

different laboratories. The best pattern is shown by *L. luteus* which also has the highest protein content. Favourable, results are given also by *L. mutabilis*. Comparison with soybeans is also favourable, both for essential amino acids and protein yield; the soybean, however, appears richer in tryptophan and lysine.

In vitro Digestibility

Pepsin, and to an even greater extent trypsin, are less active towards lupin globulins than towards casein or BSA, and the decrease differs for the various proteins. The effect is more evident for biuret-reactive, i.e. larger, peptides than for ninhydrin-reactive compounds.[21]

The decrease in activity correlates with bound sugar in the various globulins. The native conformation of the proteins seems to be a factor in the reduction of activity of trypsin since the reduced enzyme activity largely disappears after heat denaturation, whereas the increase is far less evident with casein as substrate.[21]

Denaturation probably unmasks new sites of attack for trypsin on the peptide: enzyme action is no longer hindered and the normal effect of endopeptidases to produce larger fragments prevails. This explains the different reactivity with biuret and with ninhydrin.

In conclusion, the poorer nutritional performance of lupin proteins depends not only on the amino acid composition but also on reduced digestibility by proteolytic enzymes.

In vivo Assays

Assays of PER, NPU and BV in rats on diets containing 10% protein from sweet and semi-sweet varieties of *L. albus*, as well as water and/or ethanol debittered products of *L. mutabilis*, showed that the lupin protein enriched with DL-methionine (2% in the protein) or fortified with complementary protein carriers rich in sulphur-containing amino acids, is a promising protein source for animal and human nutrition. The apparent digestibility of both raw and processed seeds, and oil cake, was 84% in *L. mutabilis* and 72·9–79·2% in *L. albus* as compared with 87% for casein. The true digestibility of the lupin protein reached 92%, which is comparable with that of casein.[45,46]

A careful review of nutritional work done with lupin up to the late 1970s has been composed by Hill.[47] Feeding trials with rats, and with poultry, indicated that methionine is the only seriously limiting amino acid. However, reported pig trials in Australia and New Zealand have given no response to methionine. In experiments in which large quantities of *L. angustifolius* testa were added to rat diets the reduced digestibility caused by the hulls was attributed to the physical entrapment of protein, and food generally, in the large mass of fibrous material, making it less accessible to enzymes.[48]

There is no doubt that lupin seed can be used successfully in pig rations as a major protein source. Nevertheless considerable differences occur between species or varieties of lupin regarding their suitability. Variation in the alkaloid content is one important determinant of their nutritive value. However, even in cases in which the alkaloid content is very low, such as the Uniwhite cultivar, the value of lupin seed meal for pigs may still be limited by factors which suppress voluntary feed consumption.[49] Lysine from lupin is less available than that from soy for growing pigs. However, while *L. angustifolius* can be freely increased in the diet, with *L. albus*, growth decreases at lupin levels higher than 20%.

The high fibre content of the food does not interfere with digestion of the protein.[58] In marked contrast to pigs, rats are more resistant to dietary lupin alkaloids.[50] Availability estimates for lysine with rats were higher than for pigs but lower as compared to broilers.[59] Further work is required on the use of lupin seed in rations based on cereals other than barley and maize, the effect of removal of testa, and heat treatment on nutritional quality and the identification of lupin species best suited for pig production.[47]

The use of lupin seed in broiler rations depends more on economic than nutritional considerations. *L. angustifolius* and *L. luteus* appear to be suitable for replacing soybean meal in wheat-based poultry rations for chicks, laying hens or intensive broiler production, provided methionine and lysine are added. There are no published results for *L. albus*, which has less testa than these two current commercial species, or for *L. consentini* which may become an important species in Western Australia.[47]

Systematic studies for using debittered lupin flour for bread and other cereal products started in 1919, with favourable results.[47] In Peru, humans fed on diets containing from 5 to 40% flour from *L. albus* did not show any abnormalities and many of the foods were better accepted than traditional foods.[51]

FUNCTIONAL PROPERTIES

Solubility

The solubility profile of the globulins shows a sharp minimum from pH 4 to pH 5·5, with two separate isoelectric regions around pH 4·65 and pH 5·60.[16] These correspond to the vicilins and the legumins respectively. Conglutin γ (globulin 1) is isoelectric in the neutral range but represents only 6% of globulins. The total extract is clearly dominated by the behaviour of the globulins: in the acid range, however, the total extract becomes significantly less soluble than the globulins. A similar decrease is described for total protein from soybeans and from haricot beans.

The ionic strength dramatically affects solubility, and by adding enough salt, more than 60% of the globulins is brought into solution at pH 5.[52] The region of minimum solubility is reduced by at least 4 pH units and, even at the point of minimum solubility, about twice as much protein remains in solution in the presence of the salt as in its absence. Sodium ions are more effective than calcium ions in promoting solubility.[16]

Water Adsorption and Desorption

Water sorption compares well with that of other new food proteins, e.g. those from alfalfa leaves or soybeans, but is less than for proteins from isolated maize germ. A minor hysteresis is observed in desorption.[54]

The emulsifying and water binding capacities of lupin protein isolates and concentrates depend on the isolation procedures used. Protein extraction by wet milling, followed by precipitation and drying at the isoelectric pH, promoted water binding capacity while dry milling and alkaline extraction enhanced the emulsifying capacity. Protein co-precipitation and ion environmental conditions are other influencing factors. They may be involved in the modification of the properties mentioned which occur on association with meat proteins.[53]

Emulsifying Properties

Some of the methods reported in the literature do not correctly represent the emulsifying potential of lupin protein because the prescribed amount of oil is too small and because, in the determination of stability, they involve heating steps which denature the proteins. When such shortcomings are avoided it appears that the emulsion capacity is good and changes with pH with a consistent decrease throughout the pH region of reduced solubility and a minimum at about the isoelectric point. Emulsion stability parallels emulsion capacity except for reduced stability around pH 6.[52]

Foaming Properties

On an equal protein basis both foaming capacity and foam stability are much higher in the defatted flour than in the globulin extract. The improved properties of the flour probably depend on other components, which appear particularly effective in the acid and alkaline ranges. At extreme pH values the effect is probably associated with the denaturation of the proteins. The sharp minima in the curves for the globulins correspond to the pH of minimum solubility; the overall shapes, however, do not correspond precisely with those of the solubility curves, and other effects related to the conformation and association of these oligomeric molecules probably come into play.[52]

Rheological Properties

Components other than proteins in lupin flour influence its rheological behaviour: isolated globulins display a much lower viscosity than the flour and require a higher temperature for gelling. Gels form only on the acid side of the isoelectric pH, whereas the flour yields gels also at neutral and alkaline pH.[55] As noted previously, globulins contain considerable amounts of free glutamic and aspartic acids. In the pH region where gels are formed carboxylic groups are less dissociated and the interaction between protein molecules and with the solvent increases, with consequent effects on gel formation and on viscosity. At the isoelectric pH, protein coagulates to form granules, and at extreme pHs no visible change is observed.

When submitted to changing shearing gradients the initial globulin gel rearranges to a less resistant structure. The overall behaviour is thixotropic, with considerable hysteresis, and is similar to that observed with soy proteins.[55] The three-dimensional network of the gel is probably altered by mechanical stress. Final values, apparently stable and reversible, reached by gels from flour, correspond to those measured in viscosity determinations of dispersions. In gels of globulins, on the other hand, the consistency decreases to much lower values. In both cases the product recovered still has gel features. Some solvent separates, corresponding to one-third of the original gel volume.[55]

ACKNOWLEDGEMENT

The work of the Author's laboratory reported in the present paper has been in part supported by grants from the Italian National Research Council (C.N.R.).

REFERENCES

1. Ritthausen, H. (1882). *Z. Prakt. Chem.*, **26**, 422–44.
2. Osborne, T. B. and Campbell, G. F. (1897). *J. Am. Chem. Soc.*, **19**, 452–72.
3. Osborne, T. B. and Harris, I. F. (1905). *Am. J. Physiol.*, **13**, 436–47.
4. Joubert, F. J. (1955). *Biochim. Biophys. Acta*, **16**, 370–6.
5. Joubert, F. J. (1956). *Biochim. Biophys. Acta*, **19**, 172–3.
6. Hudson, B. J. F. (1979). *Qual. Plant.*, **29**, 245–51.
7. Ruiz Jr., L. P. (1978). *N.Z. J. Agric. Res.*, **21**, 241–2.
8. Macrae, R. and Zand-Moghaddam, A. (1978). *J. Sci. Food Agric.*, **29**, 1083–6.
9. Pompei, C. *Proc. EEC Seminar on Perspectives for Peas and Lupins as Protein Crops, Sorrento*, 1981, in press.
10. Horn, P. E., Hill, G. D. and Porter, N. G. (1978). *Proc. Agronomy Soc. N.Z.*, **8**, 73–7.

11. Withers, N. J., King, S. and Hove, E. L. (1975). *N.Z. J. Expl. Agric.*, **3**, 331–2.
12. Hove, E. L. (1974). *J. Sci. Food Agric.*, **25**, 851–6.
13. Blagrove, R. J. and Gillespie, J. M. (1975). *Austral. J. Plant Physiol.*, **2**, 13–27.
14. Ericson, M. C. and Chrispeels, M. J. (1973). *Plant Physiol.*, **52**, 98–104.
15. Duranti, M., Restani, P., Poniatowska, M. and Cerletti, P. (1981). *Phytochemistry*, **20**, 2071–5.
16. Manrique, J., Cole, S. J. and Edwards, R. A. (1974). *Proc. 24th Annual Conf. Royal Australian Chem. Inst.*, 68–70.
17. Blagrove, R. J. and Gillespie, J. M. (1978). *Austral. J. Plant Physiol.*, **5**, 651–63.
18. Sgarbieri, V. C. and Galeazzi, M. A. M. (1978). *J. Agric. Food Chem.*, **26**, 1438–42.
19. Cerletti, P., Fumagalli, A. and Venturin, D. (1978). *J. Food Science*, **43**, 1409–11.
20. Duranti, M. and Cerletti, P. (1979). *J. Agric. Food Chem.*, **27**, 977–8.
21. Restani, P., Cerletti, P., Semino, G. and Mariani, A. 2nd Intl Lupine Congress, Torremolinos 1982, in press.
22. Duranti, M. and Cerletti, P. (1982). *Proc. EEC Seminar on Perspectives for Peas and Lupins as Protein Crops, Sorrento*, 1981, in press.
23. Gerritsan, T. (1956). *Biochim. Biophys. Acta*, **22**, 269–73.
24. Gatehouse, J. A., Croy, R. D. D., Morton, H., Tyler, M. and Boulter, D. (1981), *Eur. J. Biochem.*, **118**, 627–33.
25. Eaton Mordas, C. A. and Moore, K. A. (1978). *Phytochem.*, **17**, 619–21. *Phytochem.*, **17**, 619–21.
26. Eaton Mordas, C. A. and Moore, K. G. (1979). *Phytochem.*, **18**, 1775–9.
27. Gillespie, J. M. and Blagrove, R. J. (1975). *Austral. J. Plant Physiol.*, **2**, 29–39.
28. Restani, P., Duranti, M., Cerletti, P. and Simonetti, P. (1981). *Phytochem.*, **20**, 2077–83.
29. Blagrove, R. J., Gillespie, J. M., Lilley, G. C. and Woods, E. F. (1980). *Austral. J. Plant Physiol.*, **7**, 1–13.
30. Elleman, T. C. (1977). *Austral. J. Biol. Sci.*, **30**, 33–45.
31. Joubert, F. J. (1957). *J. S. Afr. Chem. Inst.*, **10**, 21–8.
32. Casero, P. (1981). Graduation Thesis in Food Science and Technology, University of Milan, p. 34.
33. Oram, R. N., Gillespie, J. M. and Blagrove, R. J. (1981). *Austral. J. Plant Physiol.*, **8**, 277–84.
34. Gillespie, J. M., Blagrove, R. J. and Randall, P. J. (1978). *Austral. J. Plant Physiol.*, **5**, 641–50.
35. Blagrove, R. J., Gillespie, J. M. and Randall, P. J. (1976). *Austral. J. Plant Physiol.*, **3**, 173–84.
36. Herbert, S. J. and Hill, G. D. (1976). *Proc. Agron. Soc. N.Z.*, **6**, 65–7.
37. Duranti, M., Cerletti, P., Casero, P. and Restani, P. (1981). *Proc. Int. Symp. on Seed Proteins, Versailles*, p. 13.
38. Casero, P., Duranti, M. and Cerletti, P. (1982). *Proc. 2nd Int. Conf. on Lupin, Torremolinos*, in press.

39. Duranti, M., Cucchetti, E. and Cerletti, P. (1982). *Proc. 2nd Int. Conf. on Lupin, Torremolinos*, in press.
40. Shepard, D. V. and Moore, K. G. (1978). *Eur. J. Biochem.*, **91**, 263–8.
41. Nowicki, E. and Prus-Glowacki, W. (1971). *Genetica Polonica*, **12**, 245–60.
42. Prus-Glowacki, W. (1975). *Genetica Polonica*, **16**, 38–45.
43. Prus-Glowacki, W. (1975). *Genetica Polonica*, **16**, 185–96.
44. Ney, K. H. (1974). In *The Contribution of Chemistry to Food Supplies* (Morton I. and Rhodes D. N., Eds.), p. 411, Butterworths, London.
45. Schoenberger, H., Sam, O., Gross, R., Cremer, H. D. and Elmadfa, I. (1981). *Die Nahrung*, **25**, 667–74.
46. Ballester, D., Yañez, E., Garcia, R., Erazo, S., Lopez, F., Haardt, E., Cornejo, S., López, A., Pokniak, J. and Chichester, C. O. (1980). *J. Agric. Food Chem.*, **28**, 402–5.
47. Hill, G. D. (1977). *Nutr. Abstr. Rev., Ser.* B, **47**, 511–29.
48. Bailey, R. W., Mills, S. E. and Hove, E. L. (1974). *J. Sci. Food Agric.*, **25**, 955–61.
49. Pearson, G. and Carr, J. R. (1977). *Animal Feed Sci. Technol.*, **2**, 49–58.
50. Ruiz Jr., L. P., White, S. F. and Hove, E. L. (1977). *Animal Feed Sci. Technol.*, **2**, 59–66.
51. Gross, U., Reyes, J., Gross, R. and von Baer, E. (1976). *Z. Ernährungswiss.*, **15**, 396–402.
52. Riccardi, A. and Cerletti, P. (1983). *Qual. Plant. -Pl. Fds. Hum. Nutr.*, **33**, in press.
53. Manrique, J. and Thomas, M. A. (1976). *J. Food. Technol.*, **11**, 409–22.
54. Riccardi, A., Alessandroni, A. and Quaglia, G. B. *Proc. Symp. Progress in Food Engineering, Milan*, 1981, in press.
55. Cerletti, P., Riccardi, A. and Bertolotti, G. *Proc. EEC Seminar on Perspectives for Peas and Lupins as Protein Crops, Sorrento*, 1981, in press.
56. Duranti, M. and Cerletti, P. to be published.
57. Duranti, M. and Cerletti, P. 2nd Intl. Lupine Congress, Torremolimos 1982, in press.
58. Batterham, E. S. (1979) *Austral. J. Agric. Res.*, **30**, 369–75.
59. Batterham, E. S., Murison, R. D. and Lowe, R. F. (1981) *Br. J. Nutr.*, **45**, 401–10.
60. Hudson, B. J. F., Fleetwood, J. G. and Zand-Moghaddam, A. (1976). *Plant Foods for Man*; **2**, 81–90.

Chapter 5

LEGUME PROTEIN CONCENTRATION BY AIR CLASSIFICATION

FRANK W. SOSULSKI
University of Saskatchewan, Saskatoon, Canada

SUMMARY

Compared with oilseeds and cereals, the starchy legumes present certain advantages with regard to concentration of the protein and starch constituents. The mechanical separation by air classification is much simpler than the complex technology involved in extractions with organic or aqueous solvents which are hazardous or costly processes. The double-pass procedure of pin milling and air classification is effective in isolating 95–100% of the starch in the coarse fraction, which is largely depleted in other seed constituents. The two fine fractions contain 80–90% of the flour proteins as well as a high proportion of the lipids, cell wall fiber, ash, phytic acid, sugars, oligosaccharides and antinutritive factors. Despite the crude nature of the protein fractions, they have applications as nutritional and functional ingredients in pasta, baked goods, meat patties and frankfurters, and as egg and milk substitutes. Experience with the protein fraction of field pea indicates the need for a moist heat treatment to improve flavor, and texturization to improve the functionality of the fine flour.

INTRODUCTION

Wind and air have been the traditional means for separating hull and bran, due to their lighter density, from ground seed material. Commercial air classifiers are now designed to grade endosperm or cotyledon particles in the subsieve size range of 2–60 μm into subgroups based on their differential mass, density and shape. These uniform particle size ranges may represent subcellular structures which differ substantially in physical and chemical properties.

In cereal flour milling, it was early discovered that the finest flour particles contained the highest protein percentages. This information led to the development of commercial procedures for protein displacement milling of soft wheat flours. Finely ground wheat flours are air classified into a light, fine fraction which has almost double the protein content of the original wheat flour, and a coarse starchy fraction that exhibits more uniform physicochemical characteristics than the original flour. Applications of the two fractions in bread, cake and cookie formulations, as well as industrial starch uses, has supported limited production of these refined flours in several countries.

The current interest of researchers is to adapt the concept of protein displacement milling to other cereals, root crops and legumes. The objective of the present review is to summarize the recent developments in air classification of legume flours into protein-rich and starch-rich fractions. These seed fractions represent new raw materials or commodities which can be used as ingredients in prepared and semi-processed foods, pet foods, animal feeds and in a range of industrial applications. The physical, chemical and functional properties of the crude fractions are examined with respect to potential uses as well as the possible limits on commercial development.

THE AIR CLASSIFIER AND ITS APPLICATION TO CEREALS

Spiral air classifiers such as the Mikroplex MP (Alpine AG, Augsburg, West Germany) are used commonly for the fractionation of wheat flour (Fig. 1). The inlet air blends with the particles of varying size and density, and flows in a spiral direction towards the center of the chamber.[2] The heterogeneous particles are subjected to a centrifugal force originating from a revolving rotor and an opposing centripetal force produced by the air current. The larger, denser particles are dominated primarily by the mass-dependent centrifugal forces which propel the material outward to the coarse fraction discharge worm. The small, light particles are more subject to the frictional forces of the air current, and flow through the central vortice into the air cyclone. Factors which influence the cut point between coarse and fine particles include the dimensions of the classifying chamber, peripheral forces and the spiral gradient. The cut

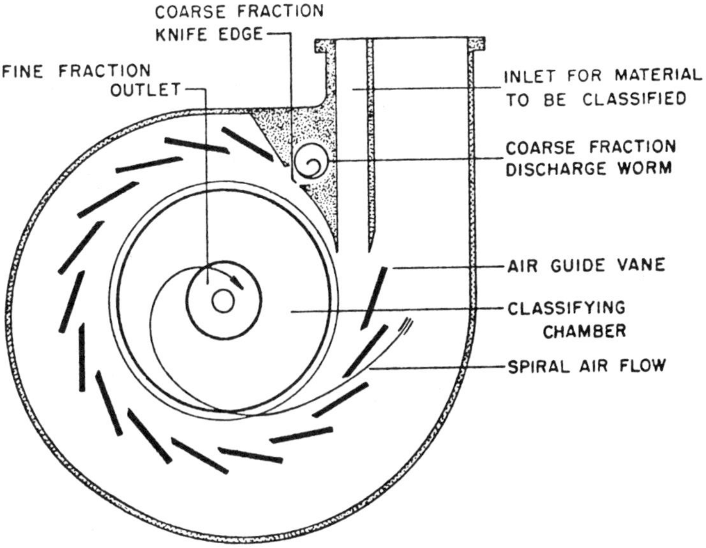

FIG. 1. Cross-section through a Mikroplex spiral air classifier.[1,2]

point is adjusted by varying the rotor speed, air velocity, vane setting and feeding rate.

When applied to the processing of grain crops, the air classifier appears to separate largely on the basis of particle size and shape. Differences in density between starch and protein are substantial but this factor has not been shown to be of major importance in effecting the air separation. Bran, hull and cell wall fragments can be of greater or lesser density than endosperm particles but it is the shape of the fragments which appears to determine their distribution between fine and coarse fractions.

Mature endosperms of most cereals are composed of thin-walled cells which are approximately 100–150 μm in diameter.[3] The cell walls are only 3–7 μm in diameter and a large proportion of the cell volume is occupied by starch granules embedded in a protein matrix. The principal starch granules are spherical or lenticular in shape with a diameter of 15–40 μm. Wheat, rye, barley and oat endosperms also contain a low proportion by weight of small, spherical granules with a diameter of 1–10 μm. The protein matrix which occurs between starch granules, often encasing the smaller granules, is usually referred to as wedge or interstitial protein.

During conventional roller milling of wheat, the endosperm particles are separated from the bran while being progressively reduced in size to pass through sieves having apertures of 100–150 μm. The resulting flours consist of particles with a range in diameter of 2–200 μm.[2] In hard wheats the majority of particles are shattered cell fragments over 50 μm in diameter. In soft wheats, the protein matrix is partially disintegrated during roller milling to yield more of the detached starch granules, wedges of protein matrix containing variable proportions of small starch granules, and cell wall fragments.

Jones *et al.*[4] classified milled wheat flours into three main categories: (a) segments or whole endosperm cells (over 40 μm diameter); (b) medium-sized starch granules with some attached protein (15–40 μm); and (c) wedge protein plus small starch granules (< 15 μm). Kent[3,5] fractionated hard and soft wheat flours using 17 and 35 μm to delineate the cut-off between fractions (Table 1). A significant yield of air-classified

TABLE 1
YIELD AND PROTEIN CONTENT OF AIR-CLASSIFIED FRACTIONS OF FLOURS WITH OR WITHOUT PINNED DISC GRINDING[3]

Flour	*Parent flour protein content (%)*[a]	*Fine (0–17 μm)*		*Medium (17–35 μm)*		*Coarse (over 35 μm)*	
		Yield (%)	*Protein content (%)*	*Yield (%)*	*Protein content (%)*	*Yield (%)*	*Protein content (%)*
Hard wheat							
Unground	13·6	1	17·1	9	9·9	90	13·8
Ground	13·4	12	18·9	41	10·0	47	14·7
Soft wheat							
Unground	7·6	7	14·5	45	5·3	48	8·9
Ground	7·7	20	15·7	71	5·0	9	9·5

[a] 14% moisture basis, N × 5·7.

fine fraction was obtained only with the soft wheat flour. Pinned disc milling increased the proportion of soft wheat fines to 20% and the protein content of this fraction was almost double that of the parent flour. Pin milling also increased the yield of intermediate starchy flour to over 70% of the final products and less than 10% remained in the coarse fraction.

Impact mills have been used extensively to reduce particle sizes of roller-milled or ground flours prior to air classification. Pin mills such as the Alpine Contraplex model 250CW (Fig. 2) have counter-rotating rows of pins which are commonly operated at 12 000 rpm on the mill side and 6000 rpm on the door side to give a differential of 18 000 rpm. This impact velocity (600 ft s^{-1}) has been found to be optimum for disintegration of the endosperm cells but is below the velocity which would damage starch extensively (800 ft s^{-1}).

FIG. 2. Illustration of a Contraplex wide-chamber mill, model 250CW.[1]

The adverse effects of starch damage on the baking properties of wheat flours, including pin-milled fractions, are well known. Principally, the water absorption properties of the flour are increased sharply and the exposed starch granules are subject to hydrolysis by amylases during dough fermentation. Thus, pin-milled flours will give higher yields of bread when short-time baking procedures are followed.[6] Longer fermentation times employed in other baking systems will result in a moist, gummy crumb in the finished product.

When applied to other cereal flours, the extent of protein shift from the coarse to the fine fractions depends on species and milling procedure.

For comparative purposes, Vose[2] calculated the degrees of protein shift based on yields of coarse/fine fractions and their protein contents relative to the parent flour (Table 2). Generally roller and turbo-milling failed to release sufficient protein particles to achieve more than a 20% protein shift by air classification. Protein shifts of 27–36% for various

TABLE 2
PROTEIN SHIFTS PRODUCED BY AIR CLASSIFICATION OF FLOURS FROM BARLEY, OATS, TRITICALE, RICE, POTATO, PEAS AND BEANS[2]

Flour	*Processing*	*% Protein shift*	*Reference*
Barley	Roller-milled	19	7
Barley	Pin-milled	28	8
Malted barley	Commercial grind	8	9
Malted barley	Pin-milled	18	8
Oats	Pin-milled	27–32	10
Triticale	Pin-milled	28–36	11
Rice	Turbo-milled	8–10	12
Potato	Pin-milled	22–25	13
White bean	Turbo-milled	22	14
Field pea	Pin-milled	42	15
Fababean	Pin-milled	45	15

pin-milled cereals may be sufficient to warrant commercial development of the process. Results with air classification of potato flour were not particularly promising. However, the protein shifts of 42 and 45% for field pea and fababean indicated that these legume seeds were more suited to processing by this technique than other seeds or storage organs.

Commercial plants for the air classification of field peas, navy and kidney beans and fababeans have been established at Saskatoon, Saskatchewan; Mississauga, Ontario; and Les Milles, France, respectively.

AIR CLASSIFICATION OF LEGUMES

Field Pea and Fababean

Field peas are particularly suited to processing into protein and starch concentrates because of their high lysine content and relatively low cost.[16] Antinutritive factors are not a problem in the consumption of raw or cooked peas. The hulls represent a small portion of the seed and are easy to remove. Smooth-seeded peas are grown for the split-pea trade and protein extraction. The wrinkled type of pea is used for canning but is not, however, suitable for protein fractionation because of the difficulty in hull removal and the susceptibility of the compound starch granules to mechanical starch damage.

Due to low and variable protein levels in commercial samples of field peas, as much as 14–27%,[17] several researchers have turned to fababean as a potential legume for component extraction. As a feed legume, the seed is inexpensive relative to other bean species and is commercially available in many countries of Europe, Africa and Asia. The protein contents are typically about 28–32% and individual values appear less variable than those of field pea. Due to the high proportion of hulls, which contain condensed tannins,[18] the removal of hulls is a necessary step in processing.

While cereal starch granules exhibit a bimodal or variable distribution in particle size, legumes have average granule dimensions of 16–21 microns in width and 23–28 microns in length.[19] After pin milling, the starch granules remain intact whereas other cell structures are largely reduced to less than 5 μm in diameter (Fig. 3). The hulls can be seen as rod-shaped fragments in the illustration of pea flour. Youngs[16] classified the pea flour at a cut point of 15 μm (800 mesh) between fine and coarse particles. The centrifugal forces appeared to isolate essentially all of the starch granules in the coarse fraction whereas the fines contained almost no starch granules or other large particles (Fig. 3).

Initial studies were done with dehulled ground peas but later it was found that pin milling and air classification worked equally well with whole, ground or dehulled seeds. When present, the hulls separated into the starch fraction, altering the yield and composition of this fraction.

Using pilot plant equipment, Youngs[16] separated a 21% protein flour (F) of field pea (S) into 25% of fines (P1) containing 60% protein and a coarse fraction (SI) containing only 8% protein and with over 85% of nitrogen-free extract. To release more of the adhering protein, the coarse fraction (SI) was remilled and reclassified to give an additional 10% of fines (PII) with 46% protein and a final starch fraction (SII) containing 2–3% protein. The schematic flowsheet for the single- and double-pass procedures is illustrated in Fig. 4.

Further work at the Prairie Regional Laboratory, Saskatoon, indicated that the yields and protein contents of PI and PII from air-classified fababean consistently exceeded those of field pea (Table 3). Even when a high protein sample of field pea was used to provide a PI fraction

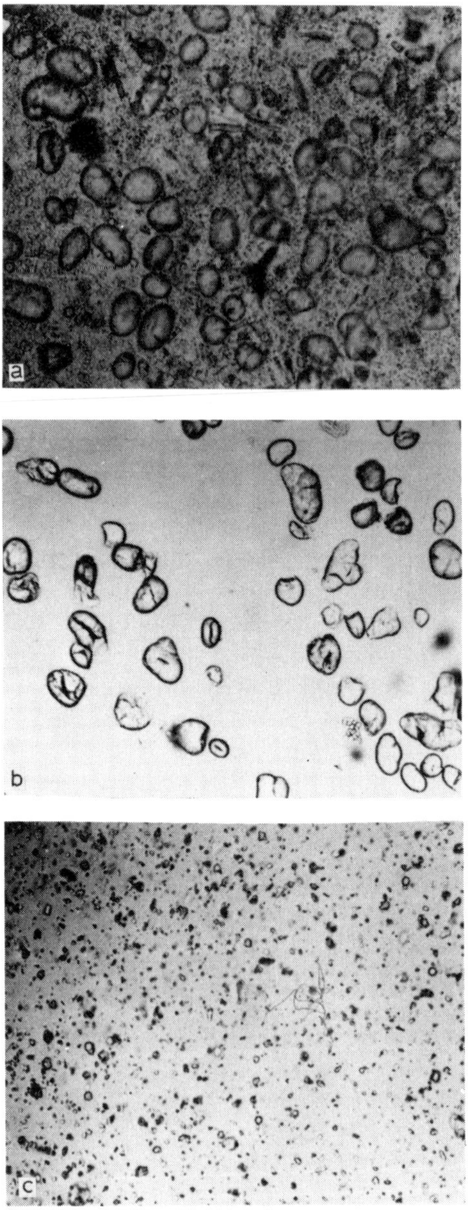

FIG. 3. Photomicrographs of pin-milled flour (a) and the air-classified starch (b) and protein fractions (c) from field pea ($\times$ 325).[16]

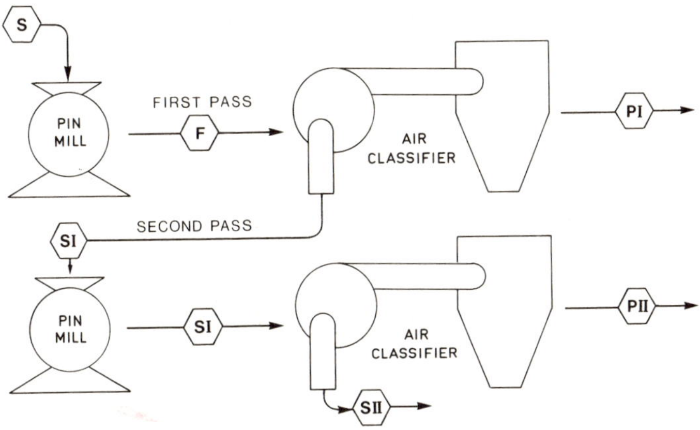

FIG. 4. Schematic flowsheet for the pin milling and air classification of field peas by the single- and double-pass procedures.[16]

containing 63·5% protein, the yield of PI in fababean was 3% greater and the protein content 2·5% higher than with field pea. Again, the second pass of SI through the pin mill and classifier gave useful yields of 19 and 14%, respectively, of field pea and fababean fine fractions containing about 50% protein. At the same time, the residual protein levels in the starch fractions were reduced from 13·3 and 15·1% to 5·6 and 5·2%, respectively. Therefore, the double-pass procedure has been adopted generally for air classification of legumes. Some investigators have passed the flour through the pin mill twice before air classification and obtained essentially the same results as for the double-pass procedure.

The same seeds were also dehulled before pin milling and air classification using a plate grinder and air aspiration (Table 3). The hulls represented about 10 and 15% of the field pea and fababean seed, respectively, and their removal increased the protein (PI):starch (SII) ratio.

Residual Protein in the Starch Fraction

To improve the functionality of the starch fraction and enhance the efficiency of protein recovery, additional pin millings and air classifications have been conducted on the starch fraction. Residual protein levels in the SIII fraction[20] and SIV fraction[21] were about 2·2–2·5%. However, Vose[22] showed that increasing the pin millings from two to

TABLE 3

YIELD AND PROTEIN CONTENTS OF WHOLE AND DEHULLED SEED OF FIELD PEA AND FABABEAN AND THEIR AIR-CLASSIFIED PROTEIN AND STARCH FRACTIONS[15]

Fractions and designation		*Whole seed*		*Dehulled seed*	
		Yield (lb)	*Protein content (%)*	*Yield (lb)*	*Protein content (%)*
			Field pea		
S	Seed	100	25·7	100	28·8
SI	Starch	75	13·3	69	14·5
PI	Protein	25	63·5	31	60·5
SII	Starch	56	5·6	52	4·8
PII	Protein	19	48·3	17	40·0
			Fababean		
S	Seed	100	27·9	100	31·9
SI	Starch	72	15·1	70	16·5
PI	Protein	28	66·1	30	69·0
SII	Starch	58	5·2	52	4·2
PII	Protein	14	51·1	18	49·6

four passes increased the starch damage in field pea from 22% to 40%. Apparently, pea starch is more susceptible to mechanical damage than corn starch which showed only 6 and 9% of shear damage when exposed to the two- and four-pass milling treatments, respectively. Recently unpublished data from the same laboratory indicated that starch damage in the SI fraction varied from 1–4% of the starch in several legumes.

The residual protein can be removed by water-washing the starch fraction from field pea to obtain a white product containing 0·06% nitrogen.[22] The nature of this water-soluble protein was investigated by Reichert and Youngs[21] by scanning electron microscopy (Fig. 5). The residual protein was found to arise from agglomerates of starch in the protein matrix, protein bodies attached to the starch granules and the presence of a residual chloroplastic membrane on the starch granule surface.

The cost of water-washing the starch fraction to remove the residual protein would probably not be justified by improved functionality of the refined starch. Therefore, product development has been directed towards utilization of the crude starch (SII) fraction directly in industrial applications or with only limited modification by heat or enzyme treatment.

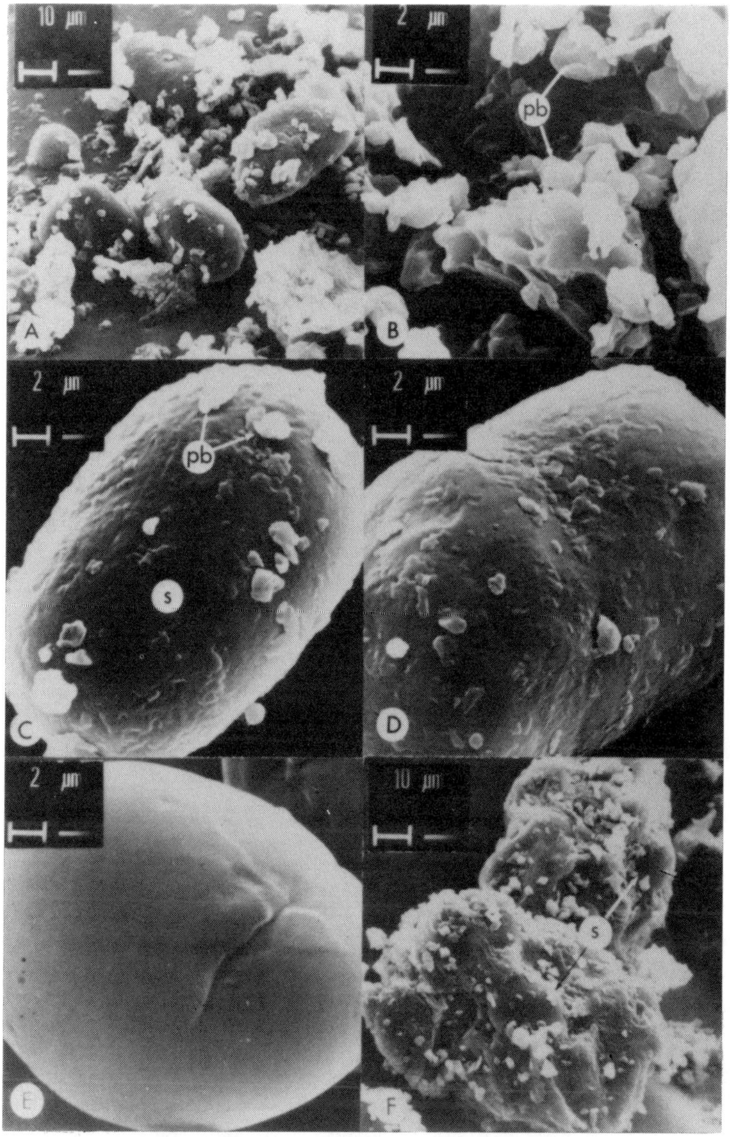

FIG. 5. Scanning electron micrographs of air-classified field pea flour fractions. (a) Flour, (b) PI; (c) SI; (d) SIV; (e) water-washed pea starch; (f) agglomerates., (pb) Protein bodies, (s) starch grains.[21]

The Hull Fraction

Dehulling before pin milling provides another seed component for food or feed uses. Pea hulls are composed primarily of cellulose and hemicellulose while the thicker hulls of fababean contained a greater proportion of lignin (Table 4). The principal sugars released on hydrolysis with acid are xylose and arabinose with smaller quantities of glucose, galactose and rhamnose. This hydrolysis pattern for the hull materials was indicative of hemicellulosic material.

TABLE 4

COMPOSITION OF HULLS OF FIELD PEA AND FABABEAN, PERCENT DRY BASIS[15]

Constituent	*Field pea*	*Fababean*
% hull in seed	8·2	13·0
Water-soluble polysaccharides	8·9	16·7
Pectin-type polysaccharides	8·2	9·0
Lignin	1·9	11·3
Hemicellulose	23·1	13·0
Cellulose	55·2	48·9
Nitrogen	0·6	0·8
Neutral lipid	0·6	0·4
Ash	2·4	2·6

Utilization of fababean hulls has not been investigated, but pea hulls have been successfully incorporated into bread products as a fiber component. One firm determined that pea hulls, when incorporated into sponge and dough bread formulations, gave the best loaf volumes, crumb scores and color ratings when compared with other natural sources of fiber (Table 5). The 20–40 mesh size of particle gave the most acceptable white bread scores. The concentration of 7·5% on a flour basis was chosen because this level gave similar fecal output characteristics to that of whole wheat bread in rat feeding trials. The bread was successfully marketed in several provinces in Canada.

Moisture Content of Seed

Seed moisture content of legumes can affect both the yield and composition of the air-classified fractions.[24] In field pea and fababean, reductions in seed moisture were accompanied by an increase in yield of protein fraction (Table 6). However, there was a corresponding decrease in protein content and an increase in starch content of the PI fraction. Seeds also became harder at the lower moisture levels which resulted in

TABLE 5
EFFECTS OF INCORPORATING 7·5% PEA HULLS ON BAKING AND PHYSIOLOGICAL FUNCTIONALITY IN RAT FEEDING TRIALS[23]

Fiber source	*Crude fiber* (%)	*Loaf volume* (cm^3)	*Fecal volume*[a] (*ml/rat*)	*Fecal density* ($g\,ml^{-1}$)	*Dry matter digestibility* (%)
Pea hulls					
on 20 mesh	41·2	2000	10·2	0·54	92·1
20–40 mesh	39·9	2100	12·0	0·44	92·4
40–80 mesh	45·2	2100	9·7	0·46	93·5
through 80 mesh	35·3	1850	8·7	0·49	93·8
Whole wheat	—	2125	18·1	0·29	92·0

[a] 5 day fecal output.

longer grinding time and slower rates of feeding to the air classifier. Tyler and Panchuk[24] concluded that the optimal protein separation efficiency was obtained at about 7% moisture in field pea and at 5% moisture for fababean.

Air Classification of Field Bean

Kon *et al.*[14] air-classified a turbo-milled sample of the small white bean to obtain a 44% protein fraction which still contained a significant level of starch. Youngs[16] also reported obtaining yields of 20–30% fines from air-classified kidney bean and northern bean flours. The protein levels in these PI fractions were 46–50%. Subsequently Patel *et al.*[25] air-classified a navy bean sample to obtain the following results:

Fraction	*% Yield*	*% Protein*
Flour	100	30·4
Protein I	19·4	63·8
Starch I	80·6	21·9
Protein II	15·3	60·0
Starch II	65·3	10·7

Due to the low yield of PI and high residual protein in SII, it appeared that the field bean was less promising as a raw material for air classification than the field pea or fababean.

A sample of navy bean has been classified by sequential removal of 5% fractions of fine material from the pin-milled flour using a laboratory air classifier.[26] The protein and starch contents of the 20 fractions are illustrated in Fig. 6. The initial three fractions contained 51–44% protein

TABLE 6

AIR CLASSIFICATION OF DEHULLES FIELD PEA AND FABABEAN AT VARIOUS MOISTURE CONTENTS[24]

Seed moisture (%)	*Protein fraction of field pea (PI)*			*Seed moisture* (%)	*Protein fraction of fababean (PI)*		
	Product yield (%)	*Protein content* (%)	*Starch content* (%)		*Product yield* (%)	*Protein content* (%)	*Starch content* (%)
14·3	18·9	51·7	7·4	14·2	24·4	74·1	3·7
12·4	27·8	49·1	8·4	11·2	29·4	71·1	3·4
10·5	29·4	46·9	8·6	8·8	32·3	71·2	3·6
6·9	29·4	46·5	9·8	—	—	—	—
5·3	29·4	46·4	9·2	5·1	33·3	69·5	5·1
3·8	30·3	45·8	9·9	3·8	33·3	69·1	5·1

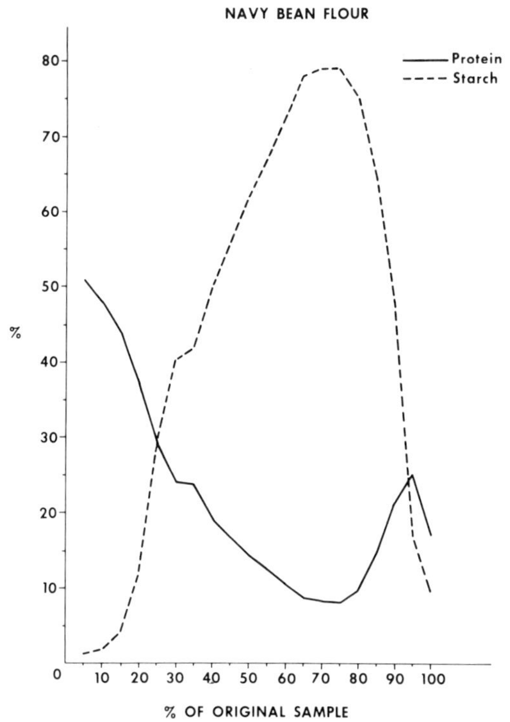

FIG. 6. Protein and starch contents of flour fractions obtained by sequential removal of 5% fines fractions from a pin-milled sample of navy bean.[26]

and essentially no starch. However, inclusion of the fourth fraction to give a PI yield of 20% would result in significant starch contamination. As the particle size increased, there were increasing proportions of flour agglomerates which contained protein. The sharp drop in both protein and starch in the final cuts was due to the presence of hull material.

The market prices for *Phaseolus* beans are quite high and the processing of the bean would not be economical under normal circumstances, especially with the rather low fractionation efficiency and starch content[27] apparent for this legume. However, one commercial plant has been operating viably in Canada due primarily to the availability of cull beans.

Air Classification of Other Legumes

The potential for preparing air-classified fractions of other major legume species was investigated by Sosulski and Youngs[28] and Tyler *et al.*[29] Of principal interest were chickpea (*Cicer arietinum*), cowpea (*Vigna unguiculata*), baby lima bean (*Phaseolus lunatus*), green lentil (*Lens culinaris*) and mung bean (*Vigna radiata*). The results obtained with these starchy legumes were compared with data obtained for smooth field pea (*Pisum sativum arvense*), small fababean (*Vicia faba minor*) and navy bean (*Phaseolus vulgaris*). Great northern bean (*Phaseolus vulgaris*) was also included in the study to assess the variability within the *Phaseolus vulgaris* biotypes. Seeds of soybean (*Glycine max*) and lupin (*Lupinus albus* or *angustifolius*) were processed by the first-pass procedure only to show that non-starchy legumes are not effectively fractionated by air classification. Flours and protein concentrates of the latter species also served as controls for chemical and functional tests.

The seed samples were equilibrated to 8% moisture before dehulling by grinding on a plate grinder followed by air aspiration. The dehulled cotyledonous material was pin-milled and air-classified by the double-pass procedure illustrated in Fig. 4. Soybean seed was ground and defatted with hexane before pin milling.

The pin-milled flours contained 20·4% (chickpea) to 30·4% (fababean) of protein as compared to 52·5% in defatted soybean flour and 41·4% in the full-fat lupin flour (Table 7). The *Phaseolus* species—navy, northern and lima beans— had 45–49% of starch while the other legumes were somewhat higher at 51–54% of starch. Fababean and lentil contained over 80% of protein + starch, a favorable combination for raw materials to be fractionated into protein and starch fractions.

Air classification of the pin-milled soybean and lupin flours gave high fine to coarse ratios in the products but there was little or no concentration of protein in either fraction (Table 7). In chickpea, the yields of fine fractions (PI and PII) were very low. Further investigation showed that the pin milling operation was satisfactory but, due to the high lipid level of 6·8% in the flour (Table 8), the flour particles agglomerated and the air classification was relatively inefficient. Subsequently the chickpea flour was defatted to show that efficient separation could be obtained on the air classifier. However, the lipid yield was too low to consider commercial extraction. The compositional data presented in the subsequent tables should be considered in the light of the poor fractionation obtained in these initial experiments.

The cut point of 15 μm used for fractionation of the remaining legume flours was effective in concentrating 95–100% of the starch in the SII fraction (Table 7). The yields of products were essentially the same for six species, namely about 38% fines (PI + PII) and 62% coarse fractions.

TABLE 7

YIELD, COMPOSITION AND SEPARATION EFFICIENCY OF LEGUME FLOURS, PROTEIN FRACTIONS (PI, PII) AND STARCH FRACTIONS (SII) PRODUCED BY AIR CLASSIFICATION OF LEGUME FLOURS, PERCENT DRY BASIS[28, 29]

Legume	*Flour composition*		*Yield of product*			*Protein content*			*Starch content*			*Separation efficiency*	
	Protein	*Starch*	*PI*	*PII*	*SII*	*PI*	*PII*	*SII*	*PI*	*PII*	*SII*	*Protein*	*Starch*
Soybean	52·5	4·7	72	—	28	54·2	—	50·5	—	—	—	—	—
Lupin	41·4	3·5	82	—	18	43·3	—	30·0	—	—	—	—	—
Chickpea	20·4	53·2	13	7	80	46·9	45·1	13·1	6·3	5·3	65·1	48·5	97·6
Navy bean	26·5	44·9	20	15	63	58·6	50·7	9·0	1·0	2·6	71·0	82·0	97·8
Northern bean	22·6	47·2	29	13	58	50·7	38·8	4·8	0·7	3·0	80·6	87·3	98·9
Cowpea	23·6	52·0	25	12	63	51·6	45·8	8·0	3·2	9·2	80·7	79·0	95·0
Lima bean	22·8	48·8	24	15	61	49·6	43·4	7·1	0·0	0·4	79·6	81·0	100·0
Field pea	23·0	54·2	22	12	66	59·2	51·8	6·2	1·6	3·7	81·5	81·3	97·9
Lentil	26·3	53·6	22	15	63	64·6	54·3	5·4	0·2	5·8	85·6	88·7	96·1
Fababean	30·4	51·3	21	16	63	71·3	63·8	7·2	0·7	4·0	81·7	86·7	96·7
Mung bean	25·8	52·5	27	12	61	61·4	52·5	4·6	2·4	11·2	82·5	89·4	96·5

TABLE 8

DISTRIBUTION OF CRUDE FAT, CRUDE FIBER AND ASH BETWEEN THE PROTEIN (PI, PII) AND STARCH (SII) FRACTIONS BY AIR CLASSIFICATION, PERCENT DRY BASIS[28, 29]

Legume	*Crude fat*				*Crude fiber*				*Ash*			
	Flour	*PI*	*PII*	*SII*	*Flour*	*PI*	*PII*	*SII*	*Flour*	*PI*	*PII*	*SII*
Soybean	0·6	—	—	—	2·6	—	—	—	6·4	—	—	—
Lupin	7·6	—	—	—	3·0	—	—	—	3·0	—	—	—
Chickpea	6·8	15·0	14·4	4·7	1·6	1·8	2·0	1·7	2·8	5·3	5·2	2·2
Navy bean	1·7	3·5	3·3	0·6	1·9	1·2	2·9	2·0	3·7	7·9	6·6	1·6
Northern bean	1·3	3·0	2·1	0·4	1·7	1·9	3·4	0·8	4·3	9·4	6·2	1·2
Cowpea	1·3	3·1	3·0	0·2	1·3	2·4	2·7	0·6	3·5	7·6	7·0	1·8
Lima bean	1·1	2·0	1·7	0·2	1·8	1·8	2·8	1·4	4·7	8·2	7·5	1·6
Field pea	1·1	3·2	2·6	0·5	1·5	1·8	2·8	1·0	2·6	5·6	5·2	1·1
Lentil	0·9	2·0	2·2	0·1	1·2	1·5	2·7	0·7	2·3	5·4	4·8	1·2
Fababean	1·6	3·4	2·8	0·5	1·3	1·9	2·6	0·7	2·6	5·0	4·8	1·2
Mung bean	0·9	1·7	1·7	0·2	1·3	2·0	1·9	1·0	3·2	7·2	6·2	0·9

Northern bean gave a 42/58 split while field pea yielded 34/66 of fine: coarse fractions.

The protein contents of the protein fractions varied between 50 and 71% for PI and 43–64% in PII (Table 7). In each case, the lowest values were for lima bean and the highest for fababean. Protein separation efficiencies were calculated to show that the lowest separation efficiencies of 79–82% were obtained for navy and lima beans, field pea and cowpea. The very high protein recoveries in PI+PII for mung bean, lentil, northern bean and fababean of 87–89% would compare favorably with any aqueous protein concentration system. Even recoveries of 80% will be rated as quite satisfactory for a dry milling process.

Another aspect of the efficiency of the process was the level of residual starch in the two protein fractions. Cowpea and mung bean had the highest starch levels in both fractions (Table 7). Further studies have demonstrated that an appropriate adjustment of cut point will reduce these starch levels substantially, although the problem cannot be eliminated entirely without adversely affecting the yield of protein fractions.

CONSTITUENTS OTHER THAN PROTEIN AND STARCH

The distribution of other cellular constituents between the protein and starch fractions will have a marked influence on their functional nutritional properties. Generally, the crude fat, fiber and ash segregated with the protein fractions (Table 8). The SII fractions were very low in lipid, except for chickpea, but the fiber split was less complete. Ash levels concentrated substantially in the protein fractions, which followed the distribution of phytic acid, as will be discussed later.

Cell Wall Material and Sum of Components

While the major constituent of legume hulls is cellulose, the cell wall fiber from dehulled pea flour is composed of 13% pectin, 25% lignin, 19% hemicellulose and 13% cellulose.[30] Fababean fiber has a somewhat different composition, the values being 38% pectin, 18% lignin, 11% hemicellulose and 8% cellulose. The neutral detergent fiber (NDF) values about 30% in each flour.

Vose[30] observed that the cell wall material was highly hygroscopic, readily absorbing 10–14 times its weight of water. Therefore, it was of interest to determine the level of cell wall material in the legume flours and air-classified fractions. Because the pin-milled pea flour was too fine for direct isolation of the cell wall component, Reichert[31] isolated the cell wall material from coarsely ground cotyledon material, obtaining a yield of about 7%. The neutral detergent fiber (NDF) of this purified cell wall material was only 44·3%, due to solubilization of pectin and hemicellulose in the detergent. Therefore Reichert extrapolated from the expression NDF × 2·26 to show that high and low protein flours contained 6·6 and 9·0% cell wall material respectively (Table 9). After air classification the corresponding values for the protein fractions were 11·7 and 16·4%, levels which would have a major influence on the physicochemical properties of these protein fractions. For example, the water absorption of the cell wall material in this study was 1816% of the sample weight as compared to 296% for pea hulls.

By obtaining a reliable estimate of the cell wall content, it was possible to account for most of the dry matter in the flours and protein fractions in field pea (Table 9). The average composition of the protein fractions was 50% protein, 14% cell wall material, 14% oligosaccharides, 8% lipid, 6% ash and 6% starch.

Oligosaccharides in Air-classified Fractions

A number of investigators have demonstrated that the oligosaccharides, especially raffinose and stachyose, are the principal causes of flatulence in man. The human digestive system lacks the enzyme α-galactosidase, and the raffinose family of oligosaccharides pass into the large intestine where they are fermented anaerobically to produce gas.

TABLE 9

TOTAL COMPONENT ANALYSIS OF FIELD PEA FLOURS AND PROTEIN FRACTIONS, PERCENT DRY BASIS[31]

Sample	*Protein*	*Starch*	*Cell wall material*	*Lipid*	*Ash*	*Sugars*[a]	*Sum of components*
High protein pea (28·5%)							
Flour	28·5	49·7	6·6	3·0	2·8	7·5	98·1
Protein fraction (PI + PII)	60·2	4·9	11·7	6·0	5·5	12·5	100·8
Low protein pea (14·5%)							
Flour	14·5	59·8	9·0	4·1	3·3	7·8	98·6
Protein fraction (PI + PII)	40·9	7·3	16·4	9·7	7·2	15·1	96·6

[a] Sum of sucrose, raffinose, stachyose and verbascose.

There are wide variations among legumes in their reported potential for formation of intestinal gas. Field bean, chickpea, lima bean, soybean and lupin are legumes which are cited as causing excessive flatus production.[32, 33]

Stachyose occurs in much higher concentrations than raffinose in mature legume seeds but certain species contain substantial quantities of the higher molecular weight α-galactoside, verbascose, and less stachyose.[34, 35] Recently, Schweizer *et al.*[36] reported that significant quantities of galactopinitol, galactinol and manninotriose occurred among the oligosaccharides extracted from several legume seeds. The role of several of these α-galactosides in flatus formation has not been elucidated. Fleming[37] has determined that stachyose concentrations in seven legumes were highly correlated with hydrogen production in rats. Raffinose gave a low positive correlation whereas verbascose contents showed a negative relation with hydrogen production.

Recently the oligosaccharide compositions of the air-classified protein and starch fractions were determined using soybean flour (Staley F-200), protein concentrate (Promosoy 100) and lupin flour (*Lupinus albus*) as controls.[38] The distribution of oligosaccharides and inositols in chickpea flour was particularly wide (Fig. 7). Seven oligosaccharides were identified and quantified, showing that the major portion of the sugars, including sucrose, was concentrated with the fines. The α-galactosides represented 9·2% of protein fraction (PI) and only 2·8% of the starch fraction (SII).

Stachyose is the principal α-galactoside in soybean, lupin, cowpea, lima bean and navy bean (Table 10). Chickpea and lentil contained high levels of both manninotriose and stachyose while verbascose was the major α-galactoside in field pea, mung bean and fababean. Soybean protein concentrate, produced by wet processing, was almost devoid of α-galactosides but the air-classified protein fractions were 40–90% higher than the original flours. The 5–7% of stachyose in cowpea and lima bean PI could present a serious flatulence problem.

Antinutritive Factors

Legume seeds contain other microconstituents which may have adverse effects on the nutritive value of the product or the health of the consumer. Trypsin inhibitors reduce protein digestibility and depress growth,[39] hemagglutinins of *Phaseolus* species will inhibit growth and even cause death of experimental animals[40] and saponins reduce feed intake.[41] The effects of phytic acid on availability of zinc, magnesium, phosphorus, calcium and iron are well known.[42]

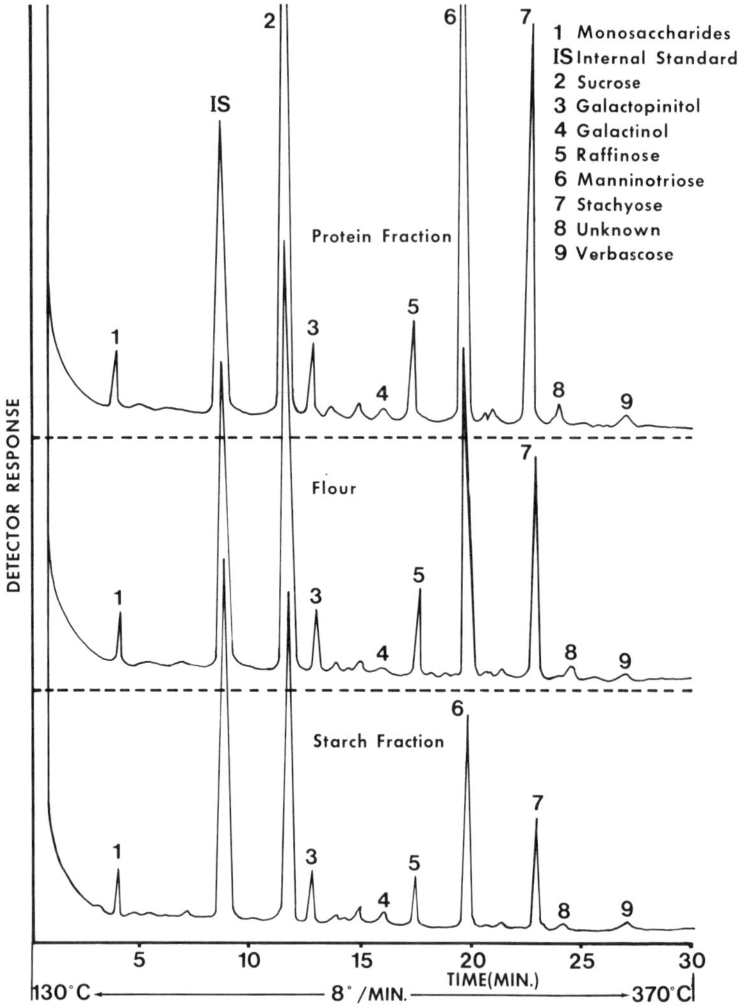

FIG. 7. GLC chromatrograms of TMS derivatives of oligosaccharides in chick-pea flour, protein fraction (PI) and starch fraction (SII).[38]

Compared with a defatted soybean flour, navy bean flour is high in trypsin inhibitor activity, hemagglutinating activity, and in saponin and phytic acid contents (Table 11). These constituents were greatly enriched in the protein fractions of navy and northern beans, and represented a serious deterrent to the use of these products in foods unless fully

TABLE 10

COMPOSITION OF α-GALACTOSIDES IN LEGUME FLOURS, PROTEIN FRACTIONS AND STARCH FRACTIONS OF NINE LEGUMES AS COMPARED WITH SOYBEAN AND LUPIN, PERCENT DRY BASIS[38]

Legume	*Hull-free flours (F)*				*Protein fractions (PI)*				*Starch fractions (SII)*			
	Raffi-nose	*Manninot-riose*	*Stachy-ose*	*Verbas-cose*	*Raffi-nose*	*Mannino-triose*	*Stachy-ose*	*Verbas-cose*	*Raffi-nose*	*Mannino-triose*	*Stachy-ose*	*Verbas-cose*
Soybean	1·15	trace	2·85	—	0·09	—	0·25	—	—	—	—	—
Lupin	0·82	0·20	4·11	0·48	—	—	—	—	—	—	—	—
Cowpea	0·41	—	4·44	0·48	0·52	—	6·85	1·27	0·29	—	2·29	—
Chickpea	0·45	2·33	1·72	0·10	0·70	4·01	2·98	0·35	0·28	1·32	0·79	—
Lentil	0·31	1·41	1·47	0·47	0·50	2·18	2·87	1·06	0·17	0·78	0·33	—
Lima bean	0·46	0·12	2·76	0·31	0·60	0·09	5·26	0·36	0·34	trace	1·55	—
Navy bean	0·37	—	2·36	0·05	0·48	trace	3·84	0·25	0·21	—	1·00	—
Northern bean	0·26	0·03	2·16	0·03	0·36	0·11	3·56	0·12	0·19	0·10	0·92	0·01
Field pea	0·60	—	1·71	2·30	1·24	—	3·76	4·64	0·42	—	0·78	0·59
Mung bean	0·23	0·12	0·95	1·83	0·32	0·21	1·49	3·13	0·25	0·08	0·45	0·60
Fababean	0·22	—	0·67	1·45	0·33	—	1·37	3·96	0·25	—	0·48	0·44

TABLE 11

TRYPSIN INHIBITOR ACTIVITY, HEMAGGLUTINATING ACTIVITY, HEMOLYTIC ACTIVITY OF SAPONINS AND PHYTIC ACID LEVELS IN LEGUME FLOURS, PROTEIN FRACTIONS AND STARCH FRACTIONS, DRY BASIS[43]

	Trypsin inhibitor activity (TUI/mg sample)			*Hemagglutination activity (HU/mg sample)*			*Saponin (HA/g sample)*			*Phytic acid (mg/g sample)*		
	Flour	*Protein*	*Starch*	*Flour*	*Protein*	*Starch*	*Flour*	*Protein*	*Starch*	*Flour*	*Protein*	*Starch*
Soybean	41·55	28·77	—	77·34	0·44	—	0	0	—	11·64	13·53	—
Lima bean	46·81	89·78	19·70	0·89	1·32	0·30	5·9	12·9	0	5·20	13·64	1·86
Navy bean	18·23	53·80	4·06	46·12	103·99	1·94	12·6	16·1	5·9	10·03	26·02	1·99
Northern bean	18·08	43·58	4·05	16·76	28·73	0·94	7·8	16·1	0·8	10·61	23·14	2·52
Chickpea	18·80	50·61	2·49	7·89	9·84	1·46	0·5	3·8	0	5·20	12·78	1·48
Cowpea	12·20	23·88	4·94	1·99	3·94	0·34	12·8	16·2	5·8	7·61	16·36	2·47
Lentil	5·12	11·69	0·99	14·61	30·47	0·69	5·6	12·6	0·5	6·39	18·74	1·49
Field pea	7·61	23·35	1·43	15·06	39·99	1·90	0	0	0	7·44	18·88	1·82
Fababean	4·75	10·71	1·36	10·55	20·15	2·98	0	0	0	3·67	8·64	1·61
Mung bean	9·96	23·03	1·78	3·42	9·02	0·26	0·8	2·4	0	5·12	14·37	1·58
Lupin	0	—	—	0	—	—	0·8	—	—	8·17	—	—

detoxified by heat or extraction. Potential nutritional problems could arise from the high levels of trypsin inhibitor in lima bean and chickpea; and saponin in lima bean, cowpea and lentil. The hemagglutinating activities of lentil, field pea and fababean have not been shown to have deleterious effects in human nutrition. The starch fractions were generally depleted in each antinutritive factor and could be used quite safely in food applications.

Fababean and related species contain pyrimidine glucosides, vicine and convicine, which are hydrolyzed during digestion into the aglucones divicine and isouramil respectively (Fig. 8). These aglucones are con-

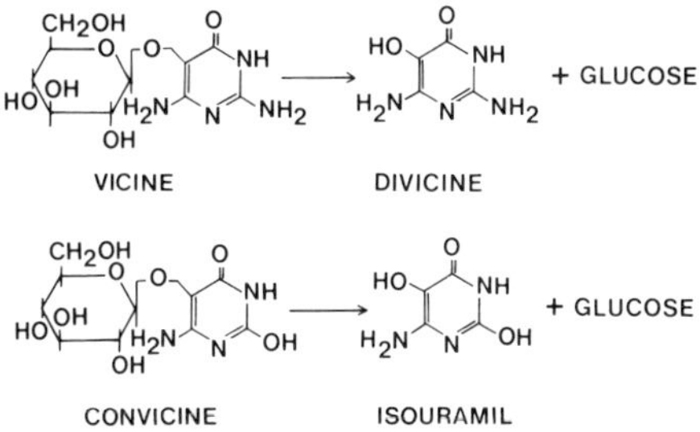

FIG. 8. Conversion of pyrimidine glucosides, vicine and convicine, to divicine and isouramil.

sidered to be the causative agents in precipitating a hemolytic disorder of the blood when fababean is consumed by susceptible individuals.[44] Like the other antinutritive factors, the pyrimidine glucosides were found to concentrate with the fine fraction during the initial air classification (Table 12).

The second pass through the pin mill and classifier was effective in reducing the total glucosides to less than 0·2% of the SII fraction. The high levels in the protein fractions would place a serious constraint on the use of these products in foods. Due to their heat stability, aqueous extraction processes are required to remove the pyrimidine glucosides from the flour or protein fractions of fababean.

PROTEIN NUTRITIVE VALUE

The legume species are similar in amino acid composition in the flour proteins (Table 13). Glutamic and aspartic acids constitute 25–30% of

TABLE 12

PERCENTAGE OF VICINE AND CONVICINE IN FABABEAN FLOUR, PROTEIN FRACTION AND STARCH FRACTIONS, DRY BASIS[44]

Fraction	*Vicine*	*Convicine*	*Total*
Flour	0·44	0·23	0·67
First pass			
Protein I	1·03	0·45	1·48
Starch I	0·24	0·15	0·39
Second pass of Starch I			
Protein II	0·89	0·40	1·29
Starch II	0·10	0·09	0·19

TABLE 13

AMINO ACID COMPOSITION OF PROTEINS IN LEGUME FLOURS IN g/16 g NITROGEN

Amino acid	*Soybean*	*Chickpea*	*Field pea*	*Lentil*	*Northern bean*	*Navy bean*	*Mung bean*	*Cowpea*	*Lima bean*	*Febabean*
Lysine	6·4	7·0	7·4	7·3	7·0	6·9	6·7	6·6	6·5	6·3
Histidine	2·2	2·5	2·3	2·3	3·1	2·8	2·6	2·9	2·9	2·4
Ammonia	1·5	1·2	1·7	1·9	2·1	1·9	1·9	2·1	1·8	2·0
Arginine	7·4	8·4	7·8	7·2	6·2	6·3	6·4	7·1	5·3	10·0
Aspartic acid	11·5	11·9	11·9	11·4	12·6	12·6	11·4	11·9	12·8	11·1
Threonine	3·7	3·8	3·7	3·5	3·9	4·3	3·1	3·5	4·1	3·3
Serine	4·5	5·2	4·6	5·0	5·5	5·7	4·8	4·6	6·3	4·6
Glutamic acid	18·3	17·0	17·3	16·7	16·4	15·6	17·6	17·8	14·10	17·2
Proline	5·1	4·3	4·0	4·0	3·6	3·7	3·9	4·1	3·6	3·6
Half-cystine	1·1	1·2	1·4	0·9	0·8	0·8	0·6	0·9	1·0	1·1
Glycine	4·2	4·0	4·5	4·0	3·8	4·0	3·5	3·7	3·9	4·0
Alanine	4·3	4·1	4·3	4·0	3·9	4·0	4·0	4·1	4·3	4·1
Valine	4·9	4·2	4·6	4·7	5·0	5·3	4·6	4·9	5·3	4·6
Methionine	1·4	1·6	0·9	0·8	1·2	1·2	1·3	1·5	1·2	0·8
Isoleucine	4·7	4·3	4·1	4·4	4·5	4·5	4·1	4·0	4·9	4·0
Leucine	7·5	7·8	7·1	7·3	8·1	8·2	7·7	7·5	8·4	7·4
Tyrosine	3·4	2·6	2·8	2·6	2·9	2·7	2·3	2·7	3·1	2·9
Phenylalanine	4·9	5·5	4·4	4·8	5·3	5·3	5·3	5·3	5·9	4·0
Tryptophan	1·3	1·2	1·0	0·9	1·0	1·1	1·1	1·0	1·3	1·0
Total	98·3	97·8	95·8	93·7	96·9	96·9	92·9	96·2	96·6	94·4

the total amino acids while lysine, arginine and leucine are also major amino acids in all species. Of nutritional significance is the high lysine in field pea and lentil when compared to fababean, soybean, lima bean and cowpea proteins. Arginine levels are high in fababean and relatively low in lima bean protein. Methionine, the first limiting amino acid, occurs in higher concentrations in chickpea, cowpea and soybean as compared with fababean, lentil and field pea.

The amino acid distributions in the air-classified proteins, which represented 79–89 % of the flour proteins, are very similar to the pattern in the flours (Table 14). In most legumes there is a slight tendency towards higher lysine, arginine and phenylalanine in the protein fractions. The data confirm the high lysine of lentil and field pea and the low methionine in fababean protein.

TABLE 14
AMINO ACID COMPOSITION OF PROTEINS IN AIR-CLASSIFIED PROTEIN FRACTIONS IN g/16 g NITROGEN

Amino acid	*Field pea*	*Lentil*	*Northern bean*	*Navy bean*	*Mung bean*	*Cowpea*	*Lima bean*	*Fababean*
Lysine	7·7	7·8	7·3	6·8	7·2	6·9	6·7	6·5
Histidine	2·4	2·3	3·0	2·7	2·9	3·4	3·1	2·5
Ammonia	1·4	1·8	1·9	1·8	1·9	1·9	1·7	2·0
Arginine	8·6	7·4	6·5	6·8	7·1	7·3	5·5	10·8
Aspartic acid	10·8	11·4	11·9	12·0	11·7	11·6	12·4	10·9
Threonine	3·5	3·6	3·9	4·4	3·2	3·7	4·3	3·3
Serine	4·7	5·1	5·9	6·1	5·2	5·1	6·7	4·8
Glutamic acid	16·7	16·3	16·6	15·8	18·1	18·2	13·9	17·4
Proline	4·1	3·9	3·8	3·9	4·3	4·2	4·1	4·2
Half-cystine	1·3	0·8	0·7	0·8	0·6	0·8	1·0	1·0
Glycine	4·0	3·8	3·8	3·9	4·0	3·8	3·9	4·0
Alanine	4·0	4·0	3·9	4·1	4·1	4·2	4·3	4·0
Valine	4·7	4·8	5·2	5·2	5·3	5·2	5·4	4·7
Methionine	1·0	0·9	1·3	1·2	1·3	1·5	1·3	0·7
Isoleucine	4·3	4·4	4·6	4·6	4·3	4·3	4·9	4·1
Leucine	7·3	7·6	8·4	8·4	8·1	8·0	8·6	7·6
Tyrosine	3·0	2·9	3·3	3·2	2·8	3·1	3·2	3·0
Phenylalanine	4·7	5·1	6·0	5·9	6·1	5·8	6·4	4·2
Total	94·2	93·9	98·0	97·6	98·2	99·0	97·4	95·7

The influence of amino acid composition on the nutritive value of the legume proteins was demonstrated in rat feeding trials in which the legume flours were autoclaved to inactivate the heat-labile antinutritive factors.[45] When incorporated as the sole source of protein in 10% protein diets, soybean, field pea and fababean supplied only 39, 36 and 30% respectively, of the sulfur amino acid requirements of the rat (Table 15). Differences in feed intake, weight gain, true protein digestibility and protein efficiency ratio paralleled the differences in methionine (Met) plus cystine (Cys) and ranked the diets in the order of casein > soybean > field pea > fababean. Methionine supplementation of the legume diets increased feed intakes to the level of the casein diet, and improved weight gains and protein efficiency ratio, especially for fababean. Protein digestibility, which varied significantly among the three legumes, appeared to be independent of methionine nutrition.

TABLE 15
RELATIONSHIP BETWEEN AMINO ACID COMPOSITION AND PROTEIN NUTRITIVE VALUE OF LEGUMES AND CEREAL-LEGUME BLENDS[45,46]

Protein and amino acid in diet	*Amino acid in diet as % of rat requirements* Lys	Met+Cys	*Feed intake (g)*	*Weight gain (g)*	*Protein digestibility (%)*	*Protein efficiency ratio*
Legumes alone						
Casein	78	48[a]	344	118·7	99	2·50
Soybean	64	39	274	68·2	87	1·81
Field pea	77	36	181	36·5	85	1·47
Fababean	61	30	140	10·0	80	0·52
Legumes plus methionine						
Casein	78	78	357	152·7	100	3·10
Soybean	64	69	378	127·0	90	2·43
Field pea	77	66	407	141·7	88	2·53
Fababean	61	60	343	100·0	83	2·11
Legume–wheat blends						
Casein	49	60	410	133·8	96	2·50
Soybean	42	56	338	83·5	92	2·04
Field pea	48	54	318	85·5	94	2·13
Field bean	45	53	360	95·9	88	1·96
Lentil	50	52	298	74·7	84	1·88
Broad bean	41	51	294	68·5	90	1·82
Fababean	40	51	293	63·4	87	1·72

[a] First limiting amino acid is underlined.

Due to their higher cystine or total sulfur amino acid levels, cereals are considered complementary to legume proteins in mixed diets. When legume flours were blended with equal proportions of protein from wheat flour in 10% protein diets,[46] lysine was found to be the first limiting amino acid for six legume species as well as for casein (Table 15). The feed intakes, weight gains and protein efficiency ratios were intermediate between those of legumes alone and methionine-supplemented legumes. Protein digestibility was improved because of the high digestibility of the wheat protein component in the diet. The weight gains and protein utilization of the field pea–wheat diet by the rats were high, due to the high lysine level and protein digestibility.

Nutritive value studies have been conducted on the protein fractions of field pea and fababean. Bell and Youngs[47] showed that protein and energy components of PI were less digestible than those of egg or casein. The biological value of field pea PI was markedly inferior to animal proteins because of methionine deficiency, and supplementation with this sulfur amino acid produced growth and biological values equal to those of egg or casein, Breads supplemented with protein fractions from field pea and fababean gave protein efficiency ratios of 1·7–1·8 as compared to the value of 1·1 for the control bread.[48]

FUNCTIONAL PROPERTIES OF LEGUME FLOUR FRACTIONS

The range of physicochemical properties which affect the behavior of proteins in food systems during processing, manufacture, storage and preparation for consumption is very broad.[49] In current applications of soybean products, emulsification and solubility are important in sausage-type processed meats and coffee whiteners; hydration and water binding affect bread dough and meat products; solubility and viscosity are important in beverages; gelation is desired in cold meat products; foaming is required in whipped toppings; cohesion is needed in textured products; and color control is desired in bread products.

Nitrogen solubility indices and extraction curves provide useful information about processing yields, heat damage to proteins and food applications of the products. Some investigators have found close correlations between solubility tests and viscosity, water absorption, emulsification and whipping properties while others have found these indices to be less than clearly indicative of functional properties.[50]

Nitrogen extraction and precipitation curves have been reported for the nine legume flours (Fig. 9). Except for narrow apparent isoelectric points, the proteins in mung bean and field pea flours are more extractable over pH 2–11 than soybean, fababean and chickpea. The solubility profiles for lupin, lentil and pea (navy) bean have broad apparent isoelectric points

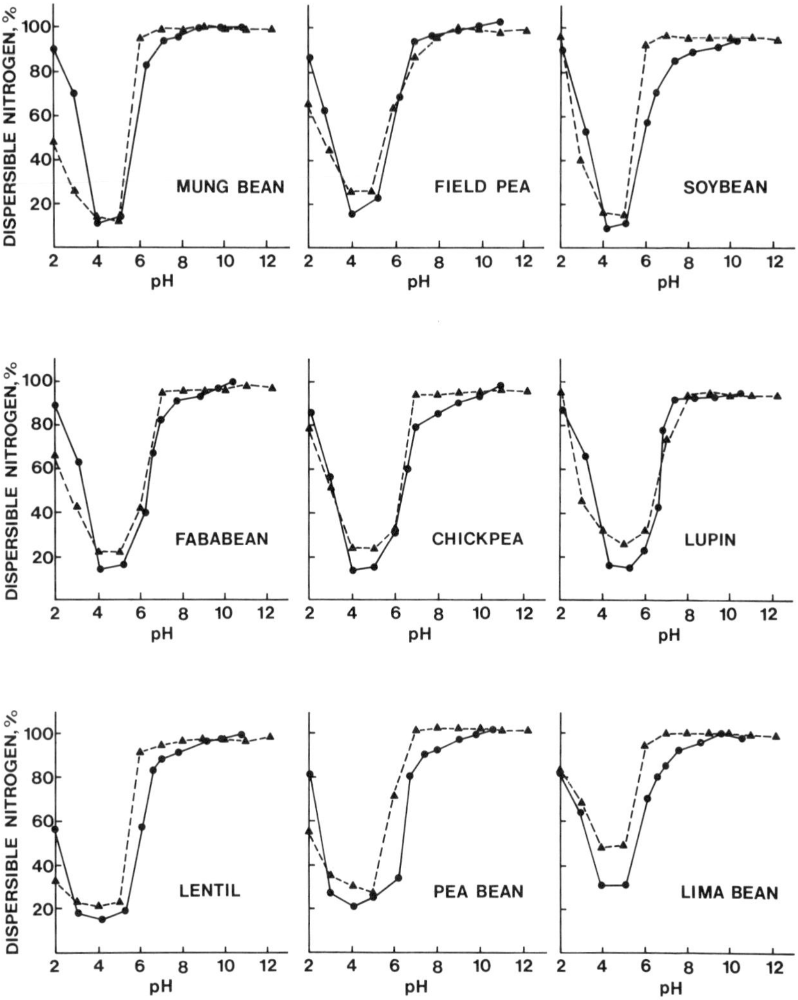

FIG. 9. Nitrogen extraction (●—●) and precipitation (▲—▲) profiles for nine legume flours.[50]

and the latter two legumes show considerable acid insolubility. The nitrogen in lima bean flour is highly soluble at the isoelectric point and the protein fraction may have specific application in acidic foods and beverages. The natural pH values for aqueous dispersions of legume flours vary over a narrow range of 6·2–6·4. Lupin has a more acid pH than soybean or the other legumes (Table 16).

TABLE 16
FUNCTIONAL PROPERTIES OF LEGUME FLOURS, PROTEIN FRACTIONS AND STARCH FRACTIONS, PERCENT DRY BASIS[28]

	pH flour	*Water absorption*			*Oil absorption*			*Oil emulsification*		
		Flour	*PI*	*SI*	*Flour*	*PI*	*SI*	*Flour*	*PI*	*SI*
Soybean	6·6	129	—	—	134	—	—	99	—	—
Lupin	5·5	173	—	—	125	—	—	90	—	—
Chickpea	6·3	75	69	78	78	85	67	94	92	79
Navy bean	6·2	89	84	93	70	83	61	64	66	20
Northern bean	6·3	122	144	113	73	85	66	92	98	57
Fababean	6·4	86	33	124	74	94	61	47	77	24
Field pea	6·4	80	72	103	72	94	61	48	48	20
Lima bean	6·3	63	32	102	63	81	54	53	70	27
Mung bean	6·3	72	41	92	65	90	49	64	64	22
Lentil	6·3	85	95	93	66	92	51	56	58	50

Most of the starchy legume flours are inferior to soybean and lupin flours in water and oil absorption, and oil emulsification (Table 16). Northern bean flour is high in water absorption and oil emulsification while chickpea shows superior oil emulsification. Generally, the starch fractions have stronger water absorptions, while the protein fractions are superior in oil absorption and emulsification.

Vegetable proteins can display excellent foam volumes when dilute aqueous slurries are aerated vigorously. For the legume flours, the navy, northern and mung beans compare favorably with soybean in their whipping characteristics (Table 17). Fractionation of the components show that the foaming property is associated with the legume proteins. Lipids disperse foams very readily so that chickpea and lupin flours develop very unstable foams.

Low initial viscosity during mixing, but high thickening properties after cooking, are desired attributes of flour ingredients in prepared

TABLE 17

PROPERTIES OF AQUEOUS SLURRIES OF LEGUME FLOURS, PROTEIN FRACTIONS AND STARCH FRACTIONS[28]

	Foam volume (ml)			*Peak viscosity (B.U.)*			*Cold paste viscosity (B.U.)*			*Color of slurry*
	Flour	*PI*	*SI*	*Flour*	*PI*	*SI*	*Flour*	*PI*	*SI*	
Soybean	960	—	—	60	—	—	60	—	—	Tan
Lupin	57	—	—	60	—	—	180	—	—	Yellow
Chickpea	7	25	5	1000	440	1000	1780	520	1820	Cream
Navy bean	747	970	415	660	140	1000	780	260	1520	Cream
Northern bean	615	1055	367	980	240	1340	860	210	1760	Cream
Fababean	220	500	117	460	60	700	1020	150	2260	Brown
Field pea	335	815	246	525	50	720	800	70	920	Cream
Lima bean	435	900	315	980	120	1040	1100	260	2000	Cream
Mung bean	832	1025	485	920	40	1160	780	180	1800	Green–brown
Lentil	435	850	400	640	80	980	840	260	2400	Tan

foods. Fababean and field pea flours gave low peak viscosities, but high cold paste viscosities were recorded on the Brabender viscoamylograph (Table 17). The thickening characteristics are essentially a property of the starch fractions.

Other investigators have considered the restricted swelling power of legume starches in water, and their stable hot paste viscosities, as being similar to those of chemically cross-linked starches.[22, 30, 35] When the starch fractions have been pasted in acid solutions, the viscosities are also very stable. These properties have been associated with the relatively high amylose contents (24–40%) of legume starches.[35, 50, 51]

Slurries of the SII fractions from field pea and fababean have been gelled under retort conditions and the viscosities measured on the Brookfield Viscometer.[22, 52] The gels were rigid, opaque and friable, and showed similar gel viscosities when heated at 95 °C (Table 18). At 120 °C

TABLE 18
GEL VISCOSITY (centipoise) OF 3% STARCH SUSPENSIONS FOLLOWING RETORTING AT 95, 120 AND 130 °C FOR 15 min AND 16 h RETROGRADATION AT 22 °C[22]

	Canning temperature (°C)		
	95	*120*	*130*
Field pea starch (SII)			
pH 6·0	600	8500	6300
pH 4·0	1000	4200	160
Commercial corn starch			
pH 6·0	600	3400	3100
pH 4·0	1000	600	200

the stability of the field pea gel greatly exceeded the viscosity of corn starch at pH 6·0 and 4·0. This phenomenon is associated with peak water uptake of legume starches at about 120 °C.[53] The property could be of value for thickening foods which require a high temperature during processing. Of concern was the gel shrinkage during storage at 5 °C, resulting in syneresis of 13–15%. Weeping losses could be prevented by acetylation to a low degree of substitution (0·1 D.S.).

There are no specific viscosity requirements for slurries of vegetable proteins but gelation after cooking can be a useful functional property in prepared and canned meats. When the apparent or cold viscosity was measured on 15% dispersions, the values obtained for field pea and fababean protein fractions were negligible compared to that of soybean protein concentrate (Table 19). After retorting and cooling, the viscosity readings for the protein products are very high and, except for soybean flour, beyond the range of the instrument. More important, both protein fractions exhibit medium gel characteristics which are inferior to the Promine D isolate gel but superior to that of soybean protein concentrate. Fleming *et al.*[54] demonstrated that isolated globulin proteins extracted from the soybean, field pea and fababean products were responsible for the gelation phenomenon.

TABLE 19

APPARENT BROOKFIELD VISCOSITY OF SEED PROTEIN SLURRIES BEFORE (INITIAL) AND AFTER (FINAL) RETORTING AT 90 °C FOR 45 min AND COOLING TO 25 °C[54]

Protein product	*Initial* (*cP*)	*Final* (*cP*)	*Characteristics of final product*
Soybean flour	1 000	45 000	smooth fluid
concentrate	6 000	166 000+	granular fluid
isolate	3 000	166 000+	firm gel
Field pea PI	40	166 000+	medium gel
Fababean PI	300	166 000+	medium gel

Undesirable dark colors have been observed in slurries containing fababean and mung bean fractions (Table 17). Soybean, lupin and lentil products also exhibit distinct colors. The cream-colored legumes would have the greatest utility in food applications.

Utilization of Protein Fraction

One of the most promising areas for use of the air-classified legume protein is in the production of protein-enriched pasta, especially for low income groups and developing countries. Nielsen *et al.*[55] prepared noodles and spaghetti from hard wheat flour fortified with 20% air-classified pea protein fraction to increase protein quantity and quality. Fortification increased the protein content up to 22·1% for raw noodles and 24·1% for raw spaghetti. The chemical score for the wheat products was only 39, the index increasing to 85–90 in the various blended products (Table 20). Noodle dough containing raw PI was sticky and sheeted poorly but precooking the protein fraction improved the handling characteristics substantially. The spaghetti doughs showed better handling characteristics and the blend containing the raw protein frac-

TABLE 20

PROTEIN AND COOKING QUALITY OF SPAGHETTI PREPARED FROM HARD WHEAT FLOUR (HRS) RAW (RPI) AND COOKED (CPI) PEA PROTEIN FRACTION I, SOYBEAN PROTEIN CONCENTRATE (SPC), DRY BASIS[55]

Spaghetti flour ingredients	*Protein (%)*		*Chemical score*	*Cooked water absorption (%)*	*Cooking loss (%)*	*Sensory evaluation*	
	Raw	*Cooked*				*Firmness* 1–8[a]	*Flavor* 1–5[b]
100% HRS	15·2	15·9	39	189	8·3	4·3	3·4
20% RPI/80% HRS	22·4	24·1	85	185	9·3	5·5	1·9
20% CPI/80% HRS	24·1	25·4	86	170	12·7	4·9	2·9
20% SPC/80% HRS	27·3	29·5	90	164	11·0	4·6	2·3

[a] 1 = very soft, 8 = very firm.
[b] 1 = poor, 5 = excellent.

tion was similar to the control wheat in water absorption, cooking losses and firmness. Due to the objectionable pea flavor of the cooked spaghetti, a 20:80 blend of cooked PI and wheat flour was tested. It was shown that the pea flavor was heat-volatile, and that a very acceptable product could be made with the pea protein fraction after a moist heat treatment.

Fleming and Sosulski[56] evaluated the breadmaking characteristics and nutritive value of air-classified protein fractions (PI) from field pea and fababean. When incorporated at 10, 15 and 20% replacement of hard wheat flour, field pea and fababean breads had lower loaf volumes than soybean bread but showed more desirable crumb grain. Samples of the breads were observed by scanning electron microscopy which showed continuous sheets of protein enveloping the starch granules in the wheat flour bread (Fig. 10). In contrast to the control, the soybean, sunflower, fababean and field pea breads had porous, broken surfaces. It appeared as though the gluten sheets were subjected to greater stress in sections

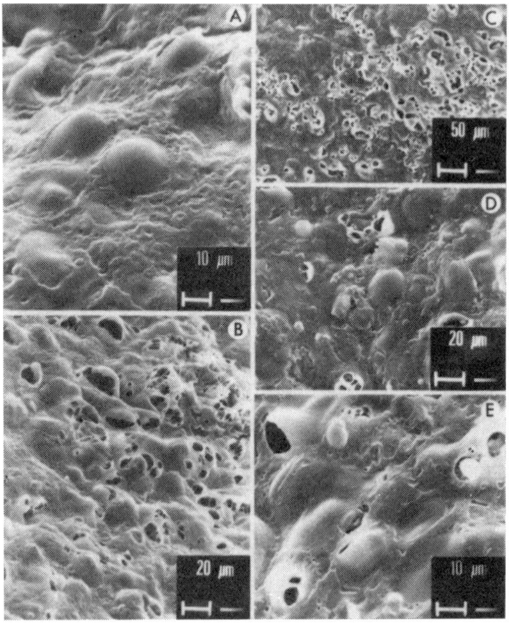

FIG. 10. Scanning electron micrographs of freeze-dried bread crumb. A = wheat flour bread; B = soybean bread; C = sunflower bread; D = fababean bread; E = field pea bread.[57]

which covered the starch granules. These pores would permit gases to escape during proofing and baking and thus reduce loaf volumes.

Acceptable breads containing 14–15% protein were obtained with composite flours containing 15% legume, 2% vital gluten and 1% polyoxyethylene-8-stearate (Table 21). However, specific volumes were substantially less in the supplemented bread, and crumb grain was adversely affected. The improvement in protein efficiency ratios due to the legume supplementation was substantial, with the high lysine field pea giving the best result.

Vaisey *et al.*[60] demonstrated that the incorporation of raw PI fraction from field pea and fababean into meat patty formulations at the 30% replacement level (legume:water = 1:2) reduced drip losses to one-half that of the control but the flavors were cereal-like, rancid and bitter. Texturizing the PI slurries by drum-drying made the broiled patties more acceptable to a panel of 90 untrained subjects. The texturization process appeared to reduce fat retention in the blended patty, and the texturized field pea and fababean gave firm patties, similar to that of the beef control.

The protein fraction from field pea has been blended with cheese whey and spray-dried to produce a replacement product for non-fat dry milk with particular application in the baking industry.[61] The heat treatments during evaporation and spray drying served to remove the ripe pea flavor which was characteristic of the air-classified fractions. Viscosity and crystallization problems encountered in spray drying the two products separately were not encountered with the blended product. The PI–whey blend had higher water hydration capacity, oil emulsification, oil absorption and greater foam stability than non-fat dry milk or a soybean–whey blend. Loaf characteristics of breads containing 6% of the blend were equivalent to those of breads containing non-fat dry milk, and sensory evaluations showed no significant differences from the control.

Utilization of the Starch Fraction

The yields of starch fraction greatly exceed that of the protein fraction and the economic viability of a commercial operation is largely dependent on markets for the starch. Industrial applications would be of prime interest because of the volumes involved. Much of the starch fraction from the Pro-Star Mills Ltd plant at Saskatoon was utilized for the production of phosphorylated derivatives of pea starch which were used in the manufacture of carbonless paper. Pea starch fraction was also

TABLE 21

BREADMAKING AND NUTRITIONAL CHARACTERISTICS OF BREADS FORTIFIED WITH CONCENTRATED PLANT PROTEINS[58, 59]

Composite flour formula[a]	*Bread protein (as is, %)*	*Loaf volume* (cm^3)	*Specific volume* ($cm^3\ g^{-1}$)	*Crumb compressibility (units)*	*Crumb grain* (0–10)	*Protein efficiency ratio*[b]
Wheat (100:0:0)	10·2	865	6·0	20	10·0	1·09
Soy (83:15:2)	14·0	810	4·7	14	7·0	1·69
Field pea (83:15:2)	14·2	750	4·9	24	8·5	1·81
Fababean (83:15:2)	14·9	800	4·6	32	9·0	1·67

[a] Wheat flour:concentrated plant protein:vital gluten.
[b] Protein efficiency ratio adjusted to 2·50 for casein.

modified for use as an adhesive, principally for the manufacture of corrugated board. Potash mines used the gelatinized starch as a clay depressant in the refining of potash.

Due to their viscosity characteristics, the starch fractions are considered to have only limited food uses.[52] Canning applications generally require heat-stable starches but unmodified pea starch exhibits syneresis, which would not be desirable in many canned foods. Pea starch could replace regular corn starch in low pH sauces and as a binder in canned luncheon meats. Most sauces and gravies require more thickening ability than is exhibited by pea starch but legume starches such as fababean and lentil could provide cold pasta viscosities which have wider application as thickeners in foods.

REFERENCES

1. Alpine Bulletin 31am and 018, Alpine American Corp., Natick, Mass.
2. Vose, J. R. (1978). *Separation and Purification Methods*, **7**, 1.
3. Kent, N. L. (1975). *Technology of Cereals with Special Reference to Wheat*, 2nd edn, Pergamon Press, Oxford.
4. Jones, C. R., Halton, P. and Stevens, D. J. (1959). *J. Biochem. Microbiol. Technol. Eng.*, **1**, 77.
5. Kent, N. L. (1965). *Cereal Chem.*, **42**, 125.
6. Tipples, K. H. and Kilborn, R. H. (1968). *Cereal Sci. Today*, **13** (9), 331.
7. Pomeranz, Y., Ke, H. and Ward, A. B. (1971). *Cereal Chem.*, **48**, 47.
8. Vose, J. R. and Youngs, C. G. (1978). *Cereal Chem.*, **55**, 280.
9. Pomeranz, Y., Finney, K. F., Bolte, L. C. and Shogren, M. D. (1976). *Cereal Chem.*, **53**, 956.
10. Wu, Y. V. and Stringfield, A. C. (1973). *Cereal Chem.*, **50**, 489.
11. Stringfield, A. C., Wall, J. S., Donaldson, G. L. and Anderson, R. A. (1976). *Cereal Chem.*, **53**, 51.
12. Houston, D. F., Mohammed, A., Wasserman, T. and Kester, E. B. (1964). *Cereal Chem.*, **41**, 514.
13. Shaw, R. and Shuey, W. C. (1977). *Nat. Symp. Food Process. Wastes Proc.*, Environmental Protection Agency, Washington, D.C.
14. Kon, S., Sanshuck, D. W., Jackson, R. and Huxsoll, C. C. (1977). *J. Food Proc. Preserv.*, **1**, 69.
15. Vose, J. R., Basterrechea, M. J., Gorin, P. A. J., Finlayson, A. J. and Youngs, C. G. (1976). *Cereal Chem.*, **53**, 928.
16. Youngs, C. G. (1975). Primary processing of pulse. *Oilseed and Pulse Crops in Western Canada — A Symposium*, Ed. Harapiak, J. T., Western Cooperative Fertilizers Ltd, Calgary, Canada.
17. Sosulski, F. W., McLean, L. A. and Austenson, H. M. (1974). *Can. J. Plant Sci.*, **54**, 247.

18. Marquard, R. R., Ward, A. T., Campbell, L. D. and Cansfield, P. E. (1977). *J. Nutr.*, **107**, 1313.
19. Biliaderis, C. G., Grant, D. R. and Vose, J. R. (1981). *Cereal Chem.*, **58**, 502.
20. Colonna, P., Gallant, D. and Mercier, C. (1980). *J. Food Sci.*, **45**, 1629.
21. Reichert, R. D. and Youngs, C. G. (1978). *Cereal Chem.*, **55**, 469.
22. Vose, J. R. (1977). *Cereal Chem.*, **54**, 1141.
23. Satin, M., McKeown, B. and Findlay, C. (1978). *Cereal Foods World*, **23**, 676.
24. Tyler, R. T. and Panchuk, B. D. (1982). *Cereal Chem.*, **59**, 31.
25. Patel, K. M., Bedford, C. L. and Youngs, C. W. (1980). *Cereal Chem.*, **57**, 123.
26. Sosulski, F. W. and Tyler, R. T. (1982). *J. Agric. Food Chem.*, **30**, in review.
27. Sosulski, F., Garrett, M. D. and Slinkard, A. E. (1976). *Can. Inst. Food Sci. Technol. J.*, **9**, 66.
28. Sosulski, F. W. and Youngs, C. G. (1979). *J. Am. Oil Chemists' Soc.*, **56**, 292.
29. Tyler, R. T., Youngs, C. G. and Sosulski, F. W. (1981). *Cereal Chem.*, **58**, 144.
30. Vose, J. R. (1980). *Cereal Chem.*, **57**, 406.
31. Reichert, R. D. (1981). *Cereal Chem.*, **58**, 266.
32. Steggerda, F. R. (1968). *An. N.Y. Acad. Sci.*, **150**, 57.
33. Srikantria, S. G. (1973). *PAG Bull.*, **3** (2), 30.
34. Cerning-Beroard, J. and Filiatre, A. (1976). *Cereal Chem.*, **53**, 968.
35. Lineback, D. R. and Ke, C. H. (1975). *Cereal Chem.*, **52**, 334.
36. Schweizer, T. F., Horman, I. and Würsch, P. (1978). *J. Sci. Food Agr.*, 1978, **29**, 148.
37. Fleming, S. E. (1981). *J. Food Sci.*, **46**, 794.
38. Sosulski, F. W., Elkowicz, K. and Reichert, R. D. (1982). *J. Food Sci.*, **47**, 498.
39. Liener, I. E. (1976). *J. Food Sci.*, **41**, 1076.
40. Honavar, P. M., Cheng-Ven, S. and Liener, I. E. (1962). *J. Nutr.*, **77**, 109.
41. Cheeke, P. R. (1976). *Nutr. Rep. Intern.*, **13**, 315.
42. Spiller, G. A. and Shipley, E. A. (1977). *Wld. Rev. Nutr. Diet.*, **27**, 105.
43. Elkowicz, K. and Sosulski, F. W. (1982). *J. Food Sci.*, **47**, 1301.
44. Pitz, W. J., Sosulski, F. W. and Hogge, L. R. (1980). *Can. Inst. Food Sci. Technol. J.*, **13**, 35.
45. Sarwar, G., Sosulski, F. W. and Bell, J. M. (1975). *Can. Inst. Food Sci. Technol. J.*, **8**, 109.
46. Sarwar, G., Sosulski, F. W. and Holt, N. W. (1975). *Can. Inst. Food Sci. Technol. J.*, **8**, 170.
47. Bell, J. M. and Youngs, C. G. (1970). *Can. J. Anim. Sci.*, **50**, 219.
48. Fleming, S. E. and Sosulski, F. W. (1977). *Cereal Chem.*, **54**, 1238.
49. Kinsella, J. E. (1979). *J. Am. Oil Chemists' Soc.*, **56**, 242.
50. Fan, T. Y. and Sosulski, F. W. (1974). *Can. Inst. Food Sci. Technol. J.*, **7**, 256.
51. Shoch, T. J. and Maywald, E. C. (1968). *Cereal Chem.*, **45**, 564.
52. Comer, F. W. and Fry, M. K. (1978). *Cereal Chem.*, **55**, 818.

53. HALLBROOK, W. U. and KURTZMAN, R. H. Jr. (1975). *Cereal Chem.*, **52**, 156.
54. FLEMING, S. E., SOSULSKI, F. W. and HAMON, N. W. (1975). *J. Food Sci.*, **40**, 805.
55. NIELSEN, M. A., SUMMER, A. K. and WHALLEY, L. L. (1980). *Cereal Chem.*, **57**, 203.
56. FLEMING, S. E. and SOSULSKI, F. W. (1977). *Cereal Chem.*, **54**, 1124.
57. FLEMING, S. E. and SOSULSKI, F. W. (1978). *Cereal Chem.*, **55**, 373.
58. SOSULSKI, F. W. and FLEMING, S. E. (1979). *Bakers' Digest*, **53** (3), 20.
59. FLEMING, S. E. and SOSULSKI, F. W. (1977). *Cereal Chem.*, **54**, 1238.
60. VAISEY, M., TASSOS, L., McDONALD, B. E. and YOUNGS, C. G. (1975). *Can. Inst. Food Sci. Technol. J.*, **8**, 74.
61. PATEL, P. R., YOUNGS, C. G. and GRANT, D. R. (1981). *Cereal Chem.*, **58**, 249.

Chapter 6

NEW APPROACHES IN THE USE OF FISH PROTEINS

I. M. MACKIE
Torry Research Station, Aberdeen, UK

SUMMARY

This chapter is a review of methods of preparing protein products from fish, with emphasis on the newer approaches that are being made to obtain products with acceptable properties for human consumption. While it has long been accepted that fish proteins are of high nutritional value it has not always been sufficiently recognised that they must be in a form that adds to the eating or aesthetic quality of the food. More recently a greater awareness of the requirements for protein preparations, in general, has led to research into means of retaining or introducing desirable functional properties in fish proteins.

The traditional fermented fish protein products of south-east Asia and the kamaboko of Japan are discussed by way of introduction to more recent developments in fish protein products. Fish protein concentrates are reviewed and their failure to be accepted is contrasted with the success of the fish gel industry of Japan. Attention is also given to protein spinning processes, enzymic fish protein hydrolysates, plasteins and the use of chemical and enzymic modifications to improve the functionality of proteins. Improved solvent extraction and extrusion processes are discussed while krill, because of its unique characteristics and particular processing problems, is treated separately.

Finally, fish protein products are considered with reference to means of making more of the world's fish resources available for direct human consumption.

INTRODUCTION

Although fish proteins have long constituted an important part of the human diet,[1-4] there has been remarkably little processing of fish into foods or components of foods in which its identity has been lost. There are many reasons for this lack of development but a major factor must be the extreme perishability of seafoods compared with other sources of proteins.[5] This is due in part to the relatively rapid bacterial spoilage of fish but other factors are the extreme sensitivity of the fish fats in general to oxidation, and associated with it, the production of rancid and other off-flavours which are notoriously difficult to remove from fish.[6] Fish proteins too are less stable than those of mammals or of vegetables and tend to lose all important functional properties on drying.[7-10] These two factors present the food scientist with formidable technical problems when the preparation of acceptable protein products from fish is considered.

On the whole the history of the development of protein products from fish has been more one of failure than success. Despite a considerable amount of effort since the early part of this century, few of the early expectations have been realised.[9] Perhaps the best known product is fish protein concentrate (FPC) which was seen in its early stages as a panacea for world malnutrition.[7-9] Although nutritious, it was completely lacking in functional properties which are now recognised as being of paramount importance if a protein product is to find acceptance. More attention is being given today to processes which will retain the valuable functional properties of fish proteins by avoiding dehydration which is known to be extremely damaging to functional properties such as solubility, emulsion or gel forming capacity.[7-10] Alternatively, biological processes which depend on the action of proteolytic enzymes to partially hydrolyse proteins, and so modify functional properties, are seen as being more likely to produce acceptable products. Greater emphasis has been given to these and other newer approaches in this review.[9-15]

FISH AS FOOD

Attitudes to Fish

While very few species of fish are inedible, many are nonetheless rejected as food for human consumption for a whole variety of reasons. Attitudes to fish vary tremendously from one region of the world to another, and even locally within countries there may be preferences for particular species of fish.[1-3] What may be a great delicacy in one country may be thrown back into the sea elsewhere or, at best, be reduced to fish meal. Squid, for example, would be rejected by many people in the UK but it is a common item of diet in many parts of the world; capelin, a small fatty fish of the northern Atlantic, is caught specifically for fish meal by European countries[16,17] yet it is highly prized as a food in Japan.[4] Viscera of fish are usually thrown back into the sea in western countries but may actually be preferred to the flesh in some eastern societies. The reasons for these widely differing attitudes to fish are deep rooted in the cultural and religious histories of peoples and, although a nutritionist might argue that all fish are nutritionally equivalent, he is not likely to convince many people of this fact, be it either in a developed or a developing country. On the whole, people still prefer to buy fish either whole or at least recognisable as fish so that they can convince themselves of the freshness and quality of the food. In developed countries such attitudes have largely changed with the acceptance of convenience frozen foods of the fish finger type. Nonetheless, attitudes to fish are conservative and impede the introduction of new species of fish, and particularly of powdered protein products made from fish.

Fish in the Diet

In the world as a whole, fish represent a major source of animal protein. The present annual landings are about 70 million tonnes, of which 50 million tonnes are caught for human food, the remainder being used for fish meal.[18] By way of comparison it is of interest that 120 million tonnes of meat are available annually (beef 40%, pork 35%, poultry 20% and sheep/goat 5%).[19]

In terms of world supplies, fisheries contribute about 6% of all proteins and 18% of animal proteins. These figures, however, conceal a wide variation in the importance of fish in the diet as in many developing countries it is the major and sometimes the sole source of animal protein. It has been quoted, for example, that as much as 59% of the population of developing countries obtains more than 30% of their 'meat' supplies from fish (Table 1). With the particular exception of Japan, and to a lesser extent Norway, Spain and Portugal, the picture is largely reversed for the developed countries. In centrally planned economies of the world too, fish is an important item in the diet.

As the world population continues to expand, further increases in

TABLE 1
CONTRIBUTION OF FISH TO TOTAL PROTEIN SUPPLIES FROM MEAT PRODUCTS (1972/74, 3-YEAR AVERAGE)

Contribution of fish	*Developed countries*	*Developing countries*	*Centrally planned economies*	*World*
	% of population	*% of population*	*% of population*	*% of population*
More than 40%	16	57	5	33
30–40%	6	2	86	30
20–30%	1	9	3	5
10–20%	38	22	3	19
Less than 10%	39	10	3	13

(Courtesy Krone.[19])

production will be required simply to maintain the present *per capita* consumption and, as there is likely to be a continuing demand (Table 2) for the readily accepted species—cod and other gadoids—in developed countries, pressure to export such fish from developing countries for valuable foreign exchange will put increasing strains on their already inadequate supplies of fish. As a consequence the problems relating to the present imbalance of distribution of fish between developed and developing countries are likely to continue and to be further compounded by the much higher rates of growth of populations in the latter countries.[19, 20]

Classification of Fish

An examination of fisheries statistics will give some indication of the range of species of fish and shellfish presently exploited by man, but the number of such species is small in relation to the many thousands of species of fish and other marine animals known to exist.

Out of 22 000 species of fish[21,22] and thousands more of shellfish and crustacea, possibly no more than several hundred are actually used for food and within these main groups the vast bulk of the supplies will be represented by no more than a few dozen species. Nonetheless the range of size and shape of those species actually exploited presents many problems in handling and processing and, as spoilage characteristics vary even among closely related species, few generalisations can be made. The term fish as used in 'fish proteins' is all-embracing only to the extent that

TABLE 2

ESTIMATED DEMAND FOR FISH 1980, 1990, 2000 (IN LIVEWEIGHT EQUIVALENT)

	Total (*million tons*)				*Feed* (*million tons*)		*Food* (*million tons*)				*Food* (*kg per capita*)			
	Con-sumption	*Projected demand*			*Con-sumption*	*Pro-jected demand*	*Con-sumption*	*Projected demand*			*Con-sumption*	*Projected demand*		
	1972–74	1980	1990	2000	1972–74	1980	1972–74	1980	1990	2000	1972–74	1980	1990	2000
World	67·2	83·4	78·8	97·1	17·8	23·0	49·9	60·4	78·8	97·1	13·1	13·9	14·9	15·5
Developing	25·5	33·1	42·6	56·7	2·6	3·8	22·8	29·3	42·6	56·7	8·4	9·3	10·6	11·5
Latin America	3·5	4·6	4·2	6·0	1·2	1·7	2·4	2·9	4·2	6·0	7·7	7·9	8·8	9·8
Africa	2·7	3·4	5·0	7·3	0·1	0·1	2·6	3·3	5·0	7·3	8·2	8·7	9·8	10·8
Near East	0·7	1·1	1·2	1·9	0·2	0·3	0·6	0·8	1·2	1·9	3·1	3·6	4·3	5·1
Far East	9·5	12·4	16·0	21·6	1·1	1·4	8·4	11·0	16·0	21·6	8·1	9·1	10·0	11·1
Asian centrally planned countries	9·1	11·5	16·0	19·6	0·2	0·3	8·9	11·2	16·0	19·6	10·1	11·4	14·2	15·4
Other developing	0·1	0·1	0·2	0·3	0·0	0·0	0·1	0·1	0·2	0·3	23·0	24·7	28·1	31·2
Developed	42·2	50·3	36·3	40·4	15·2	19·2	27·0	31·1	36·3	40·4	24·5	26·6	28·5	29·6
North America	5·5	6·2	4·9	5·7	1·8	2·0	3·7	4·2	4·9	5·7	15·7	16·7	18·0	19·4
Western Europe	13·6	15·1	8·2	9·1	6·7	7·8	6·9	7·3	8·2	9·1	19·1	19·7	21·4	23·1
EEC	8·7	9·6	4·8	5·3	4·5	5·2	4·2	4·4	4·8	5·3	16·2	16·8	18·2	19·6
Other Western Europe	4·9	5·5	3·4	3·8	2·1	2·6	2·8	2·9	3·4	3·8	26·1	26·5	28·8	30·8
Eastern Europe and USSR	11·2	14·5	11·6	13·0	3·4	4·8	7·9	9·7	11·6	13·0	22·1	25·5	28·1	29·6
Eastern Europe	3·1	4·1	2·2	2·6	1·8	2·3	1·2	1·8	2·2	2·6	11·5	15·9	18·7	20·8
Oceania	0·3	0·4	0·3	0·4	0·1	0·1	0·2	0·3	0·3	0·4	14·4	14·4	15·2	16·2
Other developed	11·7	14·1	11·2	12·2	3·3	4·5	8·4	9·6	11·2	12·2	62·1	64·3	66·4	64·8

(Courtesy Robinson.[20])

it refers to marine animals. It hides the fact that it covers the most diverse group of animals in existence. Although usually used to describe the phylum Chordata the term fish, for the purposes of this review, is taken to include the Mollusca, which comprises (a) all shellfish and the free swimming species, the squids and octopuses, and (b) the Arthropoda, the crustacea of which shrimp, scampi and lobster are typical examples.

Most fish belong to the class Pisces, the bony fishes which include the teleosts, but the older class, the Selachii or cartilaginous fish (skates, rays and sharks), is not insignificant as a source of food. The bony fish can be conveniently divided into two main groups:

(a) Pelagic—the shoaling, surface swimming fish.
(b) Demersal—bottom swimming fish.

The former are the fatty or oily species of which herring and mackerel are typical and which are readily distinguished from demersal species such as cod, hake or plaice, the white fleshed fish, by the appearance of the flesh. Pelagic species have more red muscle which is required for sustained swimming, and have fat reserves under the skin and within the flesh, giving a characteristic oily off-white appearance. Demersal species, on the other hand, deposit their reserves of fat mainly in the liver and their flesh, low in fat, is generally white. A further distinction between the two groups relates to their swimming behaviour; the shoaling pelagic species are usually caught in vast numbers by seine net and as single species of uniform size, whereas the demersal fish are caught by trawl in smaller quantities and are generally mixed both with respect to size and species. Handling and sorting of the latter group is consequently more complicated.[23]

Fish Resources Available

The total landings of fish for the world as a whole over the past few decades[18-20] show that the high rate of growth of the period between 1960 and 1970 has fallen off dramatically due largely to the collapse of the anchoveta fishery off Chile and Peru. At its peak, that industrial fishery amounted to 13 million tonnes of a world total of 67 millions.[20] The landings for food fish have, however, shown a steady increase and it is expected that by the end of the century world production will exceed 92 million tonnes,[20] most of the increase being accounted for by the increased exploitation of the fisheries off the coasts of tropical and sub-tropical countries. Although optimistic forecasts have suggested that the potential world catch is much in excess of 100 million tonnes, more recent predictions come down to figures between 80 and 90 million tonnes from conventional resources. The contribution from aquaculture and from inland fisheries should not be ignored: predictions are that contributions from these two sources will be in the region of 10–20 million tonnes by the year 2000.[20]

Further increases in the fish catch could be realised by exploiting the large stocks of cephalopods (squid), the crustacea (krill) of the southern hemisphere and mesopelagic fish (the small bony fish), each of which are large enough to yield annual catches of several tens of millions of tonnes (Table 3).

Interest in exploiting these stocks is bound to grow as pressure on the world's sources of protein increases and, although some of these new species are acceptable for human food simply as fish, others such as krill, because of their small size, are more likely to be further processed.[24,25] Today large fish-consuming countries such as Japan and the USSR are actively involved in exploratory fishing for such species and in developing acceptable products for human consumption.[24,25]

Potential Sources of Raw Material for Protein Products

In any consideration of a fish protein manufacturing process it is a logical first step to identify sources of raw material which would justify a commercial operation. It is convenient to group these sources and potential sources into four main categories:

(a) Industrial fisheries presently exploited.
(b) White fish processing waste.
(c) The by-catch.
(d) Unexploited/under-exploited stocks.

Present Industrial Fisheries

Those species of fish which are caught for fish meal represent a substantial proportion of the world's catch. In 1979,[18] in excess of 20 million tonnes of a total catch of 70 million tonnes were caught specifically for reduction to fish meal and while these species are mostly small pelagic fish which are difficult to handle and process, it must be emphasised that they are nutritionally equivalent to the food fish species. Indeed in many cases food fish species are also exploited for fish meal. In the UK, for example, over 25% of mackerel landings between 1974 and 1980 were reduced to fish meal and oil[26] and it would seem that this pattern of utilisation worldwide will continue at least for the foreseeable future.

TABLE 3

PRESENT CATCHES AND POTENTIAL FOR INCREASES FROM MAJOR GROUPS OF SPECIES

	Potential (thousand tons)	*Total world catch*[a]				*Potential increase from natural stocks (thousand tons)*			*Aquaculture*[c]
		1966	*1977*	*1966*	*1977*	*Total*	*Management*[a]	*Increased effort*	
		(thousand tons)		*(% of potential)*					
Conventional marine resources									
Salmon	650	453	480	70	74	170	170	—	+
Flounder and cods	6 700	4 692	3 786	70	57	2 500	2 914	400	c
Herring, anchovy, etc.	15 600	13 709	1 801	88	12	13 800	13 799	—	—
Shrimps and lobsters	1 670?	830	1 542	50	92	25	128	100	c
Tuna	2 260	1 031	1 592	46	70	100	668	550	c
Other demersal	37 100	9 628	17 097	26	46	4 000	20 003	16 000	—
Other pelagic	40 200	13 045	28 655	32	71	3 000	11 545	8 500	c
Unconventional marine resources									
Cephalopods	50 000?)	833	1 165	2	2	—	(50 000)	(48 000)	—
Other molluscs	Not defined	2 115	3 011	—	—	—	—	Probably large	+ +
Krill	(50 000?)	—	123	—	0·2	—	(50 000)	(50 000)	—
Mesopelagic fish	(50 000?)	—	—	—	—	—	(50 000)	(50 000)	—
Totel conventional marine resources	104 810	43 388	53 361	41	51	19 970	50 120	30 150	
Total marine[b]		48 121	62 178						

(Courtesy Krone.[19])

[a] Figures of catches exclude discards at sea. Potential increase by utilisation of these discards included under 'management'.

[b] Includes species groups (seaweeds, crabs, etc.) not considered in this Table.

[c] c denotes aquaculture possibilities limited to fattening with specially provided food. + + or + denotes opportunities to increase production from aquaculture using natural food sources.

However, as pressure on the world's existing supplies of food for human consumption increases there is growing interest in developing processes and products which will allow some at least of this catch to be diverted to human consumption. Certain of these species are readily accepted simply as fish, given adequate means of handling and processing but the less palatable and unacceptable species can reasonably be considered as raw material for products with improved functional or organoleptic properties. Processes for their manufacture can indeed be considered as being complementary to and, in some cases, alternatives to the existing fish meal manufacturing process.[27]

The White Fish Processing Waste

The preference in the UK and other western countries for bone-free fillets and products made from white bone-free flesh necessitates either the removal of the flesh as fillets or as comminuted tissue from deboning machines.[28] By either process, a substantial amount of the original weight of fish is considered as 'waste' and only 36% of the weight of cod is in fact used as fillets (Table 4).[23] A further 10% can be obtained by

TABLE 4

COMPONENT WEIGHT OF COD

Component		*Percentage whole weight*	*Percentage gutted weight*
Head		21	25
Guts	Removed	7 (range 5–8)	
Liver	at	5 (range 2–7)	
Roe	sea	4 (range 1–7)	
Backbone, etc.		14	17
Fins and lugs		10	12
Skin		3	4
Fillet skinned		36	42
		100	100

Taken from J. Waterman, Torry Advisory Note No. 33. H.M.S.O. Edinburgh.

using deboning machines. The 'waste' or residue from removal of flesh from fish contains about 10% protein and although the proportion of connective tissue is higher than in fillets the proteins are essentially equivalent nutritionally to those of the latter. At present this waste material is converted into white fish meal. In the UK last year the total landings of dermersal fish were 460 000 tonnes[29] and of this half can be assumed to be waste. If this distribution is extrapolated to the world as a whole, possibly as much as 25 million tonnes are 'wasted' in one form or another and some at least is potentially available for up-grading processes.

Although not a protein source, the corresponding waste from the crustacean processing industry, the carapace of crabs, shrimps, etc., should be mentioned. Chitin, or its deacetylated form chitosan, can be recovered in high yield but presently there is no market to justify the expense of collecting raw material from a widely scattered industry.[29,30]

The By-catch

The by-catch is generally regarded as those species of fish of low economic value which are caught incidentally in a fishery and which are either thrown back to sea or brought back for reduction to fish meal.[16,17] It is difficult to estimate the amount of this 'trash fish' but figures ranging between 5 and 30% of the annual catch have been given. In shrimp fisheries of the Gulf of Mexico it is known that a substantial proportion of the catch is white fish species for which there is little or no demand and which are considered a nuisance. However, as market trends change due in large part to a shortage of traditional species fish such as saithe, redfish and even scampi, formerly unwanted species in UK fisheries are no longer thrown overboard. Doubtless this trend towards greater utilisation of the whole catch will continue.

Unexploited and Under-exploited Stocks

While the pelagic fisheries of the world, particularly the anchoveta stocks off Chile and Peru, are already heavily exploited for reduction to fish meal, there are nonetheless substantial stocks particularly off the coasts of tropical and sub-tropical countries which could be developed further. In the world as a whole, stocks of pelagic fish far outweigh those of demersal species but their development lags behind for various reasons, not least of which is the problem of handling and processing them into acceptable products.

In our waters too, there are under-developed stocks, as for example those of blue whiting and horse mackerel.[31] The former is a demersal species the stocks of which have been estimated to have an annual sustainable yield in excess of 1 million tonnes, equivalent to more than our total annual fish landings.[31] Despite considerable effort by the Torry Research Station and the White Fish Authority in demonstrating the potential of the stock it is not exploited commercially to any significant extent by the UK.

Even larger stocks of unconventional species are known to exist in the world as a whole. Squids are widely distributed throughout the oceans and remain largely unexploited overall,[32-34] even allowing for the extensive use made of them by the Japanese and Spanish. These species account for perhaps the major untapped resource of protein in the seas. Krill, too, can be considered on this scale but its value as a food has not been established to the same extent. However, as suitable processes are developed for krill it is bound to emerge as a significant source of protein.[24, 25]

THE NATURE OF FISH

General Chemical Composition

Fish Flesh

The main constituents of fish flesh are, as would be expected for the flesh of any animal, water, protein and fat, but while the relative amounts of these components are generally within the range found for mammals, the nature and properties of both the fats and the proteins are markedly different.[35-37] Fish lipids, in general, are highly unsaturated compared with those of mammals, and are readily oxidised to peroxides or peroxy-radicals,[6] which in turn are believed to interact with the proteins to produce a loss of functional properties on frozen storage. These changes manifest themselves as a toughening and loss of succulence of the flesh.[38-40] Another theory of protein denaturation suggests that formaldehyde, one of the products of enzymic breakdown of trimethylamine oxide, interacts with free amino groups to form cross-linkages. Denaturation of proteins resulting from freezing and frozen-storage will be referred to later, in the section on wet fish protein products. The suppression of oxidation is perhaps an even greater problem in the preparation of dried protein products as even trace amounts (0·1 % fat on a dry weight basis) can be sufficient to produce unacceptable odours and flavours on subsequent storage.[7-10] The proportion of lipid in the flesh of fish varies markedly even in fatty fish species. Mackerel, for example, can have lipid contents as high as 35 % when food is most abundant but at the other extreme, levels as low as 1 % have been recorded.[41] Even in non-fatty species there is a well-defined seasonal fluctuation.[35-37] Generally, the sum of the water and fat contents amounts to 80 % with any decrease in either constituent being compensated by an increase in the other.

The protein contents (usually expressed as the nitrogen contents × 6·25) also show variations with season but the fluctuation in levels is much less than for lipids. Most protein contents of flesh lie between 15 and 19%, but it should be remembered that on average between 10 and 15% of this protein equivalent arises from non-protein sources such as free amino acids, peptides, nucleotides, trimethylamine oxide, ammonia and other volatile bases and, in the case of all fish of the Selachii, urea. For the latter class of fish, the use of conversion factors to give protein content can be highly misleading as non-protein nitrogen, mostly from urea, can be over 35% of the total nitrogen.[42] The nitrogen content of whole fish is generally 1–2% below that for flesh.[7]

Fish flesh proteins are of high nutritional value and are essentially equivalent to those from the other flesh foods such as beef and poultry (Table 5), not unexpectedly, as the edible portions of animals consist

TABLE 5
AMINO ACID COMPOSITION OF THE PROTEINS OF FLESH AND OTHER FOODS mg/g NITROGEN[43]

	Wheat (whole)	*Soya*	*Milk (cows)*	*Eggs (whole)*	*Beef (lean)*	*Poultry*	*Fish*	*Molluscs*
Ile	210	280	350	350	320	290	330	300
Leu	420	490	640	520	500	470	530	480
Lys	150	400	510	390	570	560	610	500
Met	100	80	180	200	170	150	180	170
Cys	160	100	60	110	80	80	70	100
Phe	280	310	340	320	280	280	260	260
Tyr	190	200	280	250	240	220	220	260
Thr	170	240	310	320	290	260	300	290
Trp	70	80	90	110	80	70	70	80
Val	280	300	460	470	330	300	360	390

(Courtesy *McCance & Widdowson's The Composition of Foods*, 1978. HMSO/Elsevier Press.)

primarily of muscle tissue which is structurally and functionally of a similar nature.[44,45] However, old habits die hard, it would seem, as far as eating fish is concerned, as the myth still persists that fish proteins are nutritionally inferior to meat proteins despite a welter of scientific evidence to the contrary.[46] However, the extent of usage of fish for fish meal must surely be adequate testimony to its nutritional excellence.[16,17]

Shellfish

The corresponding analytical data for shellfish and crustacea show greater variation between species.[41] The values for protein for mussels (*Mytilus edulis*) are given as 8·9–11·7 while those for shrimp (*Crangon vulgaris*) are between 22·0 and 23·2.[47] To some extent the wider range in these values can be attributed to the more varied nature of what is considered the edible portion, as parts other than the muscle are frequently eaten. While the nutritional values of shellfish and crustacea are essentially equivalent to fish in terms of nitrogen equivalence, there are differences in the concentrations of the non-protein nitrogen; in crustacea in particular the concentrations are twice as high and are believed to be associated with their characteristic sweet flavours.[47]

Fish Muscle

Structure of Muscle

As fish muscle is functionally similar to mammalian muscle, there is close correspondence in the relative amounts of the main proteins of skeletal muscle, myosin, actin and tropomyosin (see Fig. 1). Structurally, however, there are fundamental differences in the arrangement of the muscle cells and in their attachment to connective tissue; fish muscle cells are short ($\simeq$1 cm long), and interspersed between the ends of blocks of cells are sheets of connective tissue which, on heating, break down to give the characteristic fish flakes. In fish too, there are two distinct types of muscle tissue, white and red; the latter, always the minor component, is located under the lateral line and is associated with sustained or continuous swimming activity.[48] It is well developed in pelagic fish such as mackerel and tuna in which it is present as a broadly based wedge penetrating inwards from the surface. In demersal fish such as cod it is confined to the surface layers beneath the skin.

The fibres are essentially the same, in terms of the arrangement of the thick and thin filaments, as those of other animals, showing alternate arrangements of anisotropic (A) and isotropic bands (I). The I bands are divided by the thin Z lines which also delimit the sarcomeres, the regularly repeating units of the fibre. The thick filaments, comprising myosin molecules of molecular weight 470 000, and the thin filaments, double helical strings of actin molecules of molecular weight 42 000, move in relation to one another[49] to bring about contraction or relaxation of muscle. Other proteins involved in the contractile mechanism are tropomyosin and troponin, both of which are in the thin filaments.

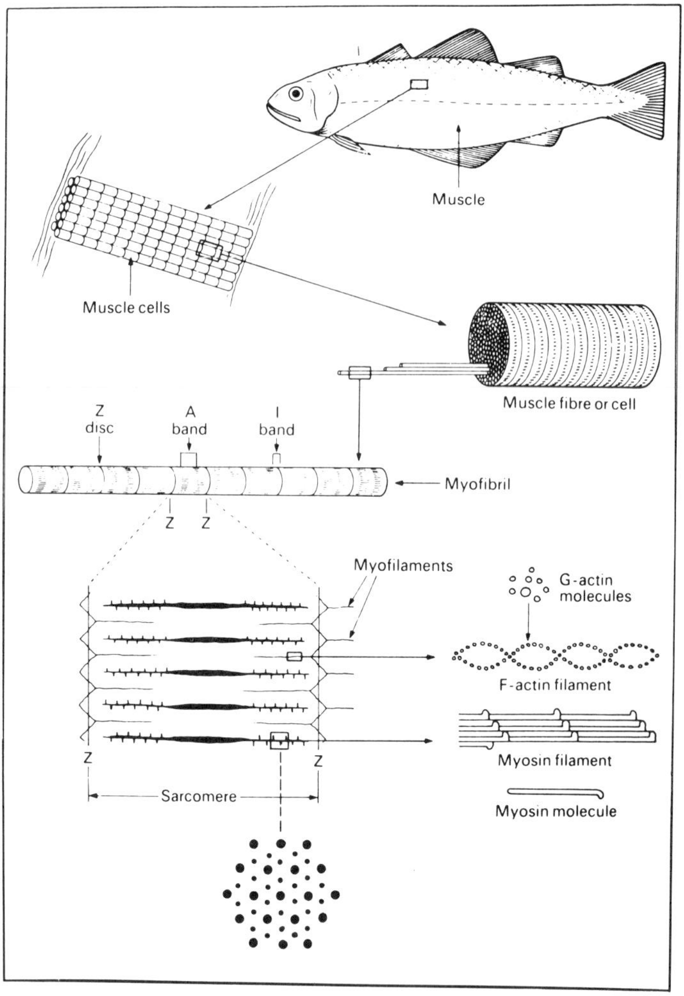

FIG. 1. Muscle cells and components. Sarcomere length (Z–Z) 2–3 μm. Myosin thick filament length 1·6 μm (spacing 35–40 nm). Thin filaments 1 μm.

Troponin, which is associated with the binding of Ca^{2+} in mammalian and fish systems, is replaced by a modified myosin in shellfish.[51]

Component Proteins

Of particular importance from aspects of protein preparations are the main myofibrillar proteins which in fish white muscle represent 65–75%

of the proteins. The interactive properties of myosin and actin are of particular importance in determining functional properties of protein preparations from fish and will be referred to later in the section on kamaboko. The sensitivity of myosin, particularly to denaturation on drying and on frozen storage of the flesh, is well established[38-40, 50] but the mechanism involved in either process is not well understood.[51] However, it is likely that during drying or dehydration on freezing the removal of the ice-like structures which mediate hydrophobic–hydrophilic interactions between and within the molecules leads to a disruption of the hydrogen bonding system and in consequence an irreversible aggregation of the proteins.[39, 40]

Other proteins of the flesh are the water-soluble or sarcoplasmic proteins which account for 20–30% of the total protein content.[38-40] They are mostly enzymes and are of importance, in their involvement in catabolic processes before significant bacterial spoilage develops, particularly in the early stages of post-mortem biochemical changes.

Although connective tissue proteins are important in relation to the appearance of fish fillets, in holding the myotomes together,[52] they account for only 3% of the protein of gadoids, for example,[38] and are of little consequence as far as their effect on the functional or nutritional properties of protein products are concerned.

Fish Spoilage

The importance of having fresh fish for further processing cannot be overemphasised as many of the flavours and odours commonly associated with fish are in fact due to spoilage—ammoniacal and amine odours are due to bacterial action and rancid flavours of fatty species can be attributed to oxidation of lipids and free fatty acids. Fish vary markedly in their rates of spoilage depending upon factors such as species, size, fishing ground, temperature of the water and availability of food. The complexity of this subject[5, 53, 54] is beyond the scope of this review but it is sufficient to mention the need for adequate measures to control these degradative changes at all stages of fish handling and processing.

Compared with mammalian flesh, that of fish spoils more rapidly, partly due to the highly active proteolytic enzymes of the gut, which invade the tissue, and also due to the nature of the spoilage bacteria themselves. Those of northern latitudes are psychrophilic and, as such, continue to grow even at temperatures near 0 °C. They are naturally present on the skin and in the intestines of fish, and during normal gutting and filleting operations they are inevitably widely dispersed over the tissue. In general, small pelagic fish, with proportionately more viscera, spoil more rapidly than large demersal fish—sprats can be kept for 1—2 days at 0 °C whereas cod can be kept for 2 weeks.[23,54]

PROTEIN PRODUCTS

Historical

The traditional methods of preserving fish are those of salting, drying and smoking,[1] all of which have the effect of reducing microbial activity and thus allowing fish to be kept in an edible form for prolonged periods of time. Over the years, well recognised products with familiar appearances and tastes have been developed but with few exceptions most processes preserve the fish in a recognisable form. The need for adequate 'ripening' or 'maturing' is, however, recognised in the preparation of salted herring, during which time various enzymic changes, both proteolytic and lipolytic, are believed to take place.[55,56] The importance of visceral organs in this and similar processes is well recognised.[55,56]

Fermented Fish Products

Products made by the action of the natural proteolytic enzymes of fish have a long history and go as far back as Roman times when liquefied fish preparations were made by adding viscera and blood of fish to salted fish. Indeed, details of actual recipes for making garum or liquamen are given in a recent translation of the cookery book of Apicius compiled during the 4th or 5th century A.D.[57] These recipes are of particular interest as they are essentially those which are still widely used in south-east Asia[50,58] and by contrast with Europe have been used there continuously for many centuries. While many variants are known, these processes are essentially the autolysis of fish, usually small pelagic species, in saturated solutions of salt over relatively long periods of time at ambient temperatures, the salt having the effect of inactivating all the spoilage bacteria but allowing proteolysis, lipolysis and other enzymic processes to take place. The products are either the clear supernatant liquid 'fish sauce' which is drained off or the whole unfractionated fermented product 'fish paste', both of which are widely consumed often as condiments and providing at the same time nutritionally important amino acids.[56,58] These fermented fish products are of particular interest as they anticipate the newer approaches now being made to hydrolyse proteins with proteolytic enzymes.

Fish Gel Products

Although quite unfamiliar to western palates, the kamaboko of Japan is perhaps the only manufactured fish protein product that can be regarded as a traditional food,[59] the manufacture of which is described in Japanese documents of the 15th century.[59] The traditional product was a means of preserving surplus fish or low grade parts of fish in a form that was still acceptable as food. The process, which is basically simple, involves grinding fish muscle with salt (2·5–4%) in a mortar, shaping with a mold, heating to near boiling point and allowing to cool. The fish meat gel which forms is called kamaboko and is savoured for its special elastic property or ASHI. This product is of particular interest as it is the basis of the most important fish protein manufacturing process in the world today,[60, 61] and which is reviewed later in this chapter under 'New Approaches'.

Dried Fish Proteins

By contrast with the successful development of fish gels and fish sauces, there has been markedly little production of dried fish proteins as traditional products. It must be assumed that wherever fish has been dried in the world, some at least of the products would be powdered, intentionally or otherwise—stockfish, for example, made in northern Norway by air-drying salted cod, is also used in a powdered form.[9, 62] In the main, however, though the concept of making protein concentrates from fish is comparatively new, various products have been tried and tested since the late 19th century and it is appropriate to review them briefly.

The first reference to a dried fish protein preparation was to a Norwegian fish flour made by A. Rosing and exhibited at the International Exhibition in the USA in 1876.[9] It was not until the 1890s, however, that an attempt was made by a Professor Waage in Norway[62] to produce a fish powder on a commercial basis. This product was essentially chopped haddock fillets which had been dried in a stream of hot air until the moisture content was reduced to 5–6%. The product could be regarded as a food grade fish meal, some modern developments of which will be discussed later.

The next most important development in fish protein preparations was the Wiking Eiweiss or egg white substitute which was manufactured in Germany during the Second World War.[63, 64] The process consisted of the following steps: treatment of the raw material with a dilute aqueous solution of acetic acid, washing and removal of water by pressing, fat extraction with 95% ethanol, solvent removal and recovery, pulverisation of the fat-free material, solubilisation of the protein with sodium hydroxide solution at 80 °C, followed by neutralisation with acetic acid and spray-drying the neutral suspension. The product, although accepted for reasons of necessity at the time, probably suffered from flavour reversion on storage. The high temperatures required for the alkaline hydrolysis stage would also have produced lysinoalanine[65] and possibly caused inversion of the configuration of amino acids,[66] two considerations which on safety grounds alone would make the process unacceptable today.

Various attempts to produce and market dried fish proteins as nutritional additives were made from the 1930s onwards in Norway, South Africa and Iceland[9] but all failed in spite of the undoubted nutritional value of the products. Unacceptable off-flavours resulting from oxidation of the lipids were undoubtedly contributory factors and in South Africa where it was intended for use as a nutritional supplement, changes in the colour of the cereal, believed to be due to Maillard-type browning reactions, led to its rejection.[9] Notwithstanding the failure of fish protein concentrates to be accepted as food additives, the whole concept was raised again in the early 1960s in the USA and Canada as a means of filling the protein gap[7-10] between the well and malnourished populations of the World. Considerable resources were expended on developing a fish protein concentrate for use primarily in developing countries. At this time too, because of the increased interest and activity in making dried fish protein products, attempts were made to classify the various concentrates being produced. As the best known product was obtained by extraction with organic solvents, for example, isopropanol (IPA), it was generally described as fish protein concentrate or FPC[7] (Fig. 2) although much broader definitions encompassing all products in which the protein concentration has been increased by the removal of water have been suggested.[7,9,10] Such a definition includes dried products made from minced flesh, whole fish and fish offal from which the fat may not have been removed, and would include dried products of proteolytic digestion. It is so broad, however, that it conveys no more in the author's view than 'dried fish protein products'. It would seem, therefore, that some additional description is required to give some indication of the process used, viz. solvent-extracted fish protein concentrates, fish protein hydrolysates, enzymically or chemically modified fish proteins, retextured fish proteins, etc. Useful definitions for fish protein concentrates were prepared by FAO in 1961[67] in an attempt to classify the various 'dried protein products[67] and can still be usefully used today (Table 6),

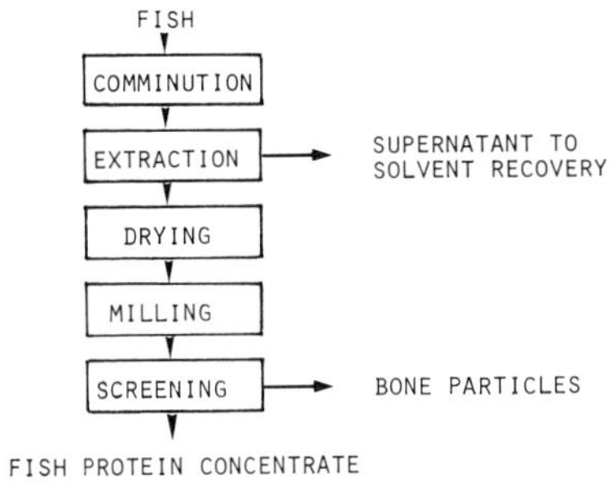

FIG. 2. Flow diagram of FPC process.

TABLE 6
SPECIFICATIONS FOR FPC[67]

Type A —	a deodorised and tasteless product containing less than 0·75% fat and a minimum of 67·5% protein
Type B —	a hygienically prepared fish meal with a maximum of 3% fat
Type C —	fish meal

although considering information now available on the stability of dried preparations, a maximum level of 0.1% would have been more appropriate for residual amounts of fat.

As solvent-extracted FPCs have been very fully reviewed and discussed by Finch,[7] Pariser,[8] Pariser *et al.*[9] and Sikorski,[10] it is unnecessary to dwell on them at any length in this review, particularly when there is now little interest in their production on an industrial scale.

The process as outlined in Fig. 2 is essentially the extraction of fat and water by organic solvents followed by removal of residual solvents by air-drying, and bones if present are removed by sieving. Over the years many variants of the basic process have been developed as, for example, the use of solvents other than IPA, such as methanol, ethanol or 1,2 dichloroethane, and repetitive extractions for varying times and at different temperatures.

FPCs have been made from a wide range of raw materials, from essentially bone-free or fat-free fillets of hake to whole fatty species such as herring and menhaden.[7,9] Generally, for products of the first type, the residual lipid is reduced to levels below 0·1% and protein contents can be as high as 96% on a moisture-free basis, if bones are removed before processing.[68] Generally, however, ash contents are considerably higher than this implies, and are more likely to be between 10 and 20% (Table 7). Not unexpectedly, the biological value is high and essentially that of the fish from which the FPC is derived[7] as the relatively mild processes cause little, if any, decrease in the availability of amino acids due to cross-linking or other reactions. FPC is thus of excellent nutritional quality but is lacking in what are now regarded as important functional properties in protein products—emulsifying properties, gel forming capacity, foam forming properties and solubility[69, 70] all of which are present to a high degree in raw fish but are lost as a result of dehydration. This complete lack of functional properties means in effect that FPC does not contribute to the acceptability of a food and that, if anything, it detracts from it because of the gritty powdery texture.

TABLE 7
COMPOSITION OF FPC MADE BY IPA EXTRACTION OF 12 SPECIES OF FISH

N × 6·25 (dry basis)	80·4–93·1%	Lysine	7·49–10·01 % N × 6·25
Ash (dry basis)	11·4–22·0%	Histidine	1·86– 3·14 % N × 6.25
Residual lipid	0·07– 0·55%	Arginine	6·14– 7·34 % N × 6·25
Calcium	2·75– 5·05%	Valine	4.42– 5·77 % N × 6·25
Phosphorus	1·87– 3·39%	Methionine	2·62– 3·33 % N × 6·25
Potassium	0·59– 1·10%	Isoleucine	3·88– 4·95 % N × 6·25
Sodium	0·34– 0·44%	Leucine	6·46– 8·09 % N × 6·25
Magnesium	0·16– 0·28%	Tyrosine	2·70– 3·66 % N × 6·25
Aluminium	67–775 ppm	Phenylalanine	3·48– 4·34 % N × 6·25
Iron	147–770 ppm	Tryptophan	0·81– 1·31 % N × 6·25
Copper	6– 21 ppm	Cystine	0·58– 0·77 % N × 6·25
Zinc	112–163 ppm		
Chromium	8– 19 ppm		
Fluorine	58–162 ppm		
Selenium	1·8– 3·8 ppm		
Arsenic	0·23– 2·55 ppm		

(Courtesy of Finch.[7])

With the benefit of hindsight, it is easy to appreciate the view of Pariser *et al.*[9] that FPC was doomed from the outset by 'conceptual myopia' as too little account was taken of the likely acceptance or indeed value of such a product in the developing countries for which it was intended. To quote Pariser *et al.*, it was 'a sophisticated and capital-intensive solution in search of a problem to which it can be applied.' Perhaps the main failing of the FPC concept, however, was that as a nutritional supplement, FPC in no way would have alleviated the primary problem of inadequate supplies of calories in developing countries. Also, the lack of interest in it as a product for human consumption in the countries in which it was developed was not likely to inspire confidence in the prospective consumers.

New Approaches

It is now recognised that new developments in fish protein products are more likely to succeed if they can be utilised in existing industrial processes[8] as, for example, alternatives to proteins currently used because of particular functional properties.

More recent research and development in the past decade has thus put greater emphasis on means of either retaining the functional properties of the native proteins or of modifying the properties of the isolated proteins by enzymic or chemical methods to give a range of products for general use in the food industry. To a large extent the successful modification of properties of soya proteins[70, 71] has stimulated renewed interest in doing similar work on fish proteins. Fish proteins in many ways should be more readily introduced into existing markets, at least in societies where fish is already widely accepted as a wholesome food. It is known to be nutritious and is free of any suggestion of being indigestible. The flavour of either soya or fish is not always acceptable, however, and although it can be greatly reduced, residual off-flavours can still lead to acceptance problems.

To overcome the problem of loss of functional properties on drying fish proteins, two basic approaches have been made:

(a) avoidance of dehydration at all stages of processing;
(b) development of suitable treatments which will permit dehydration and still give acceptable products on reconstitution.

It is recognised that drying has many attractions for long-term storage of a product and for ease of handling and distribution. Products which cannot be dried inevitably require alternative methods of preservation such as refrigeration, if they are to be stored prior to consumption. Although freezing and frozen storage is more expensive there are situations in which the acceptable qualities of the product justify the cost. Alternatively, traditional methods depending on either pH or salinity to inhibit spoilage bacteria can be used.[1-3, 55, 56, 58]

Fish Gel Products

While in some ways the fish gel products of Japan are no more than a scaling up of a traditional Japanese household process,[4, 59-61] there are features which justify their inclusion in this section. In Japan a whole new processing industry has developed over the past 20 years and today 1·5 million tonnes, equivalent to 4·5 million tonnes of wet fish, or as much as 45% of the total Japanese catch, are made into fish gel products,[8,59–61] using many species of fish once considered low grade and unsuitable for such products.[25,72–76]

The striking success of these new products contrasts markedly with the failure of FPC, the other major product concept of the late 1950s. However, an important distinction between the two concepts must be made. Well established markets for gel products already existed in Japan and the introduction of processes and products presented comparatively few problems, whereas for FPC there was neither a market nor a demand.

New developments. Until the late 1950s the production of kamaboko and related products was based on a few traditional species only and, as the special characteristics of the fish gel could only be obtained from very fresh and unfrozen fish, production was necessarily restricted. However, in the early 1960s, in response to a growing demand in Japan for kamaboko and related products, consideration was given to widening the range of species used and to developing suitable conditions for the preparation and frozen storage of the raw materials. It has been shown that kamaboko with the required ASHI can be made from a wide range of species, e.g. Alaska pollack, hake, dogfish and mackerel, among others,[74–76] and, as the process has developed, various grades of kamaboko are now produced. Factors other than species which influence the quality of the gel are removal of the fats and water-soluble components by washing, freshness, age, season and sexual maturity of the fish.

It is believed that to get good ASHI, two requirements must be met:

(a) extraction of proteins from muscle in salt solution (0·4–0·6 M) during grinding;
(b) formation of a network structure of heat-denatured proteins to form a gel with the ASHI characteristics.

To meet the first requirement proteins must be in their native state if they are to be fully extracted in salt solution.[69,77] Denaturation resulting from either prolonged storage in ice or in the frozen state leads to a loss of extractability in salt solutions and also to a loss of water-holding capacity, a related property of importance in gel formation. For the second requirement, it has been established that the most regular arrangement of the heat-denatured proteins is formed when myofibrillar proteins are washed free of the sarcoplasmic proteins[60, 73, 78] and that reagents such as cysteine and cystine[79] which cross-link protein chains, increase the elasticity of kamaboko.

The most significant development in the fish gel industry, however, was the demonstration of the effectiveness of sugars, in concentrations between 5 and 10%, in preventing denaturation of the myofibrillar proteins on frozen storage of minced fish. It then became possible to exploit the largely untapped resources of white fish species such as Alaska pollack to provide substantially increased supplies of raw material. The requirements for very fresh fish, in some cases pre-rigor, and the removal of the water-soluble proteins before freezing, led to the development of processing on factory ships. The process (Fig. 3) is the

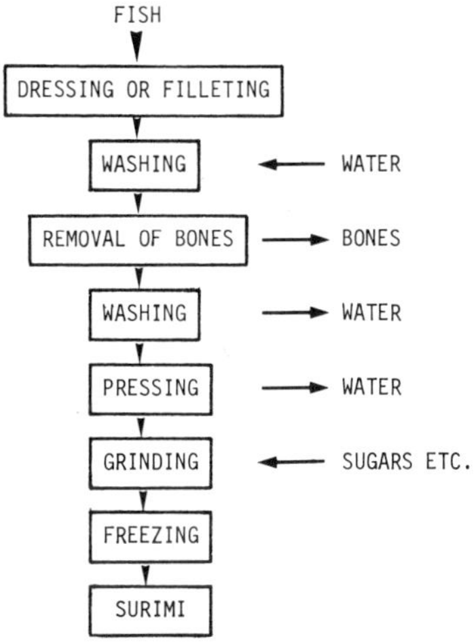

FIG. 3. Flow diagram of surimi process.

same as for the traditional kamaboko as far as the washing and dewatering stage but thereafter the cryoprotective sugars are introduced and the mince or surimi is frozen and stored at −20 °C. Under those conditions it can be used for the successful preparation of kamaboko for periods up to a year. Surimis are now available from a range of species and in various grades of quality (Table 8), depending upon the initial freshness of the fish[73] and a wide variety of fish gel products based on the kamaboko process are now manufactured in Japan. These include

TABLE 8

GRADES FOR ALASKA POLLOCK (*Theragra chalcogramma*) SURIMI ESTABLISHED BY THE HOKKAIDO SURIMI ASSOCIATION IN 1965

Grades	*Freshness of fish*	*Moisture content of washed flesh (%)*	*Additives used (%)*				*Amount of starch used in kamaboko prepared for folding test (%)*
			Sugar (sucrose)	*Sorbitol*	*Glucose*	*Poly-phosphate*	
SA (Special Grade A)	Day boat fish in rigor	79	–	5	–	0·2	0
A	Day boat fish post-rigor	80	5	–	–	0·2	3
			–	5	–	0·2	
			–	–	5	0·2	
B	1– to 3-day iced fish	82	5	–	–	0·2	7
			–	–	5	0·2	
C	3– to 4-day iced fish	83	5	–	–	0·2	10
		–	–	–	5	0·2	

(Courtesy Miyauchi *et al.*[73])

not only the many variants of kamaboko[73] but fish sausage, an entirely new and highly successful product.

Further developments in the kamaboko process. Although it has been well established that drying of fish proteins leads to a substantial loss of water-holding capacity, it is of interest that processes have been developed for spray-drying slurries of surimi, without, it is claimed, loss of kamaboko forming capacity. By adding fatty acid esters[80] or 10 mM Ca^{2+} or Mg^{2+}[81] to the slurry prior to spray-drying, the product obtained can be blended with salt and water to form a kamaboko similar to that from fresh fish. However, as fat is not removed in the process, it would be expected that off-flavours would develop on storage.

Not surprisingly, there is current interest in utilising soy protein in fish gel products in anticipation of a possible shortfall in fish supplies.[25] Recent research by Motohiro and others is aimed at developing processes which will permit the inclusion of soya protein without reducing the all important gel strength.[82] However, because of colour and textural problems, it cannot be used at levels higher than 5%.[25]

The successful use of sugars in preventing denaturation of fish muscle proteins on frozen storage has led to studies with other possible cryoprotective compounds. Amino acids[83,84] have been shown to be effective and it may be that given acceptable costs, they will extend the use of cryoprotective compounds to products which would otherwise be too sweet.

Spun Fibres

A spin-off from the technology required for the previously mentioned surimi industry, has been machines for separating the flesh from skin and bones of fish. The recovered flesh so obtained, although fully utilisable in the gel forming processes of Japan, has limited outlets in western countries. To a large extent the better quality minces obtained from bone containing parts of fillets[28] are used in fish finger production but poorer quality minces made from small whole fish such as blue whiting or skeletal filleting waste of white fish[28] have no obvious outlets and it was as a means of forming acceptable products from such raw material that the spinning process was conceived. The importance of textures simulating meat has been well recognised by proponents of the soy bean protein technology who have seen the creation of meat-like textures and flavours as one means of gaining acceptance for non-meat proteins in western diets.[71] And so it is true with fish that there is more likelihood of the poorer quality minces being accepted if they can be converted into products with more familiar textures and flavours.

Since it was first demonstrated by Boyer[85] that alkaline extracts of soya proteins could be spun into fibres suitable for making synthetic meats, or meat analogues, a number of related processes have been developed for vegetable, milk and meat proteins.[86-88] All of them are based on a wet spinning technique which is essentially the extrusion of acid or alkaline extracts of protein through narrow orifices or spinnerettes into a buffered aqueous solution near the isoelectric point of the protein. The proteins which precipitate are pulled away continuously as wet fibres which, with suitable binding and flavouring agents, are made into the appropriate meat analogues. Fibres thus prepared have superior functional properties to those prepared by the more common dry spinning techniques of the synthetic fibre industry.[86]

In a system described by Mackie[89, 90] (Fig. 4), comminuted flesh is extracted with dilute alkali and the isolate obtained by isoelectric precipitation is then redissolved in alkali at pH 11·5 to give a spinning solution with a protein content between 3 and 4%. Because of the linear

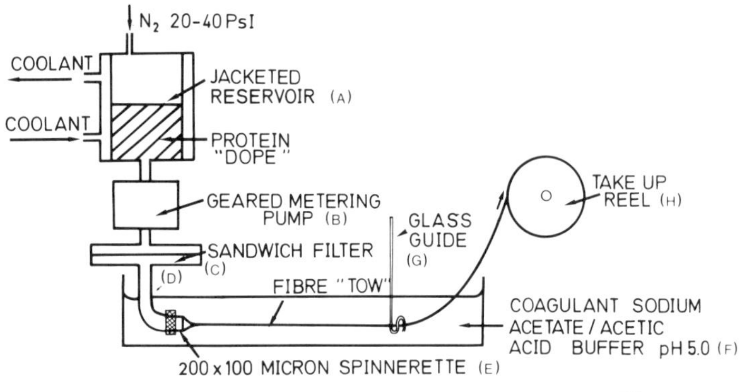

FIG. 4. Flow diagran of spinning process (Mackie[90]).

nature of flesh proteins, relatively low concentrations give highly viscous solutions of a suitable viscosity for spinning. Such alkaline extracts show a decrease in viscosity on storage but, provided the temperature is kept below 5 °C, a stable dope can be maintained for several hours. The most acceptable fibres were obtained using 1 M sodium acetate, 0·5 M acetic acid as precipitant and using 100 or 200 μm holes in the spinnerette. Fibres were white to off-white even when recovered flesh from whole fish was used (Fig. 5), with textures and composition similar to those of natural fish flesh (Table 9).

FIG. 5. Spun fibres from frame mince of cod.[90]

TABLE 9
PROXIMATE ANALYSES OF MINCES AND FIBRES (%)[90]

Material	*H_2O*	*Ash*	*N × 6·25*
Cod skeletal waste			
Unwashed mince	82·8	1·34	15·86
Washed mince	85·70	1·03	13·27
Fibre	82·31	0·20	17·50
Blue whiting			
Mince	80·25	1·41	17·56
Fibre	79·52	0·15	20·12

The improvement in whiteness is due largely to the removal of pigments with other water-soluble components in the initial extraction and precipitation steps. Variation in texture was obtained by soaking the fibre at pHs above or below the isoelectric point or by adding polyphosphates.

As presently prepared, the fibres can be stored for at least as long as natural flesh at 0 °C but on freezing and frozen storage they become unacceptably tough. The mechanism is not understood but it is likely that reactions similar to, but more rapid than, those associated with frozen storage deterioration of fish are operating. The fibres can, however, be stored very successfully in acid preservatives of the marinade type. In many ways they are highly suited for this product as the operation requires no more than a transfer from the precipitation bath to the marinading liquor. Further development work is, however, necessary before any of the fish fibre processes can be considered for commercial use.

Enzymic Fish Protein Hydrolysates (FPH)

The use of proteolytic enzymes is not in itself new, as many traditional processes of liquefying or fermenting fish have depended upon the natural enzymes of the viscera and of the flesh to digest the proteins slowly over weeks or months at ambient temperatures. However, it is only within the last decade or so that proteolytic enzymes in food grade quality have been available in sufficient quantities for industrial use. It is thus now possible to consider the application of these essentially mild processes which offer the prospect of modifying functional properties of proteins without loss of nutritional value.[91] In all of the processes, whether autolytic or accelerated by added commercial enzymes, conditions for digestion which prevent the growth of spoilage bacteria have to be selected. Salt has been used traditionally in most fermentation processes[55, 56, 58] but the high concentrations required inevitably restrict the consumption of the final product. In more recent developments in autolytic systems, acids, either mineral or organic, have been used to reduce the pH to about 4·0, as in fish silage. This process, developed originally in Finland in the 1920s, is attracting renewed interest as a means of preserving cheap, raw materials such as fish filleting waste as an animal feed and is usually considered as an alternative to the fish meal process when relatively small quantities of raw materials are available and where there are suitable liquid feed outlets, e.g. pig feeding in the local area.[92, 93] It has the advantage of simplicity and low cost as it is no more than mixing minced or comminuted fish with the appropriate volume of acid in a suitable container, but, like all autolytic processes, it is difficult to control because of the variable nature of the raw material and lack of temperature control. Silage is unlikely to be used for human consumption because of its unattractive organoleptic properties. However, by selecting higher grade material and by maintaining hygienic conditions throughout the process, the liquid autolysate could be concentrated or dried. Such a product has been made by Dagbjartsson *et al.* of Iceland who demonstrated the feasibility of making autolysates at high pH using ammonia.[94]

Notwithstanding the simplicity of the autolytic processes, the length of time required for digestion is a serious limitation for any industrial application.

Systems which are, therefore, able to utilise commercially available enzymes have the distinct advantages of short hydrolysis times and much greater control over the digestion process. Various enzymes from animal, plant and bacterial sources[95] are available today as, for example, pepsin from pig's stomach,[44] papain from the papaya tree (*Carica papaya* L.)[96] and alcalase and neutrase from the *Bacillus* sp. of bacteria.[91] These enzymes have broad specificities and, as they are all endopeptidases, they produce hydrolysates of relatively low average molecular weight. Enzymes such as papain which are optimally active at near neutral pH and at temperatures well above the survival temperature of spoilage bacteria are in many ways to be preferred, as the neutralisation step of acid or alkaline systems is avoided and with it the formation of salt. As the digestion times are short, often less than 30 min, these systems are often referred to as accelerated processes.[11–14,91,95,97–105]

The process. There are now many variants of the accelerated hydrolytic process but most include the main steps outlined in Fig. 6. The fish is minced or comminuted initially, added to an equal volume of water in the digestion vessel and, after suitable adjustment of pH and addition of activators, if required, the suspension of fish flesh is heated to the required temperature for digestion[11] and at the same time stirred vigorously. On addition of the enzyme, the viscosity drops dramatically and thereafter the hydrolysate can be treated as a free flowing liquid. On completion of the hydrolysis, the enzyme is inactivated by either heating to 100 °C (papain) or by neutralisation (pepsin or alcalase) and bones are removed by sieving. The hydrolysate, which always has a residual suspension of insoluble protein residues in a solution of amino acids and peptides, can be either spray-dried directly or separated further into its soluble and insoluble components. Fatty fish present particular problems

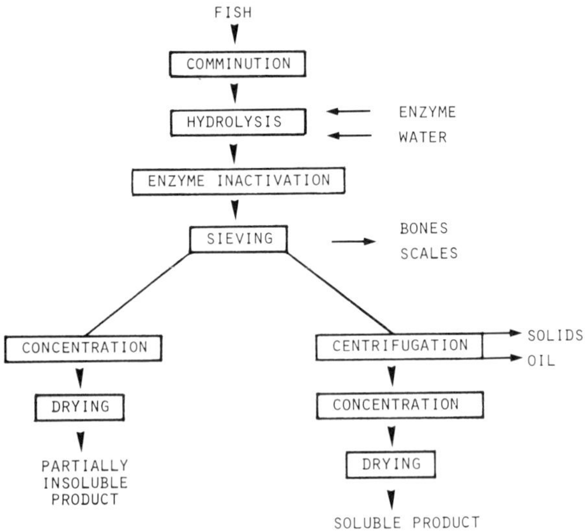

FIG. 6. Flow diagram for FPH process (Mackie[105]).

of their own as the fat must be stabilised against oxidation during the process and subsequent storage, or removed as effectively as possible by centrifugation after the hydrolysis has been completed.

Biochemical changes. Although the detailed mechanisms of such hydrolytic processes are not fully understood, it is evident that none of the added enzymes functions as the sole proteolytic enzyme, as the natural enzymes from the flesh and in some cases from the viscera, if whole fish are being digested, play a significant role. Precooked flesh is not so readily digested as raw flesh. It has been shown by Chernogortsev[100] that ungutted species with active enzymes in the viscera can be hydrolysed over 2–3 days at 40 °C if bactericides such as ethanol (7–10%) are present.[10, 100]

A typical rate curve for the solubilisation of protein is given in Fig. 7 and shown that the of hydrolysis tails off after 80% of the protein has been solubilised, and that it enters a stationary phase with no apparent hydrolysis taking place, presumably because all of the susceptible bonds have been hydrolysed.[105]

General properties of FPHs. The proximate analyses for unfractionated FPHs from non-fatty species of fish show that they are essentially all pro-

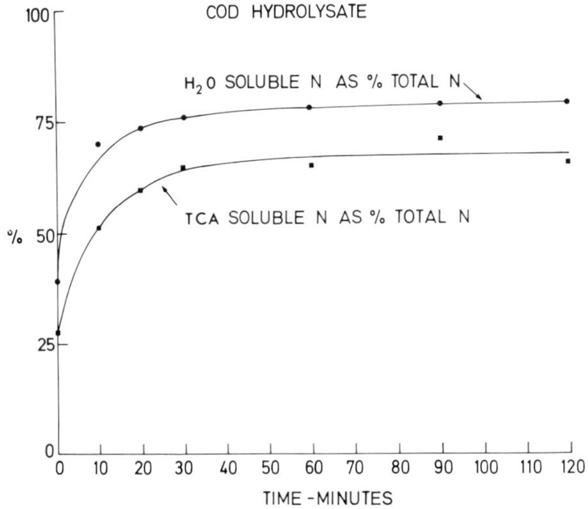

FIG. 7. Rate curves for FPH process (Ritchie and Mackie[104]).

tein (Table 10)—the bones which would still be present in FPCs are very efficiently removed in the FPH process and the products are readily dispersed in water even in the unfractionated form. The soluble component has obvious advantages when considered as a protein supplement but as a functional protein preparation it has not yet been shown to have the range of properties required as a replacement for existing proteins, and it would seem that further research on controlling the degree of hydrolysis is still necessary. In common with all enzymic hydrolysates, FPHs have a bitter taste but it is comparatively mild compared with those from casein, for example.[106] The bitterness is believed to be due to low molecular weight peptides (up to 6000 Daltons) with hydrophobic side chains.[107] This taste can be reduced in theory by controlling the

TABLE 10

APPROXIMATE ANALYSIS OF FISH PROTEIN HYDROLYSATES (UNFRACTIONATED) (MACKIE[105])

	Protein	*Lipid*	*Moisture*	*Ash*
Cod fillet	84·4	2·0	4·8	6·2
Cod filleting waste	81·8	4·2	8·5	6·9

degree of hydrolysis so that predominantly tasteless larger molecular weight peptides are produced. Alternatively, further digestion with an enzyme of different specificity, to break down the peptides to individual amino acids, can be effective.[95]

Uses of FPHs. While the long-term objective of this work is to produce FPHs acceptable for human use, more immediate outlets are for use in milk replacers for animals. FPHs are of current interest as substitutes for the casein component in milk and have advantages over FPC or fish meal in their low ash content, ready solution or dispersibility in water and good emulsion forming capacity. Feeding trials on neo-natal calves complete with FPHs from non-fatty species have shown that complete replacement is feasible, but that there is a reduction in liveweight gain.[108–111] With lambs, however, it was shown that when FPH was used as the sole source of protein it gave results for food utilisation and liveweight gains similar to casein[112] (Table 11) and indeed when hydrolysates of fatty species were used as the sole source of both protein and fat, high digestibilities for both fat and protein were obtained.[113]

As products for human consumption, they are not immediately attractive because of bitter flavours and residual fishy taints, arising from the usual

TABLE 11
FOOD UTILISATION BY LAMBS OF DIETS BASED ON FPH, LARD AND STARCH

Protein source *Carbohydrate source* *Fat source*	*Casein* *Lactose* *Butter oil*	*Casein* *Protamyl* *Butter oil*	*Casein* *Protamyl* *Lard and coconut*	*FPH* *Protamyl* *Lard and coconut*
Liveweight gain (g d^{-1})				
Weeks 2–3	151	167	128	112
Weeks 4–6	197	196	175	218
Overall	177	183	154	170
Food conversion (kg DM/kg gain)	1·02	1·06	1·08	1·00
Digestibility of:				
Fat	98·9	98·3	98·2	97·7
Starch	—	99·8	99·8	99·9
Protein	96·4	94·9	94·4	93·6

(From Soliman *et al.*[112])

spoilage changes in the fish prior to processing, and flavour changes associated with oxidation of the fats. However, the use of other flavours as masking agents is seen as one solution to the problem. It has been reported that FPHs are to be used in Chile in soups and beverages and for the fortification of cereals.[114] Recent work in the USA[14] is also encouraging to the extent that FPHs from a range of species were shown to have excellent emulsification and foaming properties. Off-flavours could easily be masked by the addition of substances containing acids, e.g. brown sugar and lemon juice.[14] However, it must be stressed that FPHs on the whole still lack important functional properties because of the preponderance of free amino acids and low molecular weight peptides and until further control over the degree of hydrolysis is achieved, their application in the food industry must be limited.[13,14,105,115,116]

Industrial development. Recent assessments of the costs of producing FPHs[14,117] indicate that products can be competitive. with skim milk proteins, given adequate supplies of raw material; and as the demand for protein additives continues to increase throughout the World,[118] it is very likely that further investment will be made in accelerated processes. In France, a soluble FPH from non-fatty species has been produced in conjunction with fish meal for over 10 years,[119] for use mainly as a milk replacer for animals. In the USA plants have been built to produce FPH from the white fish by-catch of the shrimp fishery of the Gulf of Mexico.[120] With these exceptions, however, FPH production has not gone beyond the pilot scale level of production.

Plasteins

The formation of plasteins, sometimes erroneously described as the reversal of the normal hydrolytic reaction, has been proposed as a means of removing bitterness from hydrolysates and of producing protein isolates.[121] The conditions normally required for the plastein reaction are a high concentration of substrate of amino acids and peptides (usually in excess of 30%) and a pH different from that for the hydrolytic reaction (pepsin hydrolyses at pH 2 but forms plastein at neutral pH). The product, or plastein, which forms is an insoluble gel and is variously described as a mixture of polypeptides bound by electrostatic attraction. Characteristics of these products are an increased proportion of hydrophobic amino acids and molecular weights between 2000 and 20 000. Fujimaki defined plastein formation as a polycondensation process giving rise to products of high molecular weight[121] but data of

Hofsten[122] suggest that peptide chain elongation plays only a minor role as there is no decrease in free amino groups during the formation of plastein. It would now seem, from all evidence available, that after an initial formation of low molecular weight pepetide by transpeptidation or condensation or both, the equilibrium of the reaction is favoured by hydrophobic and possibly electrostatic interactions between side chains of peptides and amino acids to yield the plasteins, as insoluble gels, a feature of which is the preponderance of hydrophobic amino acids.[123,124] Indeed, it has been shown that high yields of plasteins are strongly favoured if there is a high proportion of hydrophobic to hydrophilic amino acids in the incubation mixture.[122]

Plastein formation from proteolytic hydrolysates is of particular interest because it offers a means of removing the bitter tastes which have been attributed to low molecular weight hydrophobic peptides and hydrophobic amino acids. It is also seen as a means of enhancing the concentration of specific amino acids in a protein product by carrying out the reaction in the presence of an increased concentration of an ester of the particular amino acid. Particular applications are in the preparation of low phenylalanine proteins as dietetic foods for patients with phenylketonuria. In such a process, aromatic amino acids are preferentially removed by chromatography from a peptic hydrolysate, and prior to the plastein reaction, ethyl esters of tyrosine and tryptophan are then added to the digest to replace those removed with phenylalanine.[125]

Despite these obvious advantages of the plastein reaction, it has not been taken beyond laboratory scale experiments. It may be that the low yield of plastein normally obtained (35%),[124] the absence of functional properties and, in particular, the lack of solubility may have outweighed the positive features of the process. For the time being at least it would seem that it will remain no more than an interesting but impractical reaction associated with the preparation of protein hydrolysates.

Functional Fish Protein Preparations

Over the past decade further research into means of either retaining to some extent the natural functional properties of fish proteins or of introducing new functional properties has led to the development of several processes of potential industrial interest. The diversity of approaches has, unfortunately, led to a problem of classification and nomenclature. Terms such as 'fish protein concentrate' and 'fish protein isolate' have variable meanings in the literature, a confusion compounded by use of other terms such as 'product' and 'preparation.'

For the purpose of this review the following groupings have been used in an effort to reduce the number of terms:

(a) solvent-extracted fish proteins;
(b) chemically or enzymically modified proteins.

Solvent-extracted fish protein concentrates:

(a) Marinbeef. By utilising the known techniques for making kamaboko it has been shown by Suzuki[126, 127] that the solution of protein and salt (stage 1 of the kamaboko process) can be subsequently dehydrated by extraction with pre-cooled ethanol (at 5 °C), and drying between 25 and 30 °C. The dried product contains more than 85 % protein and 12–5 % water, and can be rehydrated by mixing with 4 volumes of water. It has been produced as a pilot scale venture in Japan and is marketed under the trade name Marinbeef®. As its name implies, it is intended as a partial replacement for beef in products such as meat loaves and hamburger stakes.

It would seem that the improvement in functional properties over previous solvent-extracted concentrates is related to the initial extraction of the protein in the salt solution (an essential requirement for satisfactory gel formation in kamaboko) and low temperature extraction of water and lipids with ethanol. However, even with these markedly improved functional properties it is apparently inadequate as the sole source of flesh protein in a product.

(b) High protein meat (HPM). A similar approach has been adopted by Stala and Krassowska[128] who developed a product essentially similar to the washed fish muscle mince used for surimi.[59] There is no prior removal of the sarcoplasmic proteins with water but lipids are removed initially by extraction of minced raw fish with 0·4 parts of organic solvent (60 % ethanol, 40 % ethylacetate). The residual mass is then treated first with 5 % tannin in glycerol in proportion of 0·66 litre kg^{-1} residue, and then with a solution of either ferric chloride or aluminium chloride. Thereafter the fish mince is subjected to pressure to reduce the water content below 40 %. The HPM is not dried further but is preserved in the frozen state. It can be used as a meat extender in sausages and hamburger-type products in which it can replace up to 50 % of the meat. Other uses are in confectionary and bakery products. HPM contains on average 30 % protein, 0·5 % lipid, less than 0·5 % ash and 70 % water and is similar to surimi. Its stability on frozen storage must be due in part at

®Niigata Engineering Co. Ltd, No. 4–1 Kasumigaseki 1-chome, Chiyoda-ku, Tokyo 100, Japan.

least to the presence of glycerol which will function similarly to sucrose and other carbohydrates in surimi. It is not clear, however, why tannin and trivalent metal salts should be involved in stabilising the protein against frozen storage deterioration other than by exerting some anti-oxidant effect.

The process was taken as far as the construction of a demonstration plant with production potential of up to 15000 tonnes of HPM per year*, but it has not been taken up industrially.

Alkali-extracted proteins. Because of the objections previously mentioned to the use of hot alkali to obtain proteins with functional properties, an alternative approach in which the proteins remained essentially undissolved was developed in the mid-1970s.[129] The process consisted of the following steps:

(a) Decolourising the comminuted fish (previously passed through a bone separator) at a pH not greater than 9·5 by mixing with alkali and hydrogen peroxide at 30–70 °C.
(b) Deodorising the fish by extraction with isopropanol.
(c) Drying the essentially fat-free protein concentrate.

While the best quality products were obtained from fillets, the white fish filleting waste previously described (p. 223) was the preferred raw material. The product had essentially the same proximate analysis as a conventionally prepared FPC but, because of the higher pH of the treatment, the final product had superior rehydration properties. In addition, its PER was only 2% below that of casein. However, even with these points in its favour, it is unlikely that a process which requires the use of hydrogen peroxide would be accepted in any product intended for human consumption. It has not been developed further.

It is of interest that as recently as 1978, a further patent[130] was awarded to an alkali process which is essentially that of the early Wiking Eiweiss[7, 63] process, but with closer control over times and temperature of hydrolysis the product has the advantage of being salt-free as the neutralisation step is carried out by eluting the alkaline solution through a cation exchange resin in the ammonium form. The ammonia is subsequently removed during the concentration step prior to drying. Any fats present in the raw material are removed by extraction with isopropanol and it is claimed that the product, which is essentially protein, may

* Navimor, Foreign Trade Enterprise, ul Matejbi 6, 80-952 Gdansk, Poland.

be used in a wide range of foods such as sausages, soups and desserts. Because of the complete solution of the proteins during preparation, the bones are readily removed by screening and centrifugation, and fluoride concentration is less than 25 ppm. Typical proximate analyses are those for the preparation from cod fillets which has 97 % protein, 2·6 % ash and 0·4 % lipid. No information on storage stability of functional properties or of possible reduction of PER in the proteins as a consequence of the extraction in alkaline solution is given. In the light of previous experience with such alkaline processes, however,[7, 65, 129, 131] it would be surprising if these problems had been resolved.

Enzymically and chemically modified proteins. Of particular interest from the standpoint of functional properties of dried fish protein preparations is the work of Spinelli *et al.* which has been aimed at modifying myofibrillar proteins in such a way that their functional properties are not only retained but improved upon. By using controlled proteolytic digestion[15, 133–135] and chemical modification by reaction with anhydrides[135, 136] they have very convincingly demonstrated the potential of these treatments used either singly or in combination. The rationale behind these approaches is that by reducing size by proteolysis, functionality is likely to be improved and by introducing new charged groups into the molecule configuration and functional properties are likely to be affected.

(a) Enzymic modification. The previously mentioned problems in modifying proteins by partial alkali hydrolysis led Spinelli *et al.* to consider the use of proteolytic enzymes to introduce major changes in functionality without at the same time breaking the proteins down to any significant extent to peptides and amino acids. The production of the latter compounds is undesirable on two counts:

(a) they remain soluble at the isoelectric point of most proteins and give poor protein recoveries.
(b) the functional properties of gel forming and emulsifying capacity are associated with large protein residues rather than small peptides.

In Spinelli's process, the myofibrillar fraction of comminuted fish flesh is washed free of low molecular weight components and sarcoplasmic proteins as in the surimi process,[59] and is then hydrolysed with the proteolytic enzyme Rhozyme P-11* under conditions which permit the

* Rohm and Haas.

formation of no more than 10 % of non-protein nitrogen. By treatment with sodium hexametaphosphate and acid to give a pH of 3·5–3·0, the modified protein complex is recovered. While still in the acidic state the complex is extracted with hot isopropanol (50–70 °C) to reduce residual lipids to less than 0·1 %. The extracted complex is then resuspended in water to remove excess isopropanol, neutralised with sodium hydroxide and spray-dried.

When freshly prepared, these enzymically modified myofibrillar proteins have an emulsifying capacity of 250 g of oil per gram of protein[135] but on storage this property deteriorates markedly. It has been shown that if a carbohydrate such as sucrose or sorbitol is co-dried with the isolate, improved stability is obtained on storage. Coincidentally, the salt solubility is increased if sugar is present; at 25 % inclusion, the solubility is in excess of 60 % compared with 20 % for the control.[133]

These proteins are envisaged as a base for artificial milk and as ingredients in cakes (30 % of egg albumen) and sausage-type products (10 % of red meat). It is stressed that effective removal of lipid is essential if off-flavour development is to be avoided on storage. In some circumstances solvent extraction can, however, be omitted if it is intended for use in food where the slight off-flavour would be masked, as in fish block prepared from under-utilised species.[15]

(b) Chemically modified proteins. The acylation reaction is of current interest as an alternative means of modifying functionality of flesh proteins. However, none of the reagents used is approved as a processing aid or as a food additive in the UK or elsewhere. Acylation of a protein involves the reaction of a nucleophilic group such as an amino group of lysine with the electrophilic anhydrides, the most commonly used being acetic and succinic anhydrides.

Acetic anhydride

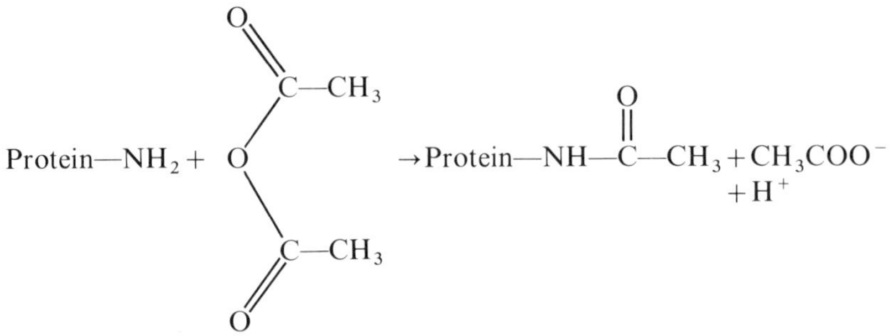

Succinic anydride

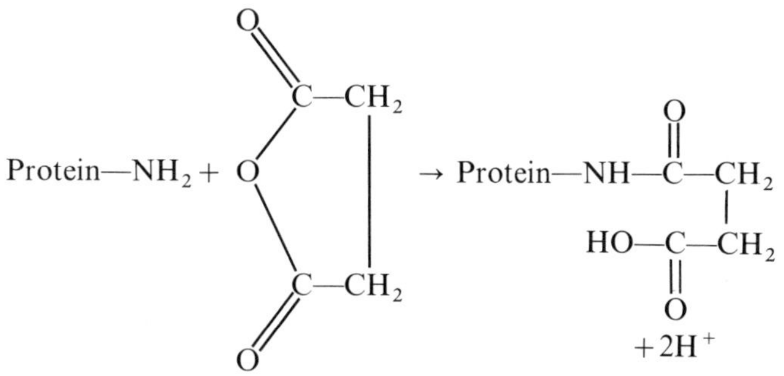

The reaction is carried out in an aqueous system usually at an alkaline pH, to form with acetic anhydride an electrically neutral residue, and with succinic anhydride a negative residue. In both cases an important change in the configuration results from the changes in charges leading often to the dissociation of aggregated or subunit proteins.[137–139] The reaction is confined mainly to the ε-amino group of lysine but other groups such as the sulphydryl group of cysteine, the phenolic group of tyrosine and the hydroxyl group of serine and threonine also react.[140] These secondary reactions are minimised if the reaction is performed near pH 9·5.

The starting material for most of the work was a 0·6 M salt extract of the washed myofibrillar proteins (Fig. 8) although a limited amount of work was also done on the washed myofibrils.

Acetylated and succinylated proteins are generally soluble at high levels of acylation but, because of their decreased nutritional value, some compromise between functionality and nutritional value must be reached.[140] Acylated fish proteins, as prepared, are so viscous that it has been found preferable to introduce partial enzymic hydrolysis by bromelain to give more amenable products. Again, by varying the degree of hydrolysis, some further control over the functional properties can be achieved. Removal of lipids from the enzyme-modified proteins is necessary, as previously, to avoid off-flavour development on storage. It is effected by extracting they acylated derivatives near neutral pH with isopropanol, particularly for incorporation into desserts, where whipping properties are important.

In general, enzymically and chemically modified proteins are heat-stable products with good emulsifying and aeration capacities. They have wide potential application in the food industry.

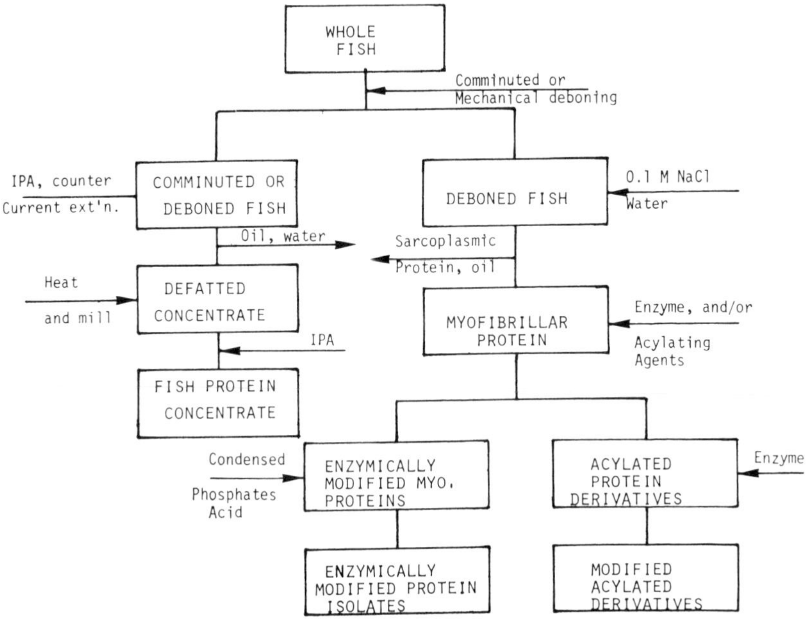

FIG. 8. Flow diagram for isolates and derivatives. (Courtesy of Spinelli *et al.*[135])

The use of reagents such as acetic and other anhydrides in a food processing system immediately raises questions of safety and, although the product of at least one anhydride is a naturally occurring compound, it is likely that full scale testing of the modified proteins would be required before they could be approved for food use.

The use of enzymes presents fewer problems from a food acceptance[141] point of view, however, as many of those referred to are already so accepted.

Other processes. While the foregoing processes represent the main areas of current and recent activity, there are others which do not readily fit into these categories and which can more conveniently be treated separately.

(a) Urea extraction process.[142] Essentially the comminuted fish muscle is dissolved in 6 M urea or guanidine hydrochloride at ambient temperature and separated from fat, bones and undissolved protein. The extracted proteins are then precipitated by dialysis and frozen. On thawing, a strongly textured protein concentrate suitable for fish mousse is obtained.

(b) Extrusion systems. The gelling properties of alginate have been the basis of an extrusion system in which highly comminuted flesh obtained from, for example, trimmings from a cod filleting operation is extruded in combination with sodium alginate into a bath containing calcium chloride, to form an alginate gel. These preparations are claimed to have potential use in products such as fish fingers.[143] Alternative extrusion systems in which minced fish and soya flour is extruded under high temperatures and pressures into moulds to give products with claimed meat-like texture on rehydration have also been developed.[144] These and similar processes utilise the heat-setting properties of both fish and soya proteins to give products with acceptable textures[144, 145] in a great variety of shapes.

(c) Fish protein concentrate type B. As a result of experience with FPC type A, FAO and the World Food Programme have included FPC type B in their list of protein products approved for aid programmes. This hygienically prepared fish meal is made in Norway mainly from capelin, mackerel, sprat and blue whiting[146] and in Sri Lanka as part of the Norwegian government's commitment to the WFP.[146, 147] There are acceptance problems[148] and while it is clearly not a solution to malnutrition, it does seem that provided sufficient research on tastes and preferences is carried out, it will make a contribution to the improved utilisation of fish resources, at least in developing countries.[148]

(d) Processes for utilisation of krill.[24,25] While some of the previously mentioned processes are also applicable to krill (*Euphausia* sp.), it is possibly more appropriate to consider potential methods for this resource separately, because of their unique characteristics and particular problems of processing. Krill are relatively small crustacea of no more than 5–6 cm in length and give proximate analyses on a dry weight basis of 53% protein, 11·3% lipid, 11·5% non-protein nitrogen and 18·9% ash.[149] Because of the highly active proteolytic enzymes in the viscera they autolyse rapidly and cannot be held for more than 3–4 h at 0–7 °C before processing.[24] Consequently, mechanical methods capable of handling and processing within the constraints of small size and rapid spoilage have had to be developed. In general, it has been found necessary to heat krill and krill products to inactivate the enzymes to stabilise them during subsequent frozen storage. A variety of products and processes have now been developed, particularly by the Russians, Japanese, Poles, West Germans and Norwegians, and are discussed fully in a recent review by Grantham.[24] Products such as pastes,[150, 151] fish protein concentrates,[152] autolysates[149] and soya–krill extruded proteins[25,153] have been prepared largely as feasibility studies and it would appear that few so far have been taken beyond this stage of development. There are many reasons for this, not least of which is the uncertainty of the market for such products[24,25] and the high costs of catching krill. Considerations of safety for human consumption also arise, however, as a recent paper on fluoride in krill by Soevik and Braekkan[154] reported levels of 7 and 24 times the permissible limit of 110 mg kg^{-1} in dried flesh and whole krill respectively.[155] Nonetheless, the development of methods of utilising krill by Japan and Russia is well advanced and will almost certainly lead to large scale exploitation provided any doubts on safety are resolved.

CONCLUSION

The problems of producing fish protein in forms that are acceptable both organoleptically and economically are now more fully appreciated than previously and it would seem that closer attention is now given to the potential markets for protein products.

To date, however, with the notable exception of the Japanese fish gel products, few protein preparations are of significance industrially. However, those processes for accelerated proteolytic digestion or for the preparation of improved solvent extracted FPCs give some indication of the varied approaches that are now being used or being considered for industrial exploitation. It is of significance too that FPC Grade B is finding acceptance in developing countries such as Sri Lanka and that new plants are being built for producing similar protein products elsewhere.[156,157] These and other essentially inexpensive processes offer one solution to the problem of supplying fish protein to those who most need it.

For developed countries where there are outlets for fish proteins in the food industry, those more sophisticated processes using proteolytic enzymes, spinning and relatively complex methods of processing have potential application. For these processes there is need for more basic research into functional properties of proteins and means of either washing or removing off-flavours associated with fish proteins. It is important too that present food labelling regulations do not restrict the development of mixed protein products as the combination of fish with vegetable proteins, for example, may well prove advantageous to both.

REFERENCES

1. CUTTING, C. L. (1955). *Fish Saving: A History of Fish Processing from Ancient to Modern Times.* Leonard Hill (Books) Ltd, London.
2. KREUZER, R. (1974). In *Fishery Products.* Ed. R. Kreuzer. Fishing News (Books), London.
3. STEINBERG, M. A. (1980). In *Advances in Fish Science and Technology.* Ed. J. J. Connell. Fishing News (Books) Ltd, London.
4. TANIKAWA, E. (1971). *Marine Products in Japan.* Koseisha-Koseikaku Company, Tokyo.
5. Connell, J. J. and Shewan, J. W. (1980). In *Advances in Fish Science and Technology.* Ed. J. J. Connell. Fishing News (Books) Ltd, London.
6. ACKMAN, R. G. (1980). In *Advances in Fish Science and Technology.* Ed. J. J. Connell. 1980. Fishing News (Books) Ltd, London; Hardy, R. (1980). Ibid.
7. FINCH, R. (1970). *CRC Crit. Rev. Food Technol.*, **1**, 519.
8. PARISER, E. R. (1973). *Adv. Food Res.*, **20**, 189.
9. PARISER, E. R., WALLERSTEIN, M. B., CORKERY, C. J. and BROWN, N. L. (1978). *Fish Protein Concentrate: Panacea for Protein Malnutrition?* MIT Press, Cambridge, Mass.
10. SIKORSKI, Z. E. and NACZK, M. (1981). *CRC Crit. Rev. Food Sci. Nutr.*, **14**(3), 201.
11. MACKIE, I. M. (1974). *Process Biochem.*, **9**(10), 12.
12. MACKIE, I. M. (1974). In *Fishery Products.* Ed. R. Kreuzer. Fishing News (Books) Ltd, London.
13. MOHR, V. (1977). In *Proc. 11th Fed. Eur. Biochem. Soc.*, **44**, 53.
14. PIGGOT, G. M., BUCOVE, G. O. and OSTRANDER, J. C. (1978). *J. Food Proc. Pres.*, **2**, 33.
15. SPINELLI, J., GRONINGER, H., KOURY, B. and MILLER, R. (1975). *Process Biochem.*, **10**(10), 31.
16. WINDSOR, M. and BARLOW, S. (1981). *Introduction to Fishery Byproducts.* Fishing News (Books) Ltd, London.
17. *The Production of Fish Meal and Oil.* (1975). FAO Fisheries Technical Paper No. 142. FAO, Rome.
18. *Yearbook of Fishery Statistics*, Vol. 48. 1979. FAO, Rome.
19. KRONE, W. (1979). *Food Policy*, **4**(4), 259.
20. ROBINSON, M. A. (1980). *Marine Policy*, **1**, 19.
21. WHEELER, A. (1975). *Fishes of the World.* Macmillan Publishing Co. Inc., New York.
22. NELSON, J. S. (1976). *Fishes of the World.* John Wiley & Sons, New York, London, Sidney and Toronto.
23. AITKEN, A., MACKIE, I. M., MERRITT, J. H. and WINDSOR, M. L. (1982). *Fish handling and Processing.* HMSO Edinburgh.
24. GRANTHAM, G. J. (1977). *The Utilization of Krill.* Southern Oceans Fisheries Survey Programme GLO/SO/77/3. FAO, Rome.
25. SUZUKI, T. (1980). *Fish and Krill Protein: Processing Technology.* Applied Science Publishers Ltd, London.
26. JOHNSON, P. O. and BAILEY, R. S. (1981). *Prospects for Fuller Utilization of UK Fish Meal Capacity.* Laboratory Leaflet No. 53. MAFF, Lowestoft.

27. CONNELL, J. J. (1978). In *5th Int. Oceans Devel. Conf.* Sept. Tokyo, Japan.
28. *Conference on the Production and Utilization of Mechanically Recovered Fish Flesh.* (1976). Ed. J. Keay. Torry Research Station, Aberdeen.
29. Scottish Sea Fisheries Statistical Tables, 1980. HMSO Edinburgh.
30. *Proceedings of the First International Conference on Chitin/Chitosan.* (1978). Eds R. A. A. Muzzarelli and E. R. Pariser. MIT Press, Cambridge, Mass.
31. BAILEY, R. S., HISLOP, J. R. G. and MASON, J. (1979). *Proc. Roy. Soc. Edinburgh*, **77B**, 479.
32. COURT, W. G. (1980). *Marine Fisheries Rev.*, **42**(7–8), 1.
33. VOSS, G. L. (1973). *Cephalopod Resources of the World.* FAO Fisheries Circular 149. FAO, Rome.
34. ARNOLD, G. P. (1979). *Squid. A Review of Their Biology and Fisheries.* Laboratory Leaflet No. 48. MAFF Direct Fish Res., Lowestoft.
35. LOVE, R. M. (1974). *The Chemical Biology of Fishes.* Academic Press, London.
36. LOVE, R. M. (1980). *The Chemical Biology of Fishes*, Vol. 2. Academic Press, London.
37. VINOGRADOV, A. P. (1953). *The Elementary Chemical Composition of Marine Organisms* (J. Efron and J. K. Setlow, translators). Sears Foundation, New Haven.
38. CONNELL, J. J. (1968). In *Low Temperature Biology of Foodstuffs.* Ed. J. Hawthorne. Pergamon Press, Oxford. p. 333.
39. SIKORSKI, Z., OLLEY, J. and KOSTUCH, S. (1976). *CRC Crit. Rev. Food Sci. Nutr.*, **8**(1), 97.
40. SHENOUDA, S. E. (1980). *Adv. Food Res.*, **26**, 275.
41. MURRAY, J. and BURT, J. R. Torry Advisory Note No. 38. HMSO Edinburgh.
42. TARR, H. L. A. (1958). *Ann. Rev. Biochem.*, **27**, 223.
43. PAUL, A. A. and SOUTHGATE, D. A. T. (1978). *McCance and Widdowson's The Composition of Foods.* HMSO–Elsevier–North Holland Biomedical Press.
44. HOWGATE, P. (1979). In *Food Microscopy.* Ed. J. G. Vaugham. Academic Press, London.
45. HOWGATE, P. (1977). In *Sensory Properties of Foods.* Ed. G. G. Birch. Applied Science Publishers Ltd, London.
46. GEIGER, E. and BORGSTROM, G. (1962). In *Fish as Food*, Vol. II. Ed. G. Borgstrom. Academic Press, London.
47. BORGSTROM, G. (1962). In *Fish as Food*, Vol. II. Ed. G. Borgstrom. Academic Press, London.
48. BONE, Q. (1966). *J. Mar. Biol. Assoc. UK*, **46**, 321.
49. OFFER, G. (1974). In *Companion to Biochemistry.* Eds A. T. Bull, J. R. Lugnado, J. O. Thomas and K. F. Tipton. Longman, London.
50. CONNELL, J. J. (1958). *Fundamental Aspects of the Dehydration of Foodstuffs.* Society of Chemical Industry, London. p. 167.
51. LAIRD, W. and MACKIE, I. M. *Bull. Int. Inst. Refrig.*, Annexe 1981–4, 349.
52. MOHR, V. (1971). On the constitution and physico-chemical properties of the connective and fish muscle skeleton. Ph.D. Thesis, University of Aberdeen.

53. CONNELL, J. J., HOWGATE, P. F., MACKIE, I. M., SANDERS, H. R. and SMITH, G. L. (1976). *J. Food Technol.*, **11**, 297.
54. CONNELL, J. J. (1980). *Control of Fish Quality*, Second Edition. Fishing News (Books) Ltd, London.
55. KIESVAARA, M. (1975). *On the Soluble Nitrogen Fraction of Barrel-salted Herring and Semi-preserves During Ripening*. Publication No. 10. Technical Research Centre of Finland, Helsinki.
56. MACKIE, I. M., Hardy, R. and HOBBS, G. (1971). *Fermented Fish Products*. FAO Fisheries Report No. 100. FAO, Rome.
57. FLOWER, B. and ROSENBAUM, E. (1958). *The Roman Cookery Book: A Critical Translation of the Art of Cookery by Apicius*. 1958. G. Harrap, London.
58. AMANO, K. (1962). In *Fish and Nutrition*. Eds E. Heen and R. Kreuzer. Fishing News (Books) Ltd, London.
59. WATANABE, T., EBINE, H. and OKADA, M. (1974). In *New Protein Foods*, Vol. 1A. Ed. A. M. Altschul. Academic Press, New York and London. p. 414.
60. OKADA, M., MIYAUCHI, D. and KUDO, G. (1973). *Marine Fisheries Rev.*, **35**(12), 1.
61. KUDO, G., OKADA, M. and MIYAUCHI, D. (1973). *Marine Fisheries Rev.*, **35**(12), 10.
62. BAKKEN, K. (1962). In *Fish in Nutrition*. Eds. E. Heen and R. Kreuzer. Fishing News (Books) Ltd, London. p. 419.
63. HILTNER, P. P., METZNER, H. and PETERS, H. (1943). German Patent 773,896.
64. SHENSTONE, F. S. (1952). *Food Pres. Quart.*, **13**(3), 45.
65. ANNAN, W. D. and MANSON, W. (1981). *Food Chem.*, **6**(3), 255.
66. TANNENBAUM, S. R., AHERNE, M. and BATES, R. P. (1970). *Food Technol.*, **24**, 96.
67. Anon. (1962). *Fishing News Int.*, **1**, 58.
68. POWER, H. E. (1962). *J. Fish. Res. Bd. Can.*, **19**(6), 1039.
69. BRISKEY, E. J. (1970). In *Evaluation of Novel Protein Products*. Eds. A. E. Bender, R. Kihlberg, B. Lofquist and L. Munck. Pergamon Press, London. p. 303.
70. KINSELLA, J. E. (1976). *CRC Crit. Rev. Food Sci. Nutr.*, **7**(3), 219.
71. HORAN, F. E. (1974). In *New Protein Foods*, Vol 1A. Ed. A. M. Altschul. Academic Press, New York and London. p. 366.
72. AMANO, K. (1965). In *Fish as Food*, Vol. III. Ed. G. Borgstrom. Academic Press, London. p. 265.
73. MIYAUCHI, D., KUDO, G. and PATASHNIK, M. (1973). *Marine Fisheries Rev.*, **35**(12), 7.
74. TAKEDA, F. (1971). In *Papers from the Symposium on the Behaviour of the Muscle Protein of Fish in Frozen Surimi. New Food Industry*, **13**(12) (Japanese).
75. IWATA, K., KARM, K. and OKADA, M. (1977). *Bull. Jap. Soc. Sci. Fish.*, **43**(2), 237.
76. OKADA, M. (1978). In FAO Fisheries Report No. 203. Supplement 1 to the Report of the Technical Consultation of the Latin American Hake Industry. FAO, Rome.

77. KAWASHIMA, T., OHBA, A. and ARAI, K. (1973). *Bull. Jap. Soc. Sci. Fish.*, **39**(11), 1201.
78. NIWA, E. and NAZAJIMA, G. (1975). *Bull. Jap. Soc. Sci. Fish.*, **41**(5), 579.
79. ITOH, Y., YOSHINAKA, R. and IKEDA, S. (1979). *Bull. Jap. Soc. Sci. Fish.*, **43**(3), 341.
80. Mitsubishi Corp., Tokyo. (1975). US Patent 3,922,372.
81. Snow Brand Milk Products Co. Ltd. (1975). US Patent 3,955,011.
82. MOTOHIRO, T. and NAKAMURA, T. (1980). *Bull. Fac. Fisheries, Hokkaido Univ.*, **31**, 115.
83. NOGUCHI, S. and MATSUMOTO, J. J. (1975). *Bull. Jap. Soc. Sci. Fish.*, **41**(2), 243.
84. EXTRACTO, A. B., SWEDEN. (1979). US Patent 4,172, 153.
85. BOYER, R. A. (1954). US Patent 2,682, 466.
86. GUTCHO, M. (1973). *Textural Foods and Allied Products.* Noyes Data Corp., New Jersey.
87. SCHMANDKE, H., PAUL, D., FRIEBE, R., ANGER, H., BARTSCH, D., PURZ, H. J., LUTHER, H., MAUNE, R. and WEBERS, V. (1976). *Nahrung*, **20**(2), 195.
88. SWINGLER, G. R. and LAWRIE, R. A. (1977). *Meat Science*, **1**, 161.
89. MACKIE, I. M. (1977). *Spectrum*, **152**, 9.
90. MACKIE, I. M. and THOMSON, B. W. (1982). *J. Food Technol.*, **17**, 243.
91. ADLER-NISSEN, J. (1977). *Process Biochem.*, **12**(6), 18.
92. WIGNALL, J. and TATTERSON, I. (1976). *Process Biochem.*, **11**(10), 17.
93. DISNEY, J. G. and JAMES, D. (Eds.) (1980). *Fish Silage Production and Its Use.* FAO Fisheries Report No. 230. FAO, Rome.
94. DAGBJARTSSON, B., ARNESEN, G., PORSTEINSSON, J., BJARNASON, J. OLAFSSON, P. and ERIKSSON, T. (1976). *Rannsóknastofnum fiskidnadarins*, 77.
95. REED, G. (1966). *Enzymes in Food Processing.* Academic Press, New York and London.
96. CAYGILL, J. C. (1979). *Enzyme Microb. Technol.*, **1**, 233.
97. HALE, M. B. (1974). *Marine Fisheries Rev.*, **36**(2), 15.
98. RUTMAN, M. and HEIMLICH, W. (1974). In *The Economics, Marketing and Technology of Fish Protein Concentrate.* Eds. S. R. Tannenbaum, B. R. Stillings and N. S. Scrimshaw. MIT Press, Cambridge, Mass. p. 135.
99. KINUMAKI, T. (1974). In *The Economics, Marketing and Technology of Fish Protein Concentrate.* Eds. S. R. Tannenbaum, B. R. Stillings and N. S. Scrimshaw. MIT Press, Cambridge, Mass. p. 125.
100. RAZUMOVSKAYA, R. G. and CHERNOGORTSEV, A. P. (1976). *Izvestiya Vysshikh Uchebnykh Zavedenii, Pishchevaya Tekhnologiya*, **5**, 921.
101. HALE, M. B. and BAUERSFIELD, P. E. (1978). *Marine Fisheries Rev.*, **40**(8), 14.
102. SEVE, B., AUMAITRE, A., JAUBERT, P. and TORD, P. (1978). *Ann. Zootech*, **27**(3), 423.
103. QUAGLEA, G. B. and MAIORANO, R. (1980). *Industrie Alimentari*, **12**, 931.
104. RITCHIE, A. H. and MACKIE, I. M. (1982). *Anim. Feed Sci. Technol.*, **7**(2), 125.
105. MACKIE, I. M. (1982). *Process Biochem.*, **17**(1), 26.
106. CLEGG, M. and MCMILLAN, A. D. (1974). *J. Food Technol.*, **9**, 21.

107. NEY, K. H. (1979). In *Food Taste Chemistry*. Ed. J. C. Boudreau. ACS Symposium Series No. 115. ACS, Washington.
108. JAUBERT, P. (1973). *L'Alim. et la Vie*, **61**(1), 32.
109. JAUBERT, P. (1973). *L'Alim. et la Vie*, **61**(1), 42.
110. GORRILL, A. D. L., NICHOLSON, J. W. G., LARMOND, E. and POWER, H. E. (1975). *Can. J. Anim. Sci.*, **55**, 269.
111. PETCHEY, A. M., OWEN, J. B., MACKIE, I. M., RITCHIE, A. H. and ØRSKOV, E. R. (1977). *Anim. Prod.*, **28**, 191.
112. SOLIMAN, H. S., ØRSKOV, E. R. and MACKIE, I. M. (1979). *J. Agric. Sci. Camb.*, **93**, 37.
113. ØRSKOV, E. R., SOLIMAN, H. S. and CLERK, C. F. S. (1982). *Anim. Feed Sci. Technol.*, **7**(2), 135.
114. YANEZ, E., BALLESTER, D., MONCKEBERG, F, HEIMLICH, W. and RUTMAN, M. (1976). *J. Food Sci.*, **41**, 1289.
115. AUTRET, M. (1978). *L'Alim. et la Vie*, **66**(1), 21.
116. THANKAMMA, R., GOPAKUMA, K., NAIR, A. L., SHENOY, A. V. and JAMES, M. A. (1979). *Fish Technol.*, **16**, 71.
117. MERRITT, J. (1982). Anim. Feed Sci. Technol., **7**(2), 147.
118. Anon. (1975). *Food Engineering*, E.F.7.
119. Anon. (1973). Fiche technique Nos. 1 and 2. 11 Production C.T.P.P. Boulogne sur mer, France.
120. Anon. (1977). *Food Engineering Int.*, **2**(12), 26.
121. FUJIMAKI, M., YAMASHITA, M., ARAI, S. and KATO, H. (1970). *Agr. Biol.*, **34**(9), 1325.
122. HOFTSEN, B. V. and LALASIDIS, G. (1976). *J. Agr. Food Chem.*, **24**(3), 460.
123. ERIKSEN, S. and FAGERSON, I. S. (1976). *J. Food Sci.*, **41**, 490.
124. ONOUE, Y. and RIDDLE, V. M. (1973). *J. Fish. Res. Bd. Can.*, **30**(11), 1745.
125. YAMASHITA, M., ARAI, S. and FUJIMAKI, M. (1976). *J. Food Sci.*, **41**, 1029.
126. SUZUKI, T., KANNA, K. and YAGI, T. (1977). Paper presented at *37th Annual Meeting Institute of Food Technologists*. Philadelphia, USA. p. 5.
127. Tokai Regional Fisheries Research Laboratory. (1977). UK Patent 1,489,268.
128. STALA, A. and KRASSOWSKA, T. (1978). UK Patent 1,526,449.
129. Astra Nutrition AB. (1975). UK Patent 1,409,876.
130. BOSUND, S. I. W., BENGTSSON, B. L. and OSTMAN, K. B. L. (1978). US Patent 4,091,003.
131. MASTERS, P. M. and FRIEDMAN, M. (1980). In *Chemical Deterioration of Proteins*. Eds. J. R. Whitaker and M. Fujimaki. ACS Symposium Series No. 123. ACS, Washington. p. 165.
132. RICHARDSON, T. (1977). In *Food Proteins, Improvement through Chemical and Enzymatic Modifications*. Eds. R. E. Feeney and J. R. Whitaker. ACS, Washington. p. 185.
133. SPINELLI, J., KOURY, B. and MILLER, R. (1972). *J. Food Sci.*, **37**, 599.
134. SPINELLI, J., KOURY, B. and MILLER, R. (1972). *J. Food Sci.*, **37**, 604.
135. GRONINGER, H. S. and MILLER, R. (1979). *J. Agr. Food Chem.*, **27**, 949.
136. BREKKI, C. J. and EISELE, T. A. (1981). *Food Technol.* **35**, 231.
137. MEANS, G. E. and FEENEY, R. E. (1971). *Chemical Modification of Proteins*. Holden-Day Inc., San Francisco.

138. CHEN, L., RICHARDSON, F. T. and AMUNDSON, C. H. (1975). *J. Milk Food Technol.*, **38**, 89.
139. HATANO, R. (1977). *New Food Ind. (Japan)*, **19**(4), 2.
140. LEE, K. H., GRONINGER, H. S. and SPINELLI, J. (1981). *Marine Fisheries Rev.*, **43**(3), 14.
141. ROLAND, J. F. (1981). *Enzyme Microb. Technol.*, **3**(2), 105.
142. Svenska Institutet för Konsiervingsforskning. (1976). UK Patent 1,423,182.
143. Unilever Ltd. (1975). UK Patent 1,414,131.
144. Ajinomoto Co. Inc., Tokyo. (1980). UK Patent 1,560,864.
145. Fuji Oil Company Ltd, Japan. (1976). US Patent 3,982,004.
146. VALAND, S. (1979). *Food Nutr.*, **5**(2), 24.
147. Anon. (1980). *Stord Bartz Review*, 6.1.
148. Anon. (1981). *Fishing News Int.*, **20**(10), 32.
149. ELLINGSEN, T. and MOHR, V. (1979). *Process Biochem.*, **14**(10), 14.
150. LAGUNOV, L. L., KRYUCHKOVA, M. I., ORDUKHANYAN, N. I. and SYSOEVA, L. V. (1974). In *Fishery Products.* Ed. R. Kreuzer. Fishing News (Books) Ltd, London, p. 247.
151. BYKOV, V. P. (1975). In *Oceanology International*, 75, 16/21. March. BPS Exhibitors Ltd, London. p. 361.
152. YANASE, M. (1979). *Bull. Tokai Reg. Fish. Res. Lab.*, **99**, 43.
153. Nisshin Oil Mills Co. (1980). Japenese Patent 5,516,628.
154. SOEVIK, T. and BRAEKKAN, O. R. (1979). *J. Fish. Res. Bd. Can.*, **36**(11), 1414.
155. Anon. (1967). *Whole Fish Protein Concentrate.* Federal Register, Washington, p. 1173.
156. HANSEN, P. (1978). In *Proc. 5th Int. Congr. Food Science and Technology.* Eds. H. Chiba, M. Fujimaki, H. Mitsuda and Y. Morita. Elsevier Publishing Co., Amsterdam. p. 58.
157. Anon. (1981). *Fishing News Int.*, **20**(10), 4.

Chapter 7

YEAST PROTEIN

M. GUZMAN-JUAREZ
Department of Food Science, University of Reading, UK

SUMMARY

Advances in protein production techniques, protein isolation, and protein refining procedures used in the removal of undesirable high levels of RNA, are bringing yeast proteins closer to use in food. More scientific and uniform approaches are being used in functionality studies for predicting protein performance in food formulations.

Both whole yeast cells and extracts of cells can be used for protein concentration. Improvements in the limited functionality of yeast proteins have been achieved by a variety of methods which include extrusion, spinning, plasmolysis and chemical modification. Some of the techniques used to improve functionality also reduce the RNA content, remove off-flavours and odours and improve relevant properties such as solubility.

The wide range of applications for which yeast proteins are envisaged includes their use as nutritional ingredients to improve the quality of food products. A more exacting role is that of a functional ingredient for the food market, and as such the nutritional characteristics occupy a less important place. However, very careful and thorough studies are still needed to evaluate safety and acceptability in use.

INTRODUCTION

During the last 15 years considerable interest has been shown in the production of food proteins by microorganisms: yeasts, fungi, bacteria and algae. The generic name adopted for such proteins is Single Cell Protein or SCP.

Studies on yeasts have occupied a great deal of the literature, at first, perhaps, due to their long association with man's food, but also because of the versatility of these organisms in their ability to grow on substrates which range from waste materials to hydrocarbon fractions. The outcome of such studies has been the establishment of a number of operational plants, with several more under construction, for the production of yeast SCP.

This chapter is concerned solely with yeast SCP for human consumption and some of the relevant active areas of research are reviewed. In the final part of the chapter the discussion is aimed at explaining the current situation regarding the restricted uses of yeast SCP in food formulations, as compared with other more conventional proteins.

HISTORICAL BACKGROUND

The use of yeast in man's diet goes back thousands of years. In Neolithic times, during the early Egyptian period, beer-making was a familiar process. Up to 40% of the grain produced in summer was used in the production of beer. Mention is made in an Egyptian papyrus of about 1400 B.C. on the ruinous effect that too much beer may have on peoples' good manners and discretion. However, fermentation in beer-making remained a spontaneous process for a long time.

The discovery of leavened bread appears to have taken place in Egypt too, although no precise date is available to mark the event. It is known, however, that by the 30th century B.C., the beginning of the Dynastic Period, in Egypt a wheat had been produced which could be threshed without the application of heat and was, therefore, suitable for bread-making. The new type of bread, although different in texture form the old bread, had no resemblance to the loaves known in present times. Despite the essential role played by yeasts in beer fermentation and bread-making, its value as a nutritious ingredient in itself, either as a 'left-over' after beer production or as a primary product of aerobic fermentation, went unrecognised for centuries.[1] Towards the end of the 19th century, in Germany left-over brewers' yeasts were added to livestock feeds with successful results. By 1920 a German team under Delbruck had developed industrial systems of yeast propagation on molasses, with improved yields and stability through the 'Zulauf' process by Hayduck in

1919. *Candida utilis* was grown on wood hydrolysates during the Second World War and used as a protein supplement.[2]

INDUSTRIAL DEVELOPMENT OF YEAST SCP

Interest in the production of protein from yeasts for human consumption developed in the early 1960s when it became apparent that effective measures were needed to counter the growing imbalance between food and people.[3,4] What was called in later years the 'protein crisis' came at a time when *per capita* food production in the less developed countries was declining and food aid was needed. The United Nations launched their programme of development and international dissemination of agricultural technology, of which the two outstanding examples are the high yielding wheat and rice varieties.[5]

The prospects of protein production from microbial sources aroused considerable interest as an attractive complement to classical methods of food production. Proteins of microbial origin, adequately supplemented, compare well with conventional protein sources such as egg or milk, in terms of their overall pattern of nutrients.[6] Microbial cell matter is particularly rich in most of the B-group vitamins and in proteins that contain essential amino acids. It therefore constitutes a potential source of enrichment for deficient diets.[7]

The term single cell protein or SCP was adopted as the generic name to describe the proteins, crude or refined, of yeasts, fungi, bacteria and algae. The initial development in this area has been thoroughly covered in the books published after the meetings at MIT in 1967 and 1973 on the prospects of the direct use of microorganisms as food.[8,9] It was there, in 1967, that the term SCP originated.

GENERAL FEATURES OF SCP PROCESSES

Among the characteristics that make SCP production attractive are the conditions which these processes require when compared with agricultural methods.[10] Such processes are independent of climatic conditions, seasonal changes, land availability, droughts, etc. Further, the organisms will grow on many different carbon substrates: natural resources such as carbohydrates, lower alcohols from the petrochemical industries or from fermentation, and hydrocarbons from the oil industry are all potential carbon sources for suitable microorganisms (Table 1).

TABLE 1
PRINCIPAL SUBSTRATES FOR SCP PRODUCTION

Industrial raw materials	*Suitable microorganisms*	*Preferred carbon substrates*
Carbohydrates		Cellulose
(i) Sugar industry	Fungi	Starch
(ii) Food effluents		Sugars
(iii) Waste timber		
(iv) Pulping industry		
Alcohols	Bacteria	Lower alcohols:
(i) Petrochemical		C_1 compounds,
(i) Fermentations		C_2 compounds.
Hydrocarbons	Yeasts	Sugars
(i) Oil refining		Lower alcohols:
		C_2 compounds.
		Paraffins:
		C_{10}–C_{23} compounds
		C_{15}–C_{30} compounds.

Other advantages of SCP processes are their applications in the areas of preservation of ecosystems and pollution control. Since growth is practicable on a variety of carbon-containing materials, microorganisms are potentially an ideal medium for the recycling of waste products, as a number of practical cases have indicated (Table 2). By these means ecosystems are preserved, the pollution problems can be controlled and protein production is feasible.[9]

The initial stages in yeast SCP production shared many features with other SCP processes. However, the outstanding development of yeast SCP plants at a pilot and industrial level in different countries to date is perhaps not surprising, considering (a) the wealth of experience in large scale production techniques amassed by the bakers' yeast industry and industrial alcohol production, and hence the knowledge of carbon and nitrogen sources, salts, aeration requirements, pH control, fermenter design and harvesting techniques, which can be applied directly to production of yeast for food purposes,[11] and (b) the extensive experience in the manufacture and use of microbial food that has been required through the centuries, mainly using yeasts, and thus establishing a link between desirable foods and the taste and smell of yeast.

A major breakthrough in yeast SCP production in the 1960s occurred when Champagnat *et al.* in 1963 reported studies on the possibility of

TABLE 2
SELECTED YEAST SCP PROCESSES

Carbon source	*Preferred microorganism*	*Process/location or reference*	*Production (tons year^{-1})*
Carbohydrates			
(i) Sugar industry	*Saccharomyces cerevisiae*	Peppler, 1971/USA (Ref. 2)	55 500
Molasses	*Candida utilis*		
(ii) Food industry effluents			
Whey	*Saccharomyces fragilis*	Delaney, 1975/USA (Ref. 19)	5 000
Potato wastes	*Endomycopsis fibuliger*	Symba/Sweden (Ref. 20)	1 000
(iii) Industrial effluents	*Candida utilis*		
Spent sulphite liquor	*Candida utilis*	Peppler, 1971/USA (Ref. 2)	
	Saccharomyces cerevisiae		
Wood hydrolysates	*Candida utilis*	*ibid*	
Lower alcohols			
Ethanol	*Candida ethanothermophilium*	Mitsubishi/Japan	100
	Candida utilis	Amoco/USA	4 500
	Not specified	Slouraft/Czechoslovakia (Ref 21)	4 000
Hydrocarbons			
(i) Gas Oil	*Candida lipolytica*	Swedt/Germany	60 000
		Roniprot/Rumania	60 000
		British Petroleum/UK	16 000[a]
		USSR (Ref. 13)	300 000
(ii) n-Paraffins	*Candida lipolytica*	Liquichimica/Italy	100 000
		British Petroleum/UK (Ref. 13)	4 000[a]

[a] Closed down

producing protein by growing yeasts on waxy alkanes, obtained as by-products during the oil refining process.[2,12] Up to that time, food and feed yeast had been grown in mineral supplemented media prepared with various sugar-containing materials, i.e. molasses, spent sulphite, waste wood hydrolysates and cheese whey. During the last 10 years, however, the industrial development of yeast SCP has moved away from the oil industry. Whereas in the 1960s the idea of using oil fractions to produce a protein was economically attractive, after the unprecedented increase in oil prices in 1973, yeast SCP based on hydrocarbons suffered a severe setback and the spread of this kind of technology is at present confined to those countries able to support such processes, i.e. oil producing or oil refining countries.[13]

Other alternatives for yeast SCP production which had been investigated primarily at a research level have been explored more extensively and as a result, some food waste carbohydrates have found industrial applications as substrates for yeast SCP alongside the more familiar carbohydrate-based processes.

This section does not attempt to cover the extensive literature on the technical and economic aspects of yeast SCP, which can be obtained elsewhere,[8,9,10] but rather to give an idea of some of the relevant aspects in the production of yeast SCP, the basic operational principles, nutritional value, and current status of these processes of food protein production.

Choice of Culture Organisms

For a culture organism with potential SCP applications some of the basic requirements are as follows:

(a) Safety—absence of pathogenicity or toxin production; absence of mutagenic character.
(b) Nutritional value—balanced source of proteins; amino acid profile as close as possible to the standard proteins such as casein or hens' egg.
(c) Other genetic features—thermotolerance; affinity for the carbon substrate; ability to grow at high cell densities; ability to withstand CO_2 during growth.

General Characteristics of the Yeasts used in Yeast SCP Production

Although a great number of yeasts exist which are important to man, only a few are interesting in terms of SCP production.

According to Phaff[14] yeasts are generally defined as fungi which do not produce distinct aerial asexual spores (conidia) and which spend at least parts of the vegetative cycle as individual single cells. Those yeasts which have been exploited as food supplements for human or as animal feeds have been strains of the genera *Saccharomyces* and *Candida*.

Genus Saccharomyces

The yeasts best known to the average person belong to the genus Saccharomyces, and are used for producing beer, wine and bread. *Saccharomyces* species are budding yeasts, a term used to define a type of asexual reproduction in which the protoplasm of the cell pushes out of the cell wall in the form of a bud and forms a daughter cell. The nucleus divides in a similar way.[15]

Saccharomyces cerevisiae, or bakers' yeast, owes its name to its ability to assimilate sugars. Besides its fermentative ability under anaerobic conditions, this yeast is available for aerobic metabolism and growth using a wider range of carbon compounds than is the case for anaerobic fermentation. It exhibits a high level of invertase activity with the result that sucrose is very rapidly split into glucose and fructose, which it can metabolise. In contrast the maltase activity is variable and can be the limiting factor in maltose fermentations. Other sugars such as mannose, acids such as lactic, tartaric, succinic and glycolic, and ethanol are utilised aerobically.[16]

S. cerevisiae is undoubtedly the most important strain of the *Saccaromyces* genus in terms of food applications. Other strains are at present used in moderate production of SCP. For example, *S. fragilis* is used in the treatment of whey. This yeast has the ability to assimilate lactose, the sugar present in whey. In 1963 *S. fragilis* was included amongst the 'food yeasts'.

The limitations of bakers' yeast appear to be related to its inability to grow without an external source of nitrogen and its need for growth factors and minerals. All these factors are of economic importance in SCP production. However, *S. cerevisiae* is able to concentrate and absorb quite large amounts of thiamine, nicotinic acid and biotin from the medium of growth, and thus form enriched products.[14] It is the favoured yeast to modify molasses, which are high in hexose sugars, vitamins of the B-group and minerals.

Genus Candida

The yeasts belonging to the genus *Candida* normally reproduce by multilateral budding, the buds originating either from single cells or from mycelial hyphae which the organism forms under certain conditions. Vegetative cells range from spheroidal to cylindrical shapes.[15]

Within the genus *Candida* two strains, *C. utilis* and *C. lipolytica*, are currently in large scale SCP production. As mentioned earlier, *C. utilis* was produced as a food supplement during the Second World War, and is presently produced as feed in various countries, and as food in the USA as Amoco yeast (Table 2).[2] It is able to metabolise hexose and pentose sugars, viz. glucose, maltose, sucrose and zylose. In addition, it assimilates a wide range of carbon and nitrogen compounds, e.g. fatty acids, such as acetic, propionic, butyric and caproic; lactic acids; pyruvic acid; ethanol; acetaldehyde; and nitrogen compounds such as urea, peptones, ammonium nitrate, nitrogen, etc.[2]

C. utilis is favoured for modifying sulphite-spent liquors and wood hydrolysates for various reasons, including its ability to assimilate not only hexose but also pentose sugars, its ability to withstand high SO_2 levels, and its ability to synthesise the necessary vitamins.

C. lipolytica has been widely used for SCP production by different companies. Its ability to utilise n-alkanes as carbon source has been the main reason for its success and potential use as a food ingredient. This yeast does not assimilate sugars, other than glucose, and lacks fermentation properties.[2]

Nutrient Requirements of a Yeast SCP Process

A block diagram of a typical SCP process is shown in Fig. 1. The detailed technique of achieving protein production varies according to each production process. However, in all yeast SCP processes and in most other SCP processes there are some basic nutritional requirements for the culture microorganisms. These characteristic needs are common to all aerobic fermentation processes and will be briefly discussed.

Carbon

This is the main energy and carbon source of the system. The role of carbon is to aid in (a) the production of energy and (b) the production of cell substance. Pasteur found that under aerobic conditions the amount of carbon used in the first function is about one-half of the total carbon, the remaining carbon being for cell substance production.

To obtain higher cell substance yields, most yeast SCP processes are operated in a continuous flow fermenter.[13,17] These vessels are operated as a chemostat, where cell growth is limited by the carbon substrate in

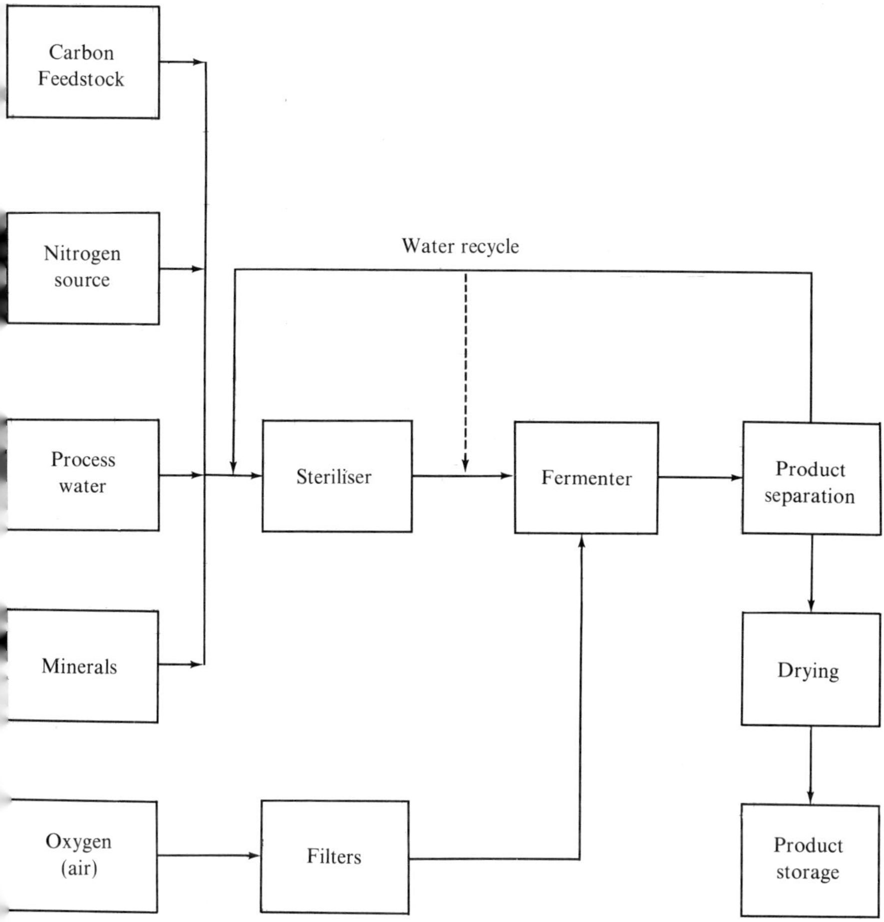

FIG. 1. SCP production process.[17]

order to enhance the protein content of the cells. If the carbon substrate is available in large amounts, certain materials, especially carbohydrates, are formed to the detriment of cell production.

Because of the difference in aerobic/anaerobic metabolism there is a great increase in available energy when cells are grown aerobically. Aerobically, 38 moles of high energy phosphate are produced per mole of glucose oxidised; anaerobically, only 2 moles of high energy phosphate are obtained per mole of glucose. Here the role of oxygen is very important and will be discussed below. Under anaerobic conditions some cells are able to continue growing, but a portion of the substrate (in this case, carbohydrate) is not completely oxidised and accumulates as ethanol, acetic acid, etc., and is lost as a potential source of energy in the cells. Under aerobic conditions the optimal use of carbon substrate is shown in the end oxidation products, CO_2+H_2O.

Oxygen

In all aerobic fermentation processes the supply of oxygen is most important for efficient use of the carbon source. It has been estimated that the log of cell mass requires an equal amount of oxygen for its production in the case of carbohydrate-produced yeast. The requirements for oxygen increase for hydrocarbon-based processes.[2,10]

The role of oxygen in SCP production is to optimise the dissimilative and assimilative functions of the carbon source. The oxygen has to be transferred from the gas phase to the liquid phase.[1] This mass transfer process is affected greatly by the degree of agitation of the air–medium–cells system and the properties of the medium.[18] Feeding of oxygen into the system is usually in the form of sterile air, which is cheaper than oxygen. High levels of oxygen, however, are inhibitory to cell growth.

Nitrogen

From the amount of nitrogen present in the cell in the form of protein, nucleic acids and enzymes, one can appreciate that a source of nitrogen is also of great importance for SCP production. The nitrogen is fed into the system as ammonium salts, amino acids, urea, etc., in incremental amounts to avoid drastic shifts in the pH of the system.

Phosphoric Acid

Soluble phosphate is essential for growth and is fed into the system as phosphoric acid or as a salt.

Minerals

Minerals such as K, Mg and S are essential for yeast production. Other elements such as Ca and Mn are not, but are incorporated into the system as a preventive measure. Essential trace elements are Cu, Fe and Zn.

General Processing Aspects of Yeast SCP Production

The choice of raw materials plays an important part in an SCP process. It is a function of the intrinsic characteristics of the carbon substrate (solid, liquid, gaseous), its composition and oxidation state. The process is designed to promote and optimise cell energy and protein production. It is not surprising to find that some of these factors are mutually opposed and they can add to the overall cost of the process, rendering it ultimately unproductive.

In terms of the carbon substrate the most important parameters to consider in a process design are as follows:

(a) yield coefficient (kg cell mass kg^{-1} substrate);
(b) productivity (kg biomass/hour-$meter^3$);
(c) solubility of the substrate in the culture medium;
(d) heat of fermentation (kcal kg^{-1} cell mass);
(e) oxygen demand (kg O_2 Kg^{-1} cell mass);
(f) resistance to contamination (pH of the process).

The yield coefficient is the most important physiological factor describing the efficiency of microbial growth and cell production. Hamer[17] defines it as the weight of cell mass produced per unit weight of carbon substrate utilised. The yield can also be expressed in terms of cell carbon per gram of substrate; depending on the degree of oxidation of the substrate, the culture organism or yeast will require different amounts of oxygen to perform its functions. Mateles[18] points out that the question of oxygen demand in highly exothermic processes like SCP production has complications both in terms of mechanical arrangements to ensure the supply of oxygen, and the need to remove large amounts of heat from the fermenter to maintain a constant temperature.

SCP processes operate as submerged cultures. In these conditions the cells are in suspension in the liquid medium and the continuous medium/cells interaction and nutrient uptake are optimised. All yeast SCP processes in operation use sterile conditions.

The industrial processes depend on continuous flow fermentation. This is done[13] (a) to ensure a continually uniform product quality, (b) to better utilise the fermenter volume and the equipment, and (c) to avoid infection more effectively.

Yeast SCP Production at an Industrial Level

Numerous processes for yeast SCP production are described in the literature, using different carbon sources as susbstrates, e.g. carbohydrates, alcohols and hydrocarbons.[6,8,9,19] Some of the industrial processes are reported in Table 2. The production of bakers' yeast, dried yeast and torula yeast, or *C. utilis*, are also included. These are the yeasts primarily grown on molasses and sulphite waste liquor specifically for human consumption and known as 'food yeasts'. Also included in Table 2 are the 'fodder' or feed yeasts—by-products of beer fermentation with lower required quality standards than the food yeasts.[2]

Carbohydrates
In addition to the traditional carbohydrate processes of yeast production, novel processes exist in which waste food products are used as carbon sources. The best known is the production of protein from whey, in which *Saccharomyces fragilis* is used to inoculate a lactose-rich growth medium. A second process is known as the Symba process. Yeast SCP is produced by the interaction of a symbiotic culture formed by the fungus *Endomycopsis fibuliger* and the yeast *C. utilis* on potato waste effluents. The function of the fungus is to degrade the starch to assimilable sugars. The sugars, in turn, serve as the growth medium for the yeast. The percentage of the fungus in the final product is of the order of 3%. The product is already on the market as a feed ingredient.[20]

Petrochemicals
The substrate used for yeast growth is ethanol, using *Candida* species (Table 2). The cost of the raw material may in the end limit the spread of this process.[3] However, up to the present the only yeast SCP product of food grade with ethanol as the raw material has been produced by the Amoco Oil Company[21]

As oil prices rise, it becomes likely that ethanol produced by fermentation of sugars or by petrochemical methods could be an alternative fuel, and its potential as substrate for yeast SCP production will be affected.

Hydrocarbons
The ability of some yeasts to utilise alkanes as carbon substrate has been industrially exploited to produce protein, as illustrated in Fig. 1. In the UK the production of hydrocarbon-based yeast was undertaken mainly by British Petroleum, starting in 1963.[12] They developed two processes of yeast SCP, both using *C. lipolytica*. One was based on the aseptic fermentation of purified n-alkanes by the yeast, and the other on the fermentation of the paraffins present in gas-oil for similar purposes, under non-aseptic conditions. In the second process a thorough solvent washing was carried out at the end of the fermentation stage to eliminate any risk of contamination from unreacted gas-oil, which, after the process, was returned to the refinery for further processing.[22] By 1975 these yeasts, under the name of 'Toprina', were already part of the feed

market for ruminants and monogastric animals, and to a lesser extent for poultry. After more than 10 years of successful production and intensive toxicological and clinical studies, and following the collapse of their gigantic SCP production venture with an Italian petrochemical concern, BP announced in July 1978 the closing down of their plants in Scotland and France, for other than technical reasons.

NUTRITIONAL VALUE AND SAFETY OF YEAST SCP

In terms of composition, the contents of protein, fat, carbohydrate and ash of different yeasts have been reviewed.[2, 6, 23] Irrespective of the kind of substrate used, the average composition of yeasts varies as follows:

	(%)
Protein	45–49
Fat	4–7
Carbohydrate	26–36
Ash	5–10

Table 3 shows the amino acid profile of different yeasts. It can be seen that yeasts are a particularly good source of lysine. Together with other SCP products, however, they have low cystine and methionine contents, which are the limiting amino acids of these and of other SCP organisms.

Feeding experiments with yeasts are widely reported in the literature.[24] The results of some of these studies have been reviewed by Waslien.[23] Diets in which the main sources of protein were yeasts were fed to rats, chicks, hens, swine and calves. From the results it seems reasonable to conclude that although toxic effects were found during the trials, they were avoided or removed by methods which indicated that yeast proteins had been either fed at an insufficient level in the diet or lacked some specific amino acid or nutrient. Nevertheless, yeast proteins are known to have a lower nutritional quality than other standard proteins such as casein or egg.

Hydrocarbon-derived yeasts may give rise to toxic effects, as a result of accumulation of paraffins in animal tissues, with possible effects on humans consuming meat from yeast-fed animals. Conclusive experiments involving long-term and multiple generation feeding studies, using dietary levels of up to 30% yeast and n-paraffin contents of up to 1500 ppm, showed that in 20 successive generations of rats and 30 of quail no evidence in either case could be found of treatment-related pathology in the yeast-fed animals.[25] Moreover, these and other studies[26] showed that

TABLE 3
AMINO ACID PROFILES OF SOME YEAST SCP AND FOOD YEASTS[a]

Amino acid	*Candida lipolytica (n-paraffins)*	*Candida utilis (sulphite liquor)*	*Saccharomyces fragilis whey*	*Saccharomyces cerevisiae molasses*	*FAO pattern*
Alanine	7·4	3·4	n.q.	n.q.	n.q.
Arginine	4·8	5·4	4·9	5·0	n.q.
Aspartic acid	0·2	4·7	n.q.	n.q.	n.q.
Cystine	1.1	0.7	n.q.	n.q.	n.q.
Glutamic acid	11·3	15·0	n.q.	n.q.	n.q.
Glycine	4·8	4·8	n.q.	n.q.	n.q.
Histidine	2·0	1·9	2·5	4·0	n.q.
Isoleucine	4·5	5·3	5·5	5·5	4·2
Leucine	7·0	7·0	4·9	7·9	4·8
Lysine	7·0	6·7	8·8	8·2	4·2
Methionine	1·8	1·2	1·5	2·5	2·2
Phenylalanine	4·4	4·3	3·9	4·5	2·8
Proline	4·4	3·5	n.q.	n.q.	n.q.
Serine	4·8	5·5	n.q.	n.q.	n.q.
Threonine	4·9	5·5	5·5	4·8	2·8
Tryptophan	1·4	1·2	1·5	1·2	1·4
Tyrosine	3·5	3·3	n.q.	n.q.	n.q.
Valine	5·4	6·3	6·6	5·5	4·2

[a] Adapted from Ref. 10.
[b] n.q. = not quoted.

the presence of n-paraffins is common in different foodstuffs (not usually exceeding 0·1% dry weight) and that they can be oxidised to fatty acids by animals and humans. Previously it had been confirmed that the odd-numbered carbon fatty acids in SCP are also submitted to metabolic turnover (Protein-Calorie Advisory Group, (PAG) Symposium, 29–30 March 1976).

Studies on carcinogenicity, using hydrocarbon-grown yeasts with mice and rats, combined toxicity–carcinogenicity using inbred and outbred mice strains, and co-carcinogenicity using rats fed on two known carcinogenic chemicals for 5 months, showed that in none of these cases did the products elicit a carcinogenic response.[25]

Conclusions

Clearly, since the consumption of significant quantities of SCP products would represent a novel addition to the human diet, either directly or indirectly via animal products consumption, SCP products must be required to pass rigorous tests aimed at assessing their safety and nutritional value before they can be cleared for sale in the market. To that end, it has been emphasised that studies in multiple species of experimental animals (PAG Guideline No. 6), production of SCP for human consumption (PAG Guideline No. 12) and clinical trials on acceptability and tolerance on human testing of supplementary food mixtures (PAG Guideline No. 7) be followed closely before attempting to incorporate any SCP product into human foods.[9] Any changes, even minor, in the substrate, culture organism or processing conditions would require these extensive tests to be satisfactorily completed before clearance of the product for human consumption could be given.

So far, yeast SCP products have had some success. By 1977, the BP yeast had been sold for animal feeding in the UK, France, West Germany, Holland, Belgium, Denmark, Luxembourg, Switzerland, Spain and Portugal. In the USSR, another yeast SCP, known as BVK, is widely used at a level of 100 000 tons per year. In Sweden, the carbohydrate-grown yeast produced by a sugar company is used at levels of up to 40% for young calves.[20]

YEAST PROTEINS FOR HUMAN CONSUMPTION

Animal feeding studies have confirmed the nutritional quality of yeast SCP. However, further improvements in yeast SCP products have been necessary, since the production of a wholesome, nutritious new protein product does not of itself guarantee a competitive place in the food market. Thus, eggs, meat and milk have not only been considered intrinsically good and nutritious, but are also very much valued for their ability to display functionality in food formulations and so contribute to desired qualities in end-products. For example, egg white is known as a foaming agent and is used in all creams, meringues and cakes whereas egg yolk is widely incorporated as a thickening and emulsifying agent in mayonnaise, cakes and soufflés. The former holds air: the latter holds otherwise immiscible phases together in emulsion systems.

The functions that proteins display in food systems are related to their ability to disperse in a given medium, to bind oil, to swell, to emulsify, to gel or to foam, all necessary qualities in the production of food products. It is therefore evident that if yeast protein is to succeed in the food industry in competition with conventional proteins, it must be capable of similar or better performance than established products.

For yeast SCP products to satisfy both the nutritional and functional aims of the prospective market several processes have been introduced with the following aims:

(a) increase the availability of the protein;
(b) to decrease ribonucleic acid (RNA) content in the products;
(c) to concentrate the protein; and
(d) to improve functionality.

Increasing Protein Availability

The complex gluco-mannan polymer, which constitutes the main component in yeast cell walls, also makes the protein unavailable to humans when ingested since digestive enzymes able to act upon the cell walls are not present in the system. Furthermore, the functionality associated with whole yeast cells is restricted to the surface of the cell wall membrane.[27]

Enzymatic methods of breaking or weakening the cell walls in yeasts have been reported, although not all of them would be practicable for industrial SCP production.[19] Enzymatic methods have received thorough investigation.[27] It would seem that the activation of endo-β-$(1\rightarrow3)$-glucanases offers an economical and efficient way of cell wall disruption. However, the control of their activity remains the crucial factor. The stomach juice from snails has been successfully used in cell wall degradation but for economic reasons extracellular microbial enzymes offer a better alternative. However, enzymatic methods present the problem of self-digestion for the yeast cell, with resulting loss in yield.

Non-mechanical methods, such as freeze–thaw cycles and treatment of cells with solvents, are impracticable for large scale operations and some of them present toxicity hazards. Other techniques, such as ultrasonication, are ineffective in the treatment of Gram-positive organisms and yeasts.

Despite heavy energy costs, cell wall disruption has been so far achieved essentially by mechanical means. Such processing is in the presence of glass beads, the apparatus design being either vertical[28] or horizontal,[29] the latter claimed to be more efficient. An industrial disintegrator in a horizontal arrangement with a capacity of 20 kg dry weight of yeast protein per hour has been reported.[30]

High pressure homogenisers (8000–15 000 psig) of the Manton-Gaulin type have been in increasing demand for SCP processes.[30] The process has been successfully used for the production of yeast glycan,[31] successive homogenisation steps ensuring the total fragmentation of yeast cells.

Reducing Ribonucleic Acid Content

Chemical Methods

The toxicity associated with a diet containing SCP products is due to the high content of nucleic acids (RNA and DNA) present in yeasts and other SCP products. Studies have shown that intakes of SCP greater than 15 g day^{-1} (equivalent to approx 2 g of nucleic acids) may result in higher uric acid levels than are considered safe (between 4 and 7 mg/100 ml plasma), and in some sensitive individuals it may cause disorders such as gout and kidney stones.[32]

In yeasts, DNA accounts for only 1–2% of the RNA/DNA content, which varies between 8 and 12% on a dry basis. Therefore efforts have been directed mostly towards the removal of RNA.

Several chemical methods are available for lowering the RNA content in yeasts. However, relatively few have been used in the treatment of yeast SCP for a variety of reasons, ranging from the impracticability of the scale of operations needed to potential toxicity of the solvents used, to risks of protein hydrolysis.[33] Most of the chemical methods of RNA removal exploit the behaviour of RNA molecules under alkaline conditions, in which the 3, 5-phosphodiester links between adjacent nucleotides in RNA molecules are able to form a cyclic triester. In the case of DNA, the deoxypentose bears no OH^- group at the C_2 carbon and therefore cannot form the essential 2′, 3′-phosphotriester. It is thus resistant to alkaline digestion. Alkaline conditions are not successful in removing RNA if temperature treatments remain low, even at high pH values.[34,35]

Robbins[36] patented a process for the manufacture of yeast protein with a reduced RNA content using alkali and/or high temperature. Slurries of food grade yeast cells, disintegrated cold at 8000 psig, were treated at pH 9·5 and 25–60 °C for about 20 min after which cell walls were recovered by centrifugation. The alkaline extract was adjusted to pH 6–8 and heated at temperatures of 110–120 °C for 2–60 min. The treatment precipitated the protein, and the intact RNA fraction was released into the supernatant. The heat treatment also sterilised the protein fraction. The final RNA level is about 2%. The process of RNA reduction can also be used to treat recovered yeast protein isolate with a full RNA content by adjusting the pH of slurries to 7 and treating them as in the previous case. The final RNA content of the isolates is of the order of 4%.

When a protein product containing cell walls is needed, the alkaline treatment is omitted and homogenates of disrupted cells at about 9% solids are treated for RNA reduction at pH 7. A final 4% RNA content is achieved.

Hedenskog and Ebbinghaus[37] used alkaline conditions for the removal of RNA from *S. cerevisiae*. The treatment involves cell disruption by disintegration in the presence of glass beads at atmospheric pressure, after which the homogenate is treated at pH 9. RNA removal is achieved by heat treatment of the homogenate at 80 °C and sudden cooling to 25 °C. The treatment precipitates the protein and leaves a high molecular weight RNA fraction in solution. The final content of RNA in the protein is 1–2%. When sodium chloride is added to the alkaline extract a better removal of RNA is achieved and lower temperatures are required to precipitate the protein.

Newell[38] used two different conditions, alternating high pH/low temperature, and low pH/high temperature treatments to reduce RNA content to under 2%. After rupturing the cells at 0–10 °C and 8000 psig the homogenate was treated at pH 9·5 at 60 °C for about 20 min. Cell walls were removed at (60 °C) and the yeast extract then treated either at pH 10–10·5 and 75–85 °C for 1–4 h or at pH 11·5–12·5 and 55–65 °C for 1–2 h. Protein precipitation was by heat using hydrochloric or phosphoric acid at the isoelectric point. The protein fractions were washed and spray-dried.

Enzymatic Methods

The use of enzymes in the removal of RNA from yeast SCP is one of the most popular techniques. Although the use of exoenzymes has been reported,[39,40] the endogenous ribonuclease present in yeasts has been most used in enzymatic studies of RNA reduction. The enzyme is temperature- and pH-dependent;[41,42] therefore most treatments involve slightly acidic conditions, at temperatures not higher than 65 °C. Ribonuclease is also affected by the presence of some metals such as Na^+ and Zn^{2+}. Mechanisms by which metals affect the enzyme have been postulated.[43]

Lindblom[44] found that 3% NaCl was necessary to initiate RNA degradation in a bakers' yeast slurry of disintegrated cells. The treatment was carried out at pH 5·6 at 50 °C: an incubation time of 20 min resulted in 85% RNA removal. A protein product containing cell walls was obtained with 1·4% RNA content.

Schuldt[45] patented a process for RNA reduction which involved the addition of Zn to yeast. The metal acts as a protective factor for RN-ase during the alkaline treatment for protein extraction and thus improves RNA reduction during the subsequent incubation step at pH 6 and 60 °C. The addition of Zn can be carried out at any of several stages, i.e. from the fermenter to just before the cell homogenate is made alkaline. At any subsequent point the RN-ase is not protected; in the absence of Zn, extraction conditions impair the action of RN-ase at the subsequent incubation step, and greater RNA reduction is achieved by activating RN-ase at a neutral pH.

Earlier, Robbins[73] had found that a holding time of 5 min at pH 9·5 was a tolerable treatment for the RN-ase present in the yeast homogenate. After cell wall removal, incubation of the yeast extract at pH 6 and 50 °C for 60–120 min allowed for the RN-ase to act upon the RNA fraction. Preparation of the protein by treatment of the sludge at 70 °C improved yield, and an RNA content of less than 2% in the end-protein product was reported.

Tannenbaum[41] developed a process for reducing the RNA content in whole cells by a heat-shock three-step process in which yeast slurry was passed through a thin steel tube for 2–20 s at 60–70 °C, after which the cells were incubated at 45–50 °C and pH 4–7, and later at 50–60 °C and pH 4·5–7. The last two steps may be combined into a single incubation period of 10–60 min at 50–55 °C at an optimum pH 6. A final RNA content of 1·5% was reported for *C. utilis*. The authors suggest that the temperature shock may have an unfolding effect upon the nucleic acids and/or on the activation of the endogenous enzymes.

Safety Considerations

In connection with RNA reduction processes, recent reports give some cause for concern. Allergic reactions specific to humans and of a gastro-intestinal or subcutaneous nature have been observed.[46] It is possible to isolate the allergenic protein fraction and although the activity of the material prevailed through temperatures of 70, 80 and 100 °C for 30 min, it eventually decreased with prolonged storage at 4 °C. These studies underline the importance of careful control and testing of new products even when their objective is the detoxification of a product designed for human use.

Concentrating the Protein

For enhancing the protein content of yeast SCP products different methods of protein concentration have been investigated. Some of them have already been mentioned in connection with the reduction of RNA, since in most cases concentration of the protein fraction involves an RNA reduction step. When the latter uses high temperature treatments, the protein fraction may be precipitated, depending on the pH of the treatment.

Most methods of protein concentration use alkaline treatments of disrupted yeast cell slurries to solubilise the protein. The use of temperatures between 40 and 60 °C is often reported: the pH of the treatments ranges from 9·5 to 12 since yields are improved by high pH extraction treatments.[34,35] Once the protein fraction is in solution the cell walls are usually removed and the protein is precipitated by heat or acid, or both, near or at the isoelectric point.

The use of high pH and temperature treatments has been found to cause undesirable effects from both nutritional and functional points of view. Essential amino acids may be lost or take part in unusual reactions, affecting their *in vivo* absorption.[37,38] Under similar conditions, heat precipitation tends to affect protein functionality due to protein denaturation.[47]

Recovery of protein fractions is affected by cell disruption, extraction and precipitation procedures.[33,48] However, Lindblom[34] found that about 10% of the nitrogen compounds in the yeast cell have high molecular weights but cannot be precipitated, and that 21% are constituents of the amino acid and nucleotide pools, together with other non-precipitable, low molecular weight compounds. She concluded that the maximum yield in a protein concentrate which could be obtained after precipitation of a suspension of disintegrated yeast would be about 70% of total nitrogen, including the cell wall fraction which accounts for about 27% of the total nitrogen. The results quoted are in agreement with other reports.[35,38]

Protein isolation procedures which involve high pH and temperature and prolonged times promote hydrolysis of protein and reduce the yield of product.[33,49]

FUNCTIONAL PROPERTIES OF YEAST PROTEINS

Functional properties are those physicochemical properties of proteins which govern their composite performance in foods during manufacture, processing, storage and consumption. They reflect the properties of the protein that are influenced by its composition and conformation, and interactions with other food components, and they are also affected by the immediate environment.[50] Some of the functional properties relevant in foods are listed in Table 4.

The above rather involved definition implies that many parameters must influence protein functionality, and gives an idea of the complexity involved in trying to predict the effect of proteins on finished food products. For this reason, interest in measuring and modifying functionality in both established and new sources of protein is widespread.[50]

TABLE 4
GENERAL CLASSES OF FUNCTIONAL PROPERTIES OF PROTEINS IMPORTANT IN FOOD APPLICATIONS[50]

General property	*Specific functional term*
Organoleptic, kinaesthetic	Colour, flavour, odour, texture, mouthfeel, smoothness, grittiness, turbidity, etc.
Hydration	Solubility, dispersibility, wettability, water absorption, swelling, thickening, gelling, rheological, water-holding capacity, syneresis, viscosity, dough formation, etc.
Surface	Emulsification, foaming, aeration, whipping, protein/lipid film formation, lipid binding, flavour binding, stabilisation, etc.
Structural, textural, rheological	Elasticity, grittiness, cohesion, chewiness, viscosity, adhesion, network cross-binding, aggregation, stickiness, gelation, dough formation, texturisability, fibre formation, extrudability, elasticity, etc.
Other	Compatibility with additives, enzymatic, inertness, modification properties

Admittedly, many reports have been of limited help to the investigator for various reasons, including lack of continuity in the studies themselves and non-standardised experimental methods. However, recently several workers have successfully used a more systematic approach to protein functionality by statistically relating the results of various functional tests, in model systems of increasing complexity, to protein performance in actual food formulations.[51] Their work has been summarised by Schoen[51] who has also advocated the measurements of basic properties of proteins and developed mathematical models to interrelate the measured variables as predictive tools in evaluating protein performance in foods. This idea reinforces the need that has become evident for a more scientific and uniform approach to functionality. Table 5 illustrates

TABLE 5
PROTEIN PROPERTIES[51]

Molecular properties	Mol. wt. determination
	Mol. wt. distribution
	Ionisation properties
	Reactive side chains
	Primary, secondary, tertiary and quarternary structure morphology
Physical properties	Hydration properties
	Solubility
	Suspendability
	Swelling
	Gelling
	Rheological properties
	Viscosity or apparent viscosity
	Yield stress
	Time dependency
	Shear dependency
	Creep and relaxation
	Thermal properties
	Gelation, coagulation
	Thermal conductivity, heat capacity
	Surface properties
	Hydrophilic and lipophilic properties
	Surface absorption and adsorption
	Foaming

some of the properties which can be measured in functional studies and subsequently used to predict protein performance in food products.

In contrast with the abundant literature available on proteins from other sources, relatively few studies on the functional properties of SCP proteins have been reported.[50] Of these few studies, however, yeast SCP products have been dominant, due to the food grade status of selected strains of the genera *Saccharomyces* and *Candida*. The patent literature recently reviewed has also provided information on the food applications of yeast SCP products.[52]

The functionality associated with whole yeast cells is determined by the surface of the cell membrane.[53–55] Not surprisingly, the protein fraction of yeasts, with or without cell walls, has provided the main material for functional tests. Processes such as extrusion, spinning, and enzymatic and chemical modification have all been used to improve the functionality of yeast SCP.

Kinsella[56] considers the solubility of proteins a crucial functional characteristic, since many other functional properties such as gelation, emulsification and foam formation depend on the initial solubilisation of the protein in the medium. The solubility of yeast SCP compares poorly with other proteins such as soy, especially at PH values common in food.[34,35,57] Lindblom found that the most important influence on yeast protein solubility was the processing steps involved in RNA reduction by endogenous RNA.[58] Other factors, such as viscosity and particle size of cellular and subcellular particles, have been considered important for quantifying protein changes due to processing conditions, especially when such changes are not reflected in solubility-pH changes.[59]

Alkali treatment during protein extraction has been reported to decrease the solubility of yeast concentrates.[34] In later studies, treatment with alkali (pH 11·8) at 70°C for 10 min of a heat-precipitated yeast concentrate was reported to improve the solubility and foam capacity over those of the original concentrate. Foam stability, however, was higher for the original concentrate.[60] Other functional studies showed that the foaming properties of yeast proteins from heat- or acid-precipitated fractions in most cases gave lower values than soy isolates, the lowest values being found in the heat-precipitated samples.[47]

Water absorption tests on yeast concentrates[34,60] gave values of 3·5–4 ml g^{-1} sample. Higher values are obtained for samples submitted to RNA reduction compared with non-RNA reduced samples.[35] In these studies yeast isolates gave water absorption values similar to soy isolates, and also similar oil binding capacities to those of their soy counterparts.

The emulsifying characteristics of yeast protein preparations have been studied. Lindblom[60] reported on the emulsifying capacity of an alkali-treated yeast concentrate obtained by heat precipitation. The original concentrate showed no emulsion forming ability, whereas the modified preparation could hold 320 g oil g^{-1} protein. Guzman-Juarez[35] studied the emulsifying properties of alkali-extracted acid-precipitated yeast protein isolates in water and sodium chloride (4% w/v). Yeast proteins showed similar behaviour to soy isolates. Emulsification values of 50–60% were obtained at levels below 5% protein. Above those levels, and up to 10%, soy isolates showed higher values but yeast preparations remained the same. Heat treatments of the emulsions to test for stability did not change the emulsifying characteristics of yeast proteins, whereas those of soy isolates were increased by a further 20%. The presence of sodium chloride had a detrimental effect upon both the activity and the stability of soy isolate emulsions. No such effect was detected in yeast isolates under similar conditions. After sodium chloride treatments, both soy and yeast isolates showed 50% emulsification for protein concentrations from 1 to 10%.

IMPROVEMENT OF FUNCTIONALITY BY PROTEIN MODIFICATION

Texturisation

The texturisation of protein from yeast has been studied as a means of improving organoleptic properties. Huang and Rha have studied apparent viscosities below the coagulation temperature for the dope formation of yeast SCP from alkali-extracted dried *Torula* yeast for the production of spinnable fibres.[61] They found that the optimum protein content of the SCP concentrate dope for spinning seemed to be in the range of 15–20%. At protein concentrations of 25% viscosity values at pH 8 and 9 were very similar (1500–4000 cps). At pH 9, protein concentrations of 20% showed viscosity values of 350–1000 cps, and were ten times greater than those of the samples at pH 8. For protein concentrations of 15%, the apparent viscosity of about 150 cps at pH 9 was approximately 2–5 times greater than the viscosity at pH 8. At pH 10, viscosity values sharply decreased and further increase in pH caused gelation. The dopes showed pseudoplastic characteristics and time dependency. Huang and Rha concluded that, since protein concentrations above 25% would cause physical difficulties in handling the dope, other ingredients could be added to manipulate the viscosity rather than increasing protein concentration to increase dope viscosity.

In later studies, Huang and Rha[62] incorporated various reagents into the dope of extracted *Torula* yeast to promote fibrous formation of the protein and investigate the potential for forming a continuous and uniform filament by spinning/extrusion. They found that, while no continuous filaments were formed using SCP/formaldehyde, SCP/glutaraldehyde or SCP/hydrogen peroxide mixtures, the use of 20% carrageen or carboxymethyl cellulose (CMC) in the SCP dope afforded superior spinnability, with an increased value for apparent viscosity at pH 9.

Further investigations,[63] in which carrageen and CMC were tested in mixtures with yeast SCP for stretchability of the filaments formed as well as continuity and uniformity, showed that CMC was a more useful additive than carrageen in yeast—SCP hydro colloid dopes. The optimum reinforcement per gram of CMC added appeared to be between 1·3 and 2%. Beyond that level, resistance to stretching increased, possibly via the resistance imparted by the increased number of bonds or by the limited availability of binding sites.

Hayakawa and Nomura[64] investigated the spinnability of yeast protein dopes using bakers' yeast. Three different processes of protein isolation were followed. In two of them, slurries of whole cells were treated with sodium hydroxide, either cold or hot, to extract the protein. In a third method, the protein was extracted by rupturing the cells in an alkaline medium. The first two were treated with acid, but the third precipitate was produced by a hot ethanol treatment after isoelectric precipitation, which also removed the lipid fraction. Using casein as control in the spinning tests, the authors found that only the ethanol-treated yeast protein could successfully be spun at 10–13% protein concentration. They found no relation between the molecular weight of casein and that of the yeast protein to account for the improved spinnability as compared with the other yeast proteins. However, viscoelastic tests based on the Markovitz equation showed the similarity of the ethanol-treated yeast protein to casein. Rha[65] comments on the size, shape, number and interactions of particles as the basic parameters affecting the rheological properties of a system, be it molecular, dispersion, suspension paste or gel.

Reverse Enzymatic Synthesis

It is possible to remove off-flavours and improve the functionality of proteins by reverse enzymatic synthesis. The process consists broadly of a controlled protein hydrolysis step, using enzymes such as papain, pepsin, etc., to obtain the desired peptides, which are then reacted with an enzyme to undergo the plastein reaction. The mixture is incubated for a suitable time and the end-product is a refined protein with improved nutritional value, colour and flavour.[66] The plastein reaction offers an effective way of refining protein.[67]

Chemical Modification

As a means of improving functionality, chemical modification is a novel approach to prospective food applications of yeast and other protein sources. The process involves chemical reactions between functional side groups in the proteins and reagents which promote nucleophilic or reductive reactions with electron-rich side chains. Kinsella[56] has reviewed relevant aspects of the mechanisms involved in such reactions.

Yeast proteins have been chemically modified by succinylation.[68] Acetylation improves thermal stability, while succinylation improves a wide range of functional properties: solubility is increased at pH 4–6 and recovery is improved. Properties such as emulsification and foaming capacities are also improved by succinylation, not surprisingly, following the improvement on solubility.

Interest in chemical modification is thus certainly justifiable. There is, however, concern about the metabolic aspects of derivatised proteins in humans exposed to large doses of chemically modified protein products. Though no studies have been reported on derivatised yeast proteins, animal studies have shown that some modified proteins impair growth in rats.[69] For example, studies on the nutritional quality of acid-soluble casein derivatives showed that rats fed on a diet containing 10% lactic casein had a PER of 2·5. Those fed on acetyl casein + 0·9% L-lysine hydrochloride had a PER of 2·0.

Acetyl whey protein gave a PER of 2·0 whilst succinyl casein allowed negligible growth (PER 0·3). Mice on toxicity trials were able to survive and reproduce when fed almost purely on acetyl casein as their dietary protein, but they were lighter and smaller than the control animals, though no degeneration of organs was detected in three generations.[70]

In vitro digestibility studies on alkylated caseins, using the enzyme chymotrypsin, have shown that alkylation lowers the initial rate of hydrolysis by the enzyme, the decrease in hydrolysis rate being in direct proportion to the size and number of the alkyl groups.[71]

Recent digestibility studies with rats, using acylated beef heart myofibrillar proteins, support the above findings.[72] These studies also indicated that the rat was able to adapt to the acetyl protein derivative in preference to the succinyl counterpart, as a dietary protein source during the 28-day trial period. No acute toxicity problems were observed and abnormalities in the organs of the rats were not detected at the end of the studies.

It would seem that chemically modified proteins are a potential alternative as functional ingredients, able to compete with more traditional food products. However, the present state of knowledge on the safety hazards of modified proteins demands more research in this area.

CONCLUSION

Yeast protein (yeast SCP) has considerable potential as a raw material for human food formulations, and there is no reason why it should not become competitive in terms of quality and functionality with similar products such as soy protein. However, being of single cell origin, special processing requirements and safety considerations arise and will place constraints, to which vegetable proteins are not subject, on its development. The use of chemically modified forms of SCP will probably have to await the satisfactory outcome of rigorous safety testing before such products come into general use.

REFERENCES

1. TANNAHIL, R. (1975). *Food in History*, Granada Publ. Ltd., St. Albans, p. 63.
2. PEPPLER, H. J. (1971). In *The Yeasts*, Vol. 3, Rose, A. H. and Harrison, J. S. (Eds.), Academic Press, London, p. 421.
3. BROWN, L. (1968). In *Single Cell Protein*, Mateles, R. I. and Tannenbaum, S. R. (Eds.), MIT Press, Cambridge, Mass., p. 11.
4. BEAN, L. H. (1978). In *New Protein Foods*, Vol. 3, Altschul, A. M. and Wilcke, H. L. (Eds.), Academic Press, London, p. 21.
5. BROWN, L. R. (1975). *Science*, **190**, 1053.
6. LIPINSKI, E. S. and LITCHFIELD, J. H. (1974). *Food Technol.*, **28**(5), 16.
7. BHATTACHARJEE, J. D. (1970). *Adv. Appl. Microbiol*, **13**, 139.
8. MATELES, R. I. and TANNENBAUM, S. R. (Eds.) (1968). *Single Cell Protein*, MIT Press, Cambridge, Mass.
9. TANNENBAUM, S. R. and WANG, D. I. C. (Eds.) (1975). *Single Cell Protein II*, MIT Press, Cambridge, Mass.
10. LIPINSKI, E. S. and LITCHFIELD, J. H. (1970). *CRC Crit. Rev. Food Technol.*, **1**(4), 581.
11. SNYDER, H. E. (1970). *Adv. Food Res.*, **18**, 18.
12. CHAMPAGNAT, A. VERNET, C. LAINE, B. and FILOSA, J. (1963). *Proc. 6th World Petroleum Cong.*, Frankfurt, **4**, 259.
13. DIMMLING, W. and SEIPENBUSCH, R. (1978). *Process Biochem.*, **13**(3), 9.
14. PHAFF, H. J., MILLER, M. W. and MRAK, E. M. (1966). *The Life of Yeasts*, Harvard University Press, Cambridge, Mass.
15. ALEXOPOULUS, C. J. (1962). *Introductory Mycology*, 2nd Edn., John Wiley & Sons, New York and Tokyo, p. 217.
16. BURROWS, S. (1971). In *The Yeasts*, Vol. 3, Rose, A. H. and HARRISON, J. S. (Eds.), Academic Press, London, p. 349.
17. HAMER, G. and HAMDAN, I. Y. (1979). *Chem. Soc. Rev.*, **8**(1), 143.
18. MATELES, R. I. (1979). In *Microbial Technology: Current State, Future Prospects*, Bull, A. T., Ellwood, D. C. and Ratledge, C. (Eds.), Cambridge University Press, Cambridge, Mass. p. 29.

19. LITCHFIELD, J. H. (1977). *Food Technol.*, **31**(5), 175.
20. SKOGMAN, H. (1976). In *Food from Waste*, Birch, C. G., Parker, K. J. and Worgan, J. T. (Eds.), Applied Science Publishers Ltd, London, p. 167.
21. LASKIN, A. I. (1977). *Biotechnol. Bioeng. Symp.*, No. 7, 91.
22. EVANS, G. H. (1968). In *Single Cell Protein*, Mateles, R. I. and Tannenbaum, S. R. (Eds.), MIT Press, Cambridge, Mass., p. 243.
23. WASLIEN, C. I. (1975). *CRC Crit. Rev. Food Sci. Nutr.*, **6**(1), 77.
24. SHACKLADY, C. A. (1970). In *Proteins as Human Food*, Lawrie, R. A. (Ed.), Butterworths, London, p. 317.
25. SHACKLADY, C. A. (1979). In *Single-Cell Protein—Safety for Animal and Human Feeding*, Garattini, S., Paglialunga, S. and Schrimshaw, N. S. (Eds.), Pergamon Press, London, p. 126.
26. VALFRE, F., BOSI, G., BELLEZZA, P., OLIVIERI, O. and MOCCA, S. (1979). In *Single-Cell Protein—Safety for Animal and Human Feeding*, Garalbini, S., Paglialunga, S. and Scrimshaw, N. S. (Eds.), Pergamon Press, London, p. 133.
27. PHAFF, H. J. (1977). In *Food Proteins—Improvement through Chemical and Enzymatic Modification*, Feeney, R. E. and Whitaker, J. R. (Eds.), Adv. Chem. Ser., ACS, Washington, D. C., p. 244.
28. REHACEK, J., BERAN, K. and BICIK, V. (1969). *Appl. Microbiol.*, **17**, 463.
29. MOGREN, H. LINDBLOM, M. and HEDENSKOG, G. (1974). *Biotechnol. Bioeng.*, **16**, 261.
30. *Cell Disruption*, Gaulin Processing Report (1974), ref. 2099, Manton, Gaulin, SA, P.O. Box 58, Hilversum, Holland.
31. SUCHER, R. W. and ROBBINS, E. A. (1975). US Patent 3,867,554.
32. EDOZIEN, J. C., UDO, U. U., YOUNG, V. R. and SCRIMSHAW, N. S. (1970). *Nature*, **228**, 180.
33. ACHOR, I. M. RICHARDSON, T. and DRAPER, N. R. (1981). *J. Agric. Food Chem.*, **29**, 27.
34. LINDBLOM, M. (1974). *Biotechnol Bioeng*, **16**, 1495.
35. GUZMAN-JUAREZ, M. to be published.
36. ROBBINS, E. A. (1976). US Patent 3,991,215.
37. HEDENSKOG, G. and EBBINGHAUS, L. (1972). *Biotechnol. Bioeng*, **14**, 447.
38. NEWELL, J. A. (1975). US Patent 3,867,555.
39. CASTRO, A. L. SINSKEY, A. J. and TANNENBAUM, S. R. (1971). *Appl. Microbiol.*, **22**, 422.
40. DALE, B. E. (1979). *Biotechnol. Bioeng.*, **21**, 1639.
41. TANNENBAUM, S. R. (1973). US Patent 3,720,585.
42. LINDBLOM, M. (1977). *Biotechnol. Bioeng.*, **19**, 199.
43. ITO, K., NAKAMURA, K., IZAKI, K. and TAKAHASHI, H. (1977). *Agric. Biol. Chem.*, **41**(2), 257.
44. LINDBLOM, M. and MOGREN, H. (1974). *Biotechnol. Bioeng*, **16**, 1123.
45. SCHULDT, E. H. (1979). US Patent 4,135,000.
46. SCRIMSHAW, N. S. and DILLON, J. C. (1979). In *Single-Cell Protein—Safety for Animal and Human Feedings*, Garattini, S., Paglialunga, S. and Scrimshaw, N. S. (Eds.), Pergamon Press, London, p. 171.
47. VANANUVAT, P. and KINSELLA, J. E. (1975). *J. Agric. Food Chem.*, **23**(4), 613.
48. VANANUVAT, P. and KINSELLA, J. E. (1975). *J. Agric. Food Chem.*, **23**(2), 216.

49. GIERHART, D. L. and POTTER, N. N. (1979). *Biotechnol. Bioeng.*, **21**, 1963.
50. KINSELLA, J. E. (1976). *Crit. Rev. Food Sci. Nutr.*, **7**, 219.
51. SHOEN, H. M. (1977). In *Food Proteins*, Whitaker, J. R. and Tannenbaum, S. R. (Eds.), Avi Publishing Co., Inc., Westport, Conn., p. 386.
52. JOHNSON, J. C. (1977). *Yeasts for Food and Other Purposes*, Noyes Data Corp., New Jersey.
53. COONEY, C. L., RHA, C. K., and TANNENBAUM, S. R. (1980). *Adv. Food Res.*, **26**, 1.
54. SCHNELL, P. G. (1975). US Patent 3,997,683.
55. CAVIT, A. (1973). US Patent 3,781,264.
56. KINSELLA, J. E. and SHETTY, K. J. (1978). In *Nutritional Improvements of Food and Feed Proteins*, Friedman, M. (Ed.), Plenum Press, New York, p. 797.
57. SCHACHTEL, A. P. (1981). *J. Food Sci.*, **46**, 377.
58. LINDBLOM, M. (1977). *Lebensm. Wiss. u.-Technol.*, **10**, 341.
59. SCHACTEL, A. P. (1981). *J. Food Sci.*, **46**, 1822.
60. LINDBLOM, M. (1974). *Lebensm. Wiss. u.-Technol.*, **7**, 295.
61. HUANG, F. and RHA, C. K. (1971). *J. Food Sci.*, **36**, 1131.
62. HUANG, F. and RHA, C. K. (1978). *J. Food Sci.*, **43**, 772.
63. HUANG, F. and RHA, C. K. (1978). *J. Food Sci.*, **43**, 780.
64. HAYAKAWA, I. and NOMURA, D. (1977). *Agric. Biol. Chem.*, **41**(1), 117.
65. RHA, K. (1978). *Food Technol.*, **32**(7), 77.
66. BEDNARSKI, W., POZNANSKI, S., JEDRYCHOWSKI, L. J., LEMAN, J. and SZCZEPANIAK, H. (1976). *J. Milk Food Technol.*, **39**(8), 521.
67. FUJIMAKI, M., ARAI, S. and YAMASHITA, M. (1977). In *Food Proteins, Improvement through Chemical and Enzymatic Modification*, Feeney, R. E. and Whitaker, J. R. (Eds.), ACS, Washington, D.C. p. 156.
68. VANANUVAT, P. and KINSELLA, J. E. (1978). *Biotechnol. Bioeng*, **20**, 1329.
69. BJORNASON, J. and CARPENTER, K. J. (1969). *Brit. J. Nutr.*, **23**, 859.
70. CREAMER, L. K., ROEPER, J. and LOHREY, E. H. (1971). *J. Dairy Sci. Technol.*, **6**(3), 107.
71. SEN, L. C., LEE, H. S., FEENEY, R. E. and WHITAKER, J. R. (1981). *J. Agric. Food Chem.*, **29**, 348.
72. EISELE, T. A., BREKKE, C. J. and MCCURDY, S. M. (1981). *J. Food Sci.*, **47**, 43.
73. ROBBINS, E. A. (1975). US Patent 3887431.

Chapter 8

THE ESTIMATION OF PROTEIN QUALITY

Ann F. Walker
Department of Food Science, University of Reading, UK

SUMMARY

Protein quality of a foodstuff is not easily definable, as not only does it depend on the amino acid pattern and digestibility of a protein, but also on the physiological status of the consumer. As it is not practical for large numbers of samples to be assayed using humans, rat bioassay, and in particular the protein efficiency ratio (PER), is used as a reference method for protein quality evaluation. However, rat assay is time-consuming, costly and subject to methodological and interpretative difficulties. Recent research has been applied to the development of indices of protein quality based on amino acid profiles and in vitro *digestibility. These measurements are relatively easy to make, and, in contrast to bioassay data, are additive, so that predictions for the protein quality of actual diets can be made.*

INTRODUCTION

Protein quality is the capacity of a protein to meet animal or human nutritional requirements for non-essential nitrogen and essential amino acids. It can be judged partly by the amino acid pattern and its relation to human amino acid requirements, and partly by protein digestibility. Methodologies for the determination of the quality of proteins, based on nitrogen balance studies or rat growth, were developed early in this century, although significantly useful advances in amino acid analysis were not made until the 1940s. After the Second World War the determination of protein quality of foods became a subject of considerable concern, as, with the assumption that protein deficiency was world-wide, there was considerable interest in the treatment of malnutrition with protein-rich diets. During this period much research effort was orientated towards finding new sources of protein and the development of techniques for protein quality evaluation.

Towards the end of the 1960s the relationship between energy and protein intake in man was re-examined[1] and human protein requirements were re-appraised.[2] The outcome has been that many nutritionists now support the view that it is not the shortage of protein per se which is responsible for undernutrition in the less-developed countries, but a shortage of dietary energy. The basis for this view, which has been summarised by Waterlow and Payne,[3] is that most staples in the diet, such as cereals, contain adequate protein to fulfil the protein requirements[2] of adults and even children, as long as enough food is eaten to provide energy requirements.

Some researchers now consider[4] that protein quality may have been overemphasised for its importance in human nutrition. To some extent the methodology for the determination of protein quality in humans and animals may have contributed to this, as these bioassays are generally carried out at inadequate levels of protein, to exaggerate the differences between proteins so that measurements can be taken. However, at the moment, the importance of protein quality to human nutrition remains a contentious issue. Scrimshaw[6] has argued that the recommended daily allowances (RDA) for protein are too low and do not allow for proper maintenance of nitrogen balance. Indeed, the RDA for protein in developed countries such as the United Kingdom or the USA are higher than those in the international tables, being based on protein which is actually consumed, rather than on physiological requirements. In a report of the Advisory Committee of the United Nations University World Hunger Program[7] the point is made that marginally fed or malnourished people are rarely free from disease, and little is known about increased requirements for protein and other nutrients in relation to energy requirements under these conditions. People likely to succumb to protein deficiency are those marginally nourished, 'at risk' groups such as infants. While most recent data suggest that a protein:energy ratio of greater than 5% is adequate for children gaining weight normally, discussion still continues with respect to protein requirements of children during 'catch up' growth.[8,9] Infant and pre-school children constitute the human group with the highest nutrient requirements. Not only are these nutrients needed to support optimal growth and mental development, but also to foster health and well-being, part of which is the maintenance of the body's defence against invasion by micro-organisms.

The importance of protein quality to farm animals is a far less controversial subject. In practice, the food intake of farm animals is carefully monitored, so that the animals are fed adequate, but not excess, protein, as this is the most costly item of the feed. There will continue to be a need within this industry for rapid methods for estimating protein quality.

More recently, interest in the development of rapid methods of protein quality evaluation has been coming from a different direction. Regulatory agencies in the USA, convinced of the importance of protein quality to human nutrition, have insisted that a statement of protein quality be made along with other items in the nutritional labelling of food products. The protein efficiency ratio (PER) method, based on the growth of young rats, has become the 'official' method in the USA for protein quality bioassay. However, as it is time-consuming and costly, a rapid method for protein quality evaluation with a high correlation to the PER bioassay method would be particularly welcome.[5]

BIOASSAY PROCEDURES

Methods of estimation of protein quality of a foodstuff may be broadly categorised into *in vivo* and *in vitro* methods. Clearly for the evaluation of the protein quality of foods, methods involving the use of humans would be ideal, but cost and time prohibit this. *In vivo* methods using the rat are well established,[10] although there is debate over which is the most suitable. In the USA, use of the PER of Osborne *et al.*[11] is favoured, whereas in Europe, preference is given to the net protein utilisation (NPU) assay of Miller and Bender.[12] *In vitro* methods have advantages over *in vivo* ones, as they are quicker and more precise. However, their relationship to *in vivo* methods must be clearly understood, and they should be applicable to a wide range of proteins in food and food products. Recently a number of rapid methods have been published, but so far none of these is sufficiently advanced to be used by the food industry.

In this chapter no attempt has been made to reference the literature completely, as so much published material now exists in the field of protein quality evaluation. Instead, more recent research is discussed, together with comments on problems of interpretation of data for human nutrition. References to reviews of the older literature are given whenever appropriate.

Bioassays Using Humans

Protein quality is difficult to define precisely as it is dependent not only on the amino acid composition and digestibility of the protein itself, but also on the composition and adequacy of the diet as a whole and on the physiological, nutritional and health status of the individual consuming it (Table 1[14]). Methods for the evaluation of protein quality in humans are discussed extensively by eminent authors in Ref. 13.

TABLE 1

SOME FACTORS WHICH CAN AFFECT PROTEIN UTILISATION AND HENCE CAN INFLUENCE PROTEIN QUALITY[14]

Diet:	Total protein, total food energy, amino acid composition (both deficiency and excess), digestibility, fibre, and other dietary constituents.
Consumer:	Age, sex, physiological status (growth, pregnancy, lactation), activity, infection, injury, and emotion.
External:	Food frequency, social, economic, hygiene, and sanitation.

Most bioassays for protein quality using humans are based on nitrogen balance studies, and data are most often used as a basis for standardisation of rat bioassay. As well as the amino acid composition, digestibility of a protein is an important factor in determining its protein quality, the digestibility of a protein being influenced by its structure or its association with other food components such as lipids. Digestibility of a protein may be measured as 'true' digestibility, in which case allowance is made for the faecal nitrogen loss on a protein-free diet (See Table 5), or as 'apparent' digestibility for which no such correction is made. Despite the difficulties of protein quality assay using humans, values for true digestibility obtained by different authors on the same foods show little variation (Table 2). However, there are considerable differences between true and apparent protein digestibility values as the latter measurement is particularly susceptible to differences in the protein content of the diet.

One of the factors which influences the number of studies which can be

TABLE 2
COMPARISONS OF APPARENT AND TRUE DIGESTIBILITY OF PLANT PROTEIN SOURCES AS ESTIMATED IN DIFFERENT STUDIES WITH ADULT HUMAN SUBJECTS[a]

Protein source	*Reference*	*Intake level (g protein kg^{-1} body weight day^{-1})*	*Apparent digestibility (%)*	*True digestibility (%)*
Peanut flour	42	0·4	76	91,96[b]
Peanut flour	98	0·46		99
Peanut butter	99	1·0	87[c]	95
Soy isolate	42	0·4	81	95,100[b]
Soy flour	100		70	92
Wheat gluten	42	0·4	82	96,100[b]
Wheat gluten	98	0·50		104
Wheat gluten	101	0·27		99[c]
		0·73		100[b]
Wheat gluten	102	0·1	21[c]	105[c]
		0·2	52	98
		0·4	80	102
		0·6	84	97
		1·0	88	96

[a] Adapted from Bodwell *et al.*[42]
[b] Values obtained by use of 9 and 12 mg faecal obligatory N losses kg^{-1} body weight day^{-1}, respectively.
[c] Calculated from data given by authors.

carried out using humans is the time necessary for nitrogen balance studies, 10–14 days for each feeding period. Therefore, several workers have attempted to develop shorter methods. Bressani *et al.*[15] have reported on a short-term, 2–3 days for each feeding period, assay. Clearly more information will be needed but initial results are encouraging as values obtained for the short-term assay were similar to those obtained using more conventional methods. Bodwell *et al.*[16] investigated a short-term, 4–6 day nitrogen balance in humans and found that urinary nitrogen is particularly useful as a direct predictor of protein quality.

A basic criticism of nitrogen balance studies is that protein quality tends to be overestimated, as protein is fed at inadequate levels. The efficiency of utilisation of proteins of the highest quality when consumed by adults at the requirement level is only 65–70%.[2] Young *et al.*[17] have further criticised the nitrogen balance method. They comment that adults retain nitrogen beyond that expected when receiving high nitrogen intakes, which cannot be explained on the basis of methodological error.

They suggested that nitrogen gas may be produced in the body from the activity of intestinal microflora on substances such as nitrite.

Young *et al.*[17] used ^{15}N- and ^{13}C-labelled amino acids for the study of nitrogen turnover in humans. Using this technique they found that the amount of protein synthesis associated with net body protein gain for babies was only a small proportion of the total daily protein synthesis (Table 3), and that this varied considerably from nitrogen turnover in the rat. Therefore, at least in this respect, the rat may not be an adequate model for the study of protein metabolism in humans.

TABLE 3

BODY PROTEIN SYNTHESIS FOR THE RAT AND THE HUMAN COMPARED WITH ESTIMATED PROTEIN ALLOWANCES[a]

Subject	*Protein synthesis* ($g\ kg^{-1}\ day^{-1}$)	*Protein allowance* ($g\ kg^{-1}\ day^{-1}$)	*Ratio of protein synthesis to protein allowance*
Growing rat[103]	40·0	10·0	4·0
Human newborn[104]	17·5	3·2	5·5
Child[104]	7·0	1·4	5·0
Young adult[104]	3·5	0·6	5·8
Elderly[104]	3·0	0·5	6·0

[a] Compiled by Pellett.[14]

Bioassays Using Rats

There are many reviews of the use of laboratory animals for the determination of protein quality, including those of Porter and Rolls,[18] Friedman,[19] Bender[20] and Pellett.[14]

When protein quality data are to be interpreted in terms of human nutrition, the rat is normally chosen for *in vivo* assay. However, the use of the rat may lead to an underestimate of the quality of a protein for man. In general, man requires less protein for growth and more for maintenance than does the rat.[20] Also, the rat's requirements for the sulphur-containing amino acids and lysine are higher than man's[21] (Table 4).[21] Differences in requirements between the two species may be due to the faster growth rate of rats, and also the greater amount of hair of the rat and therefore a higher requirement for the sulphur amino acids.

Despite the inadequacies of the rat model, rat bioassay continues to be

TABLE 4
COMPARISON OF AMINO ACID REQUIREMENT PATTERNS

	Amino acids as % of protein		
	Rat[105]	*Human*	
		NAS/NRC (1974)[63]	*FAO (1973)*[2]
Histidine	2·5	1·7	1·4[a]
Isoleucine	4·6	4·2	4·0
Leucine	6·2	7·0	7·0
Lysine	7·5	5·1	5·5
Methionine and cystine	5·0	2·6	3·5
Phenylalanine and tyrosine	6·7	7·3	6·0
Threonine	4·2	3·5	4·0
Tryptophan	1·25	1·1	1·0
Valine	5·0	4·8	5·0
Arginine	5·0	—	—

[a]Infants only.

the principal laboratory technique for protein quality assay. Table 5 summarises the main assay methods, while Table 6 indicates some of the inherent errors of rat bioassay. As far as the PER is concerned, the methodology is standardised, at least in the USA.[22] It is carried out using weanling rats, fed *ad lib.* with 10% protein in the diet for a period of 28 days. Values are adjusted so that weight gained per unit of protein ingested by animals fed the test diets is ranked against the same response criterion for a control group of animals fed a diet containing casein. Values of the test protein are adjusted to that which would be expected if the PER of 2·5 were obtained for casein. The comparison with casein is made to decrease the variability in PER measurements between laboratories, although Steinke[23] has criticised the use of casein for this purpose, as it does not have a composition corresponding to the amino acid requirements of humans. The popularity of the PER method for the determination of protein quality in the USA is not only a result of its 'official' status, but also because it is relatively easy to carry out compared with other rat assay methods, requiring, in the main, only weights of diets and rats to be recorded.

The net protein ratio (NPR[24]) is similar to the PER, but also involves a measurement of weight loss of a group of rats fed a protein-free diet. In this way, not only is protein quality for growth determined but also that required for maintenance (Table 5). The biological value (BV) of Mitchell[25] is determined by nitrogen balance studies using rats. Apparent or true digestibility can also be determined by this method,

TABLE 5
RAT BIOASSAYS FOR MEASURING PROTEIN QUALITY[a]

Name of assay	*Computation*	*Reference*
Protein efficiency ratio	$\text{PER} = \dfrac{\text{weight gain}}{\text{protein consumed}}$	22
Net protein ratio	$\text{NPR} = \dfrac{\text{weight gain of test group} + \text{weight loss of group fed non-protein diet}}{\text{protein consumed}}$	24
Net protein utilisation	$\text{NPU} = \dfrac{\text{body N of test group} - \text{body N of group fed non-protein diet}}{\text{N consumed}}$	12
Biological value	$\text{BV} = \text{absorbed N from food} - (\text{urinary N} - \text{endogenous urinary N})$	25
True digestibility	$\text{TD} = \dfrac{\text{N intake} - (\text{faecal N} - \text{metabolic faecal N})}{\text{N intake}}$	25
Relative nutritive value (slope ratio)	$\text{RNV} = \dfrac{\text{slope of response curve of animals fed test protein}}{\text{slope of response curve of animals fed standard protein}}$ (all response lines have common intercepts)	106
Relative protein values (slope ratio)	$\text{RPV} = \dfrac{\text{slope of response curve of animals fed test protein}}{\text{slope of response curve of animals fed standard protein}}$ (response curves do not have common intercept)	107

[a]Adapted from Hopkins and Steinke.[29]

which is an advantage over the PER method, which does not include a value for digestibility. According to the method of Mitchell,[25] the NPU can be calculated from the product of BV and true digestibility.

The NPU method of Miller and Bender[12] is still the preferred method in Europe. This method involves analyses of carcass nitrogen of groups of rats on a test diet and on a nitrogen-free diet. A value for protein digestibility can also be obtained at the same time. Both the NPR and

TABLE 6
SOME INHERENT ERRORS OF RAT BIOASSAY FOR EVALUATION OF PROTEIN QUALITY[a]

1.	Calculation of protein as nitrogen × 6·25
2.	Use of experimental animal other than man
3.	Use of single index for multiplicity of purposes
4.	Rat's needs for growth are greater than for maintenance; the reverse applies to man.
5.	Results vary with level of dietary protein, other nutrients and previous diet
6.	PER reduced by low food intake
7.	At low levels of intake, some amino acids are conserved
8.	All methods measure only limiting amino acid
9.	Quality differs with type of assay
10.	Accuracy unknown, since there is no criterion

[a] Adapted from Bender.[20]

the NPU[12] methods are two-dose assays, the test protein being fed at zero and one other level, whereas the PER is a single-dose assay. The slope ratio assay, which is a multiple-dose procedure, includes a non-protein diet and three or more dietary levels of protein (Table 5). Only levels falling on the linear portion of the response curve are used for computing the slope. The slope of the test protein is expressed as a percentage of the slope of lactalbumin, which is used as a reference protein. The main problem in this procedure is that it is very expensive to carry out and requires large amounts of dietary materials. McLaughlan[10] showed that it yielded erroneous results for some proteins, especially those low in threonine. He also concluded from his studies that if only the linear portion of the response curve were used, then data at the mid-point of the curve would be wasted. The extra expense involved in the multiple-dose assays was not justified as the information obtained showed little improvement over the NPR or NPU[12] methods for protein quality evaluation.

All methods of rat bioassay for the determination of protein quality yield data subject to wide variation. Therefore, recommended procedures must be followed closely and a sufficiently large group of animals used to provide statistically significant results. Because of its widespread use, the PER method has been the most thoroughly investigated of the rat bioassays for protein quality. The method has been severely criticised.[14,27] Hurt *et al.*[28] showed that a number of factors influenced the PER values even when the assay was carried out by the standard method recommended by AOAC.[22] Apart from the level of protein in the diet, these included the amount of water, dietary fibre and mineral mix and the proportions of saturated to unsaturated fatty acids in the diet, as well as the strain, sex and age of the rats. However, the major criticisms of the PER method are that the results are positively correlated to food intake,[24,26] and results are not directly proportional to the quality of the protein.[10]

The NPU and NPR bioassays are essentially the same assay, and it is therefore not unexpected that agreement between the two methods is good.[29] Of the two methods, the NPR is the easier to carry out. The NPU bioassay is not without its critics, as all animals, including man, have the ability in some degree to re-use amino acids for the synthesis of protein. Anomalies related to this are well known. Bender[30] showed that a protein containing no lysine (Score 0) has an NPU of about 40. However, an absence of valine or total sulphur amino acids gives the expected zero response for NPU, while the other essential amino acids behave in an intermediate manner.

More recently there has been relatively little progress in the development of new techniques for rat bioassay. Phillips[32] compared multiple-dose bioassays calculated using a linear relationship with a non-linear mathematical model. Proteins such as peanut protein and casein, for which protein quality was difficult to integrate using a linear model, were not more easily interpreted using the non-linear model. Farina *et al.*[31] have published a rapid method in which relative weight changes of protein-depleted rats are measured after feeding them with protein diets. The weight lost after 4 h fasting subsequent to a 24 h feeding period, expressed as a percentage of the previously gained weight (WL%), showed close agreement to several indicators of protein quality. These authors concluded that with careful control of the protein content of the diet, WL% would be a good indirect indicator of protein utilisation. New bioassay methods for protein quality evaluation require considerable study for validity and repeatability before they can replace established methods. As this type of work is costly and tedious it is unlikely that enough resources will be allocated to it to enable a new method to become sufficiently established to be accepted for routine analysis.

Of obvious interest in the interpretation of protein quality data published in the literature is the manner in which the various rat bioassays compare. Bender[33] reported a high correlation between the PER and NPU[12] methods for a large range of foodstuffs. Henry and Toothill[34] found a high correlation between the method of Mitchell[25] and the NPU method (see Table 7). They concluded that although the

TABLE 7
RELATIVE PER, NPU AND NPR VALUES FOR SOME PROTEIN SOURCES (CASEIN = 100)[34]

Protein source[a,b]	*PER*	*NPU*	*NPR*
Casein	100	100	100
Fishmeal (16)	109	100	103
Dried skim milk (13)	100	105	109
Soya (5)	83	83	91
Cottonseed meal (49)	76	94	94
Yeast	63	78	79
Sunflower seed meal (2)	63	85	82
Groundnut (12)	48	66	69
Wheat gluten	19	64	71

[a] Values expressed as a percentage of value obtained for casein.
[b] Figures in parentheses refer to authors' sample numbers.

Mitchell method gave more fundamental information and was more precise, it was also more time-consuming. Wolzak *et al.*[35] found that the correlation between PER and NPR values for groups of similar foods was high. However, for a wide range of foods this may not be the case, as the NPR tends to overestimate the protein quality of lysine-deficient proteins, while the PER strongly underestimates them.[37] As far as the multiple-dose assays are concerned, McLaughlan[10] showed that, except for high quality proteins, the PER gave lower relative values than those derived by this method, but agreement between NPR and the multiple-dose assay was excellent except for casein, white flour and wheat gluten.

Although any of the bioassays can be used successfully to rank proteins in order of nutritive value and to detect deterioration of protein quality during processing, none of them can be used to predict the supplementary or complementary value of proteins in mixed diets, because protein quality data obtained using these methods are not additive. Also, none of the methods permits an accurate prediction of the amount of protein required to meet animal or human requirements. Despite this, Bender[20] and Young and Scrimshaw[38] conclude that there is, in general, approximate agreement between rat and human assay, which would justify continued use. Hopkins and Steinke,[29] however, cautioned that for the future it is important that the measurement of protein quality be carried out using humans, so that procedures used by the food industry and by regulatory agencies reflect the requirements of the human and not that of the test animals. Only limited data are available from studies in which the same food protein has been evaluated in both rats and humans.[36] More comparative data of this nature are therefore needed, as, at present, data derived from rat assay methods cannot be used to predict protein quality accurately for humans.

IN VITRO PROCEDURES

Although rat assay methods such as the PER may be established practice, there are considerable advantages in using *in vitro* methods for protein quality evaluation. In general, *in vitro* methods are less time-consuming and require less sample, so that more samples can be assayed at one time. In this way processing can be monitored conveniently at various stages during food production. However, it is necessary that *in vitro* methods are adequately correlated with existing rat assay methods, or better with protein quality assays carried out with humans.[29] Although *in vitro* methods have the advantage that no special premises or licence is needed to undertake them, less information about the food material will be obtained. In particular, *in vivo* procedures are valuable for the detection of unexpected toxic components of foods. Data from *in vitro* methods may be less variable than *in vivo* data, but caution is required in the interpretation of this as precision may not necessarily be of biological significance.

In recent years there has been renewed interest in *in vitro* methods of protein quality determination, particularly as these methods are convenient for use by food technologists in industry. An ideal chemical method would be rapid, precise, related well to established or accepted *in vivo* methods, require only a small amount of material and be easy to carry out by relatively unskilled technicians using inexpensive chemicals and equipment. At present no method complies with all these criteria. Thus, for example, amino acid analysis requires skilled operation and expensive chemicals, and is time-consuming as well as incurring high equipment maintenance costs.

In vitro Digestibility

All *in vitro* methods for the determination of digestibility of human foods are based on the use of enzymes, usually those of the digestive system, and many simulate physiological conditions. On the whole there are few satisfactory *in vivo* data published alongside these methods.

Among published *in vitro* assay methods for protein digestibility is that of Saunders *et al.*[39] who studied a number of leaf protein concentrates processed under different conditions. These authors obtained a high correlation for apparent digestibility in rats with *in vitro* digestibility obtained by a stage pepsin/trypsin incubation technique. However, total incubation was 40 h and the method is by no means simple. A promising, rapid technique is that of Hsu *et al.*,[40] in which a number of isolated, commercially available pancreatic enzymes were added to a protein suspension in water at pH 8·0 and the change in pH after a period of time (usually 10 min) was measured. In their hands this method showed good agreement with rat nitrogen digestibility for 23 proteins, mainly of plant origin.

A number of workers have published work continuing that of Hsu *et al.*[40] Thus, Marshall *et al.*[41] investigated the possibility of using human, porcine and rat pancreatin in place of the expensive isolated enzyme of Hsu *et al.*[40] They assayed 33 foods and food ingredients and calculated regression equations, for each pancreatin, of the pH drop against apparent protein digestibility obtained in the rat. Resulting correlations were low, but when the regression equations were recalculated specifically for food protein categories, such as plant, combination (plant and animal) or animal proteins, the correlation coefficients were improved.

The different behaviour of animal and plant proteins when assayed by this type of *in vitro* digestibility method was also noted by Bodwell *et al.*,[42] who correlated *in vitro* digestibility obtained by three enzymic procedures, using (a) a three-enzyme method,[40] (b) a modified, three-enzyme method,[43] and (c) a four-enzyme method,[43] with apparent and true protein digestibility determined in humans and rats. Although not statistically significant, correlations were much higher when the data for the animal and plant proteins were analysed separately (Fig. 1). On the whole, the enzymic methods tended to underestimate the true digestibility for humans of both plant and animal proteins. For plant protein, correlations between values for the three-enzyme and four-enzyme methods and apparent rat digestibility were similar to those published by Hsu *et al.*[40] and Satterlee *et al.*[43]

There may be problems in the interpretation of *in vitro* digestibility data when heat-processed samples are analysed. Pronczuk *et al.*[44] found, using casein autoclaved with glucose for various time intervals, that *in vitro* digestibility (with papain and pancreatin) was more closely associated with NPU than with digestibility determined with rats. The longer the time of autoclaving, the greater the resistance to *in vitro* digestion, but this was not reflected in a decrease in the *in vivo* digestibility even after 135 min of autoclaving. They found that urine from rats

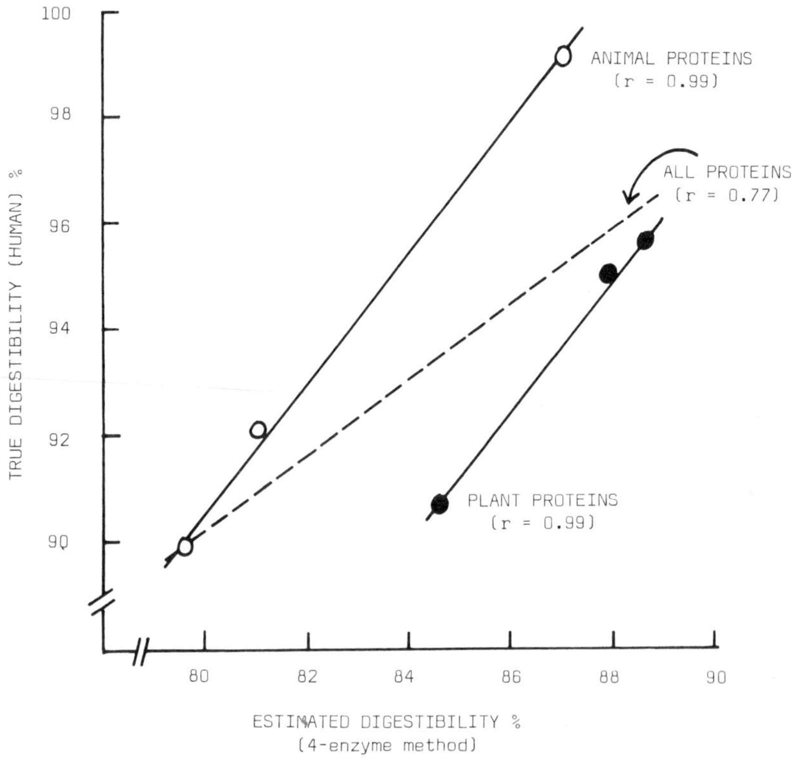

FIG. 1. Relationship between digestibility as estimated by the four-enzyme *in vitro* digestion method and true digestibility. Source: Bodwell *et al.*[42]

fed heat-damaged casein contained protein and high molecular weight peptides, which were not found in the urine of rats fed non-heat-damaged casein. Ford and Shorrock[45] obtained similar results for other types of heated foods. These reports indicate that large heat-damaged peptides are absorbed by the rat, but, being unsuitable for utilisation in the body, are excreted in the urine.

For reasons of lack of comparable animal data or procedural impracticality, *in vitro* procedures to estimate protein digestibility have not found wide acceptance as alternatives to rat bioassay. More research is therefore needed in this area, particularly as *in vitro* digestibility data in combination with amino acid composition data are now needed to calculate combined indices, as is shown below.

AMINO ACID ANALYSIS AND SCORING PATTERS

Amino acid analysis, based on the method of Spackman *et al.*,[46] using ion-exchange chromatography (IEC) of acid hydrolysates of proteins is one of the most important chemical techniques for predicting protein quality. The shortcomings of the technique are that there is complete destruction of tryptophan, and partial destruction of some other amino-acids during acid hydrolysis, and discrepancy may exist between the content of amino acids in a protein obtained by this method and their nutritional availability to man. Problems of analysis are greatest with some of the essential amino acids, particularly threonine, lysine, tryptophan and the sulphur-containing amino acids, methionine and cystine, i.e. those amino acids which are likely to be limiting in the human diet. Analytical problems of amino acid analysis have been reviewed by a number of authors.[47-49] Some of these problems are summarised in Table 8. While there are extensive data in the literature on amino acid composition of individual foods and diets, the reliability of this data is variable.

During acid hydrolysis, amino acids with polar side chains, such as threonine and serine, are progressively destroyed, and some amino acids

TABLE 8

RELATIONSHIP OF AMINO ACID VALUES OBTAINED BY ION-EXCHANGE CHROMATOGRAPHY OF ACID HYDROLYSATES OF PROTEINS TO BIOLOGICAL VALUE

Values[a]	*Amino acid*	*Cause*
High	Lys, All	No account of bioavailability, including digestibility[14]
Low	Try	Destroyed by acid hydrolysis[49]
	Met, Cys	Partially destroyed by acid hydrolysis[49]
	Thr, Ser	Increasingly destroyed with increase in hydrolysis time[49]
	Val, Ile	Increasingly released with increase in hydrolysis time[49]
Variable	All	Reproducibility of individual amino acid depends on protein source[b] [51]
	All	Equipment choice
	All	Operator error
	All	Units used for expressing data[47,53]

[a] High, low or variable in relation to biological value.
[b] See Table 9.

(e.g. valine and isoleucine) are released from a protein only slowly.[49] Therefore, the time normally employed (22–24 h) for acid hydrolysis of proteins prior to amino acid analysis may not result in amino acid patterns which are fully representative of the protein. Downs and Pigman[50] attempted to correct for loss or gain of amino acids during acid hydrolysis by hydrolysing them for different times and extrapolating data to zero time. However, this work was carried out with free amino acids, and Davies and Thomas[48] have shown that correction for incomplete hydrolysis and/or destruction of amino acids cannot be made for proteins in food materials on this basis, as results for individual proteins are so variable.

The results of several collaborative trials have shown that differences in the reproducibility of analysis of specific amino acids of proteins varies considerably depending on the source of the protein.[51,52] Table 9 shows data obtained in the author's laboratory for essential amino acids of a pure protein, β-lactoglobulin A, and a protein concentrate extracted from leaves. Data for all amino acids analysed gave approximately the same mean coefficient of variation (CV), i.e. 4·3 and 4·2% respectively, but the CV of the individual amino acids differed considerably (Table 9). The

TABLE 9

REPRODUCIBILITY OF ESSENTIAL AMINO ACID VALUES FROM ACID HYDROLYSATES OF TWO PROTEINS (AMINO ACID EXPRESSED AS RESIDUES % OF TOTAL RESIDUES RECOVERED)

	β-Lactoglobulin A		*Leaf protein concentrate*	
Amino-acid[10]	*Mean*[a]	*CV %*[b]	*Mean*[a]	*CV %*
Lys	10·2	3·7	6·5	0·8
Thr	4·5	1·1	4·9	1·6
Ser	3·3	2·1	4·2	1·9
Cys	1·9	4·7	0·6	18·2
Val	5·1	2·4	6·0	2·2
Met	3·3	4·5	2·5	6·3
Ile	5·6	3·2	5·1	2·0
Leu	14·3	1·4	9·5	1·5
Tyr	3·1	6·4	4·5	3·8
Phe	3·0	1·3	6·4	1·3
Overall[c]		4·3		4·2

[a] Mean of 10 separate analyses.
[b] CV = coefficient of variation.
[c] Mean CV % for all amino acids analysed

reproducibility of amino acid analysis depends on a number of factors, including the type of ion-exchange chromatography equipment, and the skill of the operator. Particularly when large discrepancies occur in nitrogen recovery, normalisation of amino acid data is necessary to achieve the reproducibility necessary for comparisons to be made between different food samples. The usual method is to calculate values to a total of 100 g of recovered amino acids.[53] However, Cavins *et al.*[47] preferred to reject data from hydrolysates showing nitrogen recoveries outside the range of 90–100 %.

A separate hydrolysis of the protein after performic acid oxidation is often recommended to determine the sulphur-containing amino acids.[54] Thus for a complete analysis of all amino acids other than tryptophan, two separate hydrolyses would be necessary: the oxidised hydrolysate itself cannot be used for the determination of all amino acids, as phenylalanine, histidine and tyrosine would be greatly underestimated.[55] However, for the determination of methionine, the performic acid oxidation may offer no advantage. The addition of the methionine sulphoxides peak to that for methionine to give a value for total methionine greatly improved the precision of the determination of methionine from unoxidised acid hydrolysates of leaf protein concentrates, giving a CV of 6·3 % compared with 75·0 %.[56] For leaf protein concentrate, and β-lactoglobulin A, calculation of methionine in this way gave as good an estimate of methionine as that obtained by the performic acid oxidation method (Table 10). However, cystine was undervalued with respect to the performic acid oxidation method by approximately 40 % for the two materials. With the increasing interest in the nutritional properties of the oxidation products of methionine, the performic acid method may be less useful, as it does not distinguish between the different forms of methionine in a foodstuff. In particular, methionine sulphone, which has no biological activity, would be included in an overall value for methionine by this method.

Various reagents have been investigated with a view to devising a hydrolytic method which will give reliable information on all amino acids, including tryptophan. One of the most promising is that of Liu and Chang[57] using *p*-toluene sulphonic acid. These authors obtained reliable values for the tryptophan content of pure protein, but indicated that carbohydrate would interfere. In addition, hydrolysates prepared in this manner underestimated the sulphur-containing amino acids. New methods of hydrolysis such as this are generally developed using pure proteins, so the effects of other food components such as starch are not

TABLE 10

SULPHUR-CONTAINING AMINO ACID CONTENTS (g/16g N) OF TWO PROTEINS DETERMINED BY TWO METHODS[56]

	Methionine		*Cystine*	
	Mean[a]	*CV %*	*Mean*[a]	*CV %*
Leaf protein concentrate				
Acid hydrolysis[b] +IEC (no prior oxidation)	2·46	6·5	0·64	18·75
Performic acid oxidation+acid hydrolysis[c]+IEC	2·45	7·4	1·13	6·2
β-Lactoglobulin A				
Acid hydrolysis[b]+IEC (no prior oxidation)	3·51	5·4	2·14	5·1
Performic acid oxidation+acid hydrolysis[c]+IEC	3·53[d]	ND	3·38[d]	ND
From known amino acid sequence[e]	3·3		3·4	

[a] Mean value of 10 separate analyses.
[b] 6 M HCl, 110°C, sealed tube, 24 h.
[c] 6 M HCl, 110°C, open flask, 24 h, adaptation of method of Moore.[108]
[d] Triplicate analysis.[109]
[e] Calculated from data of Frank and Braunitzer.[110]
ND = not determined.

known. It has taken considerable time and effort, since automated procedures for IEC of amino acids were first introduced by Spackman *et al.*,[46] to ascertain the reproducibility of the now standard laboratory procedure for the preparation of proteins for amino acid analysis (i.e. hydrolysis with 6 M HCl), so it is unlikely that a new hydrolytic reagent would be readily accepted for general use. Even so, despite the considerable knowledge we have on the hydrolytic effects of 6 M HCl on proteins, the interpretation of data must be carried out cautiously. Paddock *et al.*[58] found that hydrolysis of protein-free mixtures of nucleotides and nucleic acids with 6 M HCl led to the formation of a variety of artefacts whose peaks separated by IEC, the predominant one eluting at the same time as glycine.

Other methods for obtaining amino acid compositional data for a protein are based on gas–liquid chromatographic methods. However, it is necessary to form volatile derivatives of the amino acids after acid hydrolysis, and reproducibility is no better than by IEC. In addition, using this method certain amino acids such as histidine, arginine and tyrosine are not easily resolved.[59]

Harper[62] has suggested that it is only necessary to analyse proteins for lysine, total sulphur-containing amino acids, tryptophan and possibly threonine, for, if requirements for these are met, it is unlikely that the requirements for the other amino acids will not be met. However, it is just these amino acids which require most analytical time and expertise. Indeed, tryptophan values are usually unreported in the literature, as a separate alkaline hydrolysis of the protein is necessary for their determination.

Combined Indices of Amino Acid Composition and *in vitro* Digestibility

Amino acid scores are usually directly correlated with bioassay values, no correction being made for protein digestibility. For the majority of foods and diets this may not be important, as nitrogen digestibility is high. For amino acid data to be useful for the interpretation of the protein quality of poorer proteins or processed foods, it would be necessary to include an estimate of protein digestibility. In recent years, rapid *in vitro* methods for the determination of protein digestibility have been used for this purpose. In the C-PER (computed PER) method of Hsu *et al.*,[64] which includes a modification of multi-enzyme *in vitro* digestibility method of Hsu *et al.*,[40] the PER can be estimated using a computer program based on correlations from experimental data. The calculation of the C-PER involves expressing the essential amino acids for the test protein and reference casein as a percentage of the provisional amino acid scoring pattern.[2] Each essential amino acid is adjusted for *in vitro* protein digestibility, and the essential amino acid score of the test protein is expressed as a ratio of the essential amino acid score of the casein reference. A correcting quadratic equation is then used to compute the C-PER from an established model derived from experimental data. C-PER values were found to relate well to PER values for most proteins (Table 13), although the method tended to overestimate the PER of low quality proteins and to underestimate high quality ones. Wolzak *et al.*[35] determined the PER and C-PER for a number of foods and showed that a high correlation existed for certain groups of foods such as cereal grains and oilseeds. The correlation coefficient rose when certain foods such as cornflakes, cornmeal and white wheat flour were excluded. For leguminous seeds alone, the correlation coefficient between PER and C-PER was not significant. Thus protein quality of leguminous seeds and processed samples was overestimated while that of mixtures supplemented with animal protein was underestimated. The data suggest that for calculation of the C-PER,

TABLE 13

COMPARISON OF C-PER AND RAT PER OF VARIOUS FOODS

Sample	*Rat PER*	*C-PER*
Cornmeal	0·7	1·1
Whole corn	1·4	1·4
White wheat flour	0·7	0·8
Durum wheat flour	0·9	0·7
Bran flour	1·2	1·3
Non-fat dry milk	2·7	2·3
Whey	2·9	2·7
Soy isolate	1·3	1·3
Soy flour	1·6	1·3
Yeast protein concentrate	1·8	2·0
Whole egg	3·2	2·7
Egg white	2·5	2·3
Lactalbumin	2·4	2·4
Turkey meat	2·6	2·7

it would be preferable to divide foods into groups prior to computer analysis.

The DC-PER (discriminant computed PER) was devised by Jewell *et al.*[65] with the object of finding a method to compute a chemical PER without prior knowledge of the protein source. The DC-PER model was developed from a broad range of plant and animal proteins having corrected rat PERs ranging from 0·30 to 3·22. From the amino acid profile of each protein, digestibility was estimated and this estimate was then used with the amino acid composition data to calculate the DC-PER. From these data separate regression equations were developed for four groups of proteins. Group 1 consisted primarily of low quality proteins, Group 4 of the high quality proteins with Groups 2 and 3 in the middle of the range. Having established the model for the determination of the DC-PER, the amino acid composition data for any new protein can be fed into a computer and is automatically assigned to one of the four food groups for calculation. Further research is now needed to ascertain how well a wide range of food materials fit the model. So far Jewell *et al.*[65] have found that the method overestimates the protein quality of products with respect to rat assay, but less so than the C-PER of Hsu *et al.*[64] Anomalous data for three food materials was obtained, even in setting up the DC-PER model: therefore it is likely that other proteins will also be found to give anomalous data when using the model.

SPECIFIC CHEMICAL PROCEDURES FOR PROTEIN QUALITY ASSAY

Assay of the availability of amino acids by chemical techniques has been discussed by a number of authors such as Carpenter,[66] Porter and Rolls,[18] Friedman,[19] Mauron[67] and Evans and Witty.[68]

Lysine, as well as being easier to measure in its available, reactive, form than other amino acids, is reported to be the first essential amino acid to be rendered unavailable during processing,[66] and therefore its availability can be used to monitor processing damage. The best established method is the fluorodinitrobenzene (FDNB) method of Carpenter.[69] This method has been correlated with chick assay,[70] although it does not correlate so well with rat assay. Boctor and Harper[71] compared the FDNB-reactive lysine value with the rat assay and showed that available lysine values for heat-treated foods varied greatly with the method of assay used. However, the number of samples that they investigated was limited, owing to the laborious nature of both the FDNB and rat assay methods. Boctor and Harper postulated that FDNB-reactive lysine values may underestimate the extent of lysine availability. If this is so, it may account for the reports of FDBN-reactive lysine being found in rat faeces.[72] Despite the drawbacks, the FDNB method remains a good indicator of processing damage, and there is information based on this method from many laboratories for a wide range of foods.

Even with established methods such as Carpenter's method, subsequent modification may lead to interpretative difficulties. For example, the FDNB method originally involved the direct estimation of the FDNB-reactive lysine.[69] Roach *et al.*[73] modified the method to determine unreacted lysine by IEC. This value was then subtracted from a total lysine value also obtained by IEC.[73] These authors reported differences in the results obtained by the two methods. The presence of large quantities of carbohydrates in a food may also interfere with the reaction of FDNB with lysine, leading to an underestimate of lysine availability of a protein.[74] Another difficulty is that the Carpenter[69] method does not always give satisfactory data for samples which have been processed under mild conditions.[75]

More recently, dye-binding methods have been modified to estimate available lysine in foodstuffs. The dye-binding method, using C.I. Acid Orange 12, involves the blocking of the ε-amino group of lysine, and various reagents have been used for this in the past, e.g. ethyl chloroformate[76] and trinitrobenzene sulphonic acid.[77] A modification of the propionylation method of Hurrell *et al.*[78] using standard laboratory equipment was useful for the detection of heat damage of leaf protein concentrates.[79, 80] Dye-binding methods have the disadvantage of requiring rigid standardisation of procedure and establishment of suitable conditions for each class of food material.

There are no well-established methods for the determination of the availability of other essential amino acids. It is known that there is a loss of availability of methionine during processing, leading to the formation of methionine sulphoxides and methionine sulphone. Methionine sulphoxides have only part of the activity of methionine for rats, whereas the other oxidation product, methionine sulphone, has none.[81] One tentative approach to the determination of methionine and its oxidation products has been the use of gas–liquid chromatography.[82]

MICROBIOLOGICAL METHODS

Microbiological assays for protein quality have several advantages over animal assays. Although not rapid by chemical standards, they are by those of rat bioassay, requiring also little space, cost and equipment. There are two main types of microbiological methods, those for the determination of protein quality, which give a relative nutritive value (RNV) compared with a standard such as casein, and those for the determination of the availability of certain essential amino acids.[83] The RNV includes an estimate of both digestibility and amino acid composition and can therefore be compared with rat bioassay. Organisms which have been used for both types of determination are bacteria and the protozoan *Tetrahymena pyriformis* W. *T. pyriformis* was considered suitable as a test organism as it requires the same amino acids as man and ingests the particulate matter of which food materials are comprised.

Relative Nutritive Value

Poor reproducibility of date due to erratic growth of the test organisms are the main problems with this method. In earlier work, Fernell and Rosen[84] found that RNVs were in general agreement with chemical scores and in even better agreement with rat PERs for a range of proteins. In an attempt to improve the agreement between the available lysine determination of foodstuffs using *T. pyriformis*, and rat bioassay,

Shorrock and Ford[85] introduced an enzymic predigestion stage during the preparation of the test material. Others subsequently introduced this step into the methodology for RNV. Landers[86] and Dryden *et al.*[87] used pepsin and Shepherd *et al.*[88] used pronase for predigestion. Satisfactory reproducibility of RNV has now been reported both by workers who have used no predigestion[89,90] and by those who have used enzyme predigestion.[86,91]

Many problems exist in the use of the *T. pyriformis* RNV assay. Smith and Pena[89] rejected poor growth of *T. pyriformis* on leaf protein concentrates (LPCs) which had been oven-heated. They also found that all LPCs assayed by this method showed a high response not only to methionine but also to lysine, although lysine has been shown by IEC not to be a limiting amino acid in LPC. They suggested that autoclaving the protein, which was necessary for the preparation of aseptic media before inoculation of the organism, may have rendered lysine less available to enzymic digestion, possibly by promoting the Maillard reaction. They also suggested that food components other than protein may interfere with the assay. Indeed, Dryden and Satterlee[92] showed that chlorogenic acid bound to casein gave a much lower RNV with *T. pyriformis* than the casein control.

Despite the reported close correlation between RNV determined by *T. pyriformis* with rat bioassay, all authors have mentioned the shortcomings of the method. So far no attempt has been made to compare the various procedures which have been published in the literature. These procedures differ considerably with respect to sample preparation, in particular the use of enzymic predigestion, and basal media. The method is highly empirical and therefore differences of this nature are likely to influence results greatly. On the whole, *T. pyriformis* assays for protein quality are quite involved, requiring skilled technical ability. The assay is not yet at the stage where it could be considered for routine use.

More recently, Sutton[93] shortened the *Tetrahymena* assay to 66 h by the use of a closely related organism, *T. thermophila* WH_{14}, and determined growth of the organism using a Coulter counter. Hsu *et al.*[64] incorporated this method into a T-PER (*Tetrahymena* – PER) assay for protein quality. Greater variation in results was obtained using *Tetrahymena* than from the use of indices based on amino acid composition (the C-PER), due to erratic growth of the organisms. Wang *et al.*,[94] however, reported that for 11 commercially prepared food products and eight heat-treated complete diet mixtures, results with *T. thermophila* correlated better with rat assay than with *T. pyriformis* W. Since the growth of *T. thermophila* is greatly affected by various food additives and spices, caution should be exercised if this method is to be used for the measurement of protein quality of foods for which the composition is not known. Although *T. thermophila* has not been studied so extensively as *T. pyriformis*, it is likely that its growth will be subject to variability for similar reasons.

Other organisms that have been used in RNV assays of proteins include *Leuconostoc mesenteroids*, *Streptococcus faecalis* and *Clostridium perfringens*, but none of them was entirely satisfactory, for various reasons, including pathogenicity. Ford[95] had limited success with *Streptococcus zymogenes* but RNVs were high in relation to rat NPU, which may have been due to the lack of lysine requirement of this organisms.

Availability of Amino Acids

Microbiological methods at best give a crude ranking order for protein quality but may be more useful in the future for the determination of the availability of amino acids.[83] Methodology for the determination of available methionine using *Streptococcus zymogenes*[96] is now well advanced and the technique is relatively straightforward. So far there is no suitable microbiological method for available lysine determination, as *S. zymogenes* has no requirement for this amino acid. Shorrock[97] determined available lysine using *T. pyriformis*, but results were subject to wide variation.

CONCLUSION

The importance of protein quality to human nutrition may have been overemphasised in the past, but we still do not know enough about the protein requirements of certain groups of the population to be certain of this. As far as the animal feeds industry is concerned, a knowledge of protein quality is necessary to maximise profits.

The requirement in the USA for a statement of protein quality to be made alongside other nutrient information in the labelling of food and food products has led to the rat PER bioassay becoming accepted methodology for protein quality estimation by regulatory agencies. This method, like all rat bioassays, is time-consuming and costly. So, recently, there has been renewed interest in finding rapid *in vitro* methods for protein quality evaluation, which compare favourably with the PER bioassay. Unfortunately, the PER method is subject to a number of methodological and interpretative difficulties, but at the moment there is no suitable bioassay to replace it.

The main problem in protein quality evaluation is that protein quality is not a fixed entity, being dependent on the physiological constitution of the consumer as well as the protein digestibility and amino acid profile. Therefore, it is not possible to obtain an absolute value for protein quality. A number of research workers are now of the opinion that perhaps we are wrong in choosing a bioassay as the yardstick for comparison of *in vitro* data. Instead we should concentrate on more easily measurable indices of protein quality, such as the content of essential amino acids, and put more research effort into finding satisfactory *in vitro* methods for the determination of bioavailability of amino acids. After all, data on amino acid composition have the advantage over bioassay data of being additive, and therefore predictions of the protein quality of whole diets can be made.

REFERENCES

1. McLaren, D. S. (1975). *Lancet* (ii), 93.
2. FAD (1973). *Energy and Protein Requirements.* Nutrition Meetings Report Series No. 52, FAO, Rome.
3. Waterlow, J. C. and Payne, P. R. (1975). *Nature*, **258**, 113.
4. Kies, C. (1981). *J. Agric. Food Chem.*, **29**, 435.
5. Satterlee, L. D. (1977). *New Concepts for the Rapid Develbpment of Protein Quality.* Univ. of Nebraska and Nat. Sci. Found., Washington, D.C.
6. Scrimshaw, N. S. (1976). *New England J. Med.*, **294**, 136.
7. Advisory Committee of the United Nations University World Hunger Program (1979). *Protein–energy Requirements under Conditions Prevailing in Developing Countries: Current Knowledge and Research Needs.* Food Nutr. Bull. (U.N.) Supplement No. 1.
8. Hegsted, D. M. (1977). *Am. J. Clin. Nutr.*, **30**, 465.
9. Scrimshaw, N. S., Young, V. R. and Garza, C. (1977). *Am. J. Clin. Nutr.*, **30**, 466.
10. McLaughlan, J. M. (1979). In *Soy Protein and Human Nutrition.* Wilcke, H. L., Hopkins, D. T. and Waggle, D. H. (Eds.). Academic Press, New York.
11. Osborne, T. B., Mendel, L. B. and Ferry, E. L. (1919). *J. Biol. Chem.*, **37**, 223.
12. Miller, D. S. and Bender, A. E. (1955). *Br. J. Nutr.*, **9**, 382.
13. Bodwell, C. E. (Ed.) (1977). *Evaluation of Proteins for Humans.* Avi Publishing Co. Inc., Westport, Conn.
14. Pellett, P. L. (1978). *Food Technol.*, **32**, 60.
15. Bressani, R., Navarrete, D. A., Elias, L. G. and Braham, J. E. (1979). In

Soy Protein and Human Nutrition. Wilcke, H. L., Hopkins, D. T. and Waggle, D. H. (Eds.). Academic Press, New York.
16. Bodwell, C. E., Kyle, E. M., Schuster, E. M., Vaughan, D. A., Womack, M. and Hackler, L. R. (1979). *Nutr. Rep. Int.*, **19**, 703.
17. Young, V. R., Scrimshaw, N. S. and Bier, D. M. (1981). *J. Agric. Food Chem.*, **29**, 440.
18. Porter, J. W. G. and Rolls, B. A. (Eds.) (1973). *Protein in Human Nutrition.* Academic Press, London and New York.
19. Friedman, M. (Ed.) (1975). *Protein Nutritional Quality of Foods and Feeds*, Vol. 1, Parts 1 and 2. Marcel Dekker Inc., New York.
20. Bender, A. E. (1975). In *Clinical and Biological Evaluation of Protein Quality.* Proc. 9th Int. Congr. Nutr., Mexico, 1972. Vol. 3. Karger, Basel.
21. Bodwell, C. E. (1977). *Cereal Chem.*, **54**, 958.
22. AOAC (1975). *Official Methods of Analysis*, 12th edn, 48.183. p. 857. Assn. Offic. Anal. Chem., Washington, D.C.
23. Steinke, F. H. (1979). In *Soy Protein and Human Nutrition.* Wilcke, H. L., Hopkins, D. T. and Waggle, D. H. (Eds.). Academic Press, New York.
24. Bender, A. E. and Doell, B. H. (1957). *Br. J. Nutr.*, **11**, 140.
25. Mitchell, H. H. (1948). In *Proteins and Amino Acids in Nutrition.* Sayhuan, M. (Ed.). Reinhold Publishing Co. Ltd., New York.
26. Hegsted, D. M. and Chang, Y. (1965). *J. Nutr.*, **85**, 159.
27. Steinke, F. H. (1977). *Cereal Chem.*, **54**, 949.
28. Hurt, H. D., Forsythe, R. H. and Krieger, C. H. (1975). In *Protein Nutritional Quality of Foods and Feeds*, Vol. 1, Part 1. Friedman, M. (Ed.). Marcel Dekker Inc., New York.
29. Hopkins, D. T. and Steinke, F. H. (1978). *Cereal Food World*, **23**, 539.
30. Bender, A. E. (1961). In *Progress in Meeting Protein Needs of Infants and Preschool Children.* Voris, L. (Ed.), Pub. 843, Nat. Acad. of Sciences – Nat. Res. Council, Washington, D.C.
31. Farina, R., Rio, M. E. and Sanahuja, J. C. (1979). *Nutr. Rep. Int.*, **20**, 693.
32. Phillips, R. D. (1981). *J. Nutr.*, **111**, 1058.
33. Bender, A. E. (1956). *Br. J. Nutr.*, **10**, 135.
34. Henry, K. M. and Toothill, J. (1962). *Br. J. Nutr.*, **16**, 125.
35. Wolzak, A., Élias, L. G. and Bressani, R. (1981). *J. Agric. Food Chem.*, **29**, 1063.
36. Bodwell, C. E. (1979). *J. Am. Oil Chem. Soc.*, **56**, 156.
37. McLaughlan, J. M. and Keith, Q. M. (1975). In *Protein Nutritional Quality of Foods and Feeds*, Vol. 1, Part 1. Friedman, M. (Ed.). Marcel Dekker Inc., New York.
38. Young, V. R. and Scrimshaw, N. S. (1974). In *Nutrients in Processed Foods – Proteins.* White, P. L. and Fletcher, D. C. (Eds.). Publishing Sciences Group Inc., Acton, Mass.
39. Saunders, R. M., Connor, M. A., Booth, A. N., Bickoff, E. M. and Kohler, G. O. (1973). *J. Nutr.*, **103**, 530.
40. Hsu, H. W., Vavak, D. L., Satterlee, L. D. and Miller, G. A. (1977). *J. Food Sci.*, **42**, 1269.
41. Marshall, F. F., Jr, Wallace, G. W. and Satterlee, L. D. (1979). *Nutr. Rep. Int.*, **19**, 901.

42. Bodwell, C. E., Satterlee, L. D. and Hackler, L. R. (1980). *Am. J. Clin. Nutr.*, **33**, 677.
43. Satterlee, L. D., Marshall, H. F. and Tennyson, J. M. (1979). *J. Am. Oil Chem. Soc.*, **56**, 103.
44. Pronczuk, A., Pawlowska, D. and Bartnik, J. (1973). *Nutr. Metabol.*, **15**, 171.
45. Ford, J. E. and Shorrock, C. (1971). *Br. J. Nutr.*, **26**, 311.
46. Spackman, D. H., Stein, W. H. and Moore, S. (1958). *Analyt. Chem.*, **30**, 1190.
47. Cavins, J. F., Kwolek, W. F., Inslett, G. E. and Cowan, J. L. (1972). *J. Ass. Off. Agric. Chem.*, **55**, 686.
48. Davies, M. G. and Thomas, A. J. (1973). *J. Sci. Food Agric.*, **24**, 1525.
49. Blackburn, S. (Ed.) (1978). *Amino Acid Determination: Methods and Techniques*, 2nd edn. Marcel Dekker Inc., New York and Basel.
50. Downs, F. and Pigman, W. (1969). *Int. J. Protein Res.*, **1**, 181.
51. Knipfel, J. E., Aitken, J. R., Hill, D. C., McDonald, B. E. and Owen, B. D. (1971). *J. Ass. Off. Agric. Chem.*, **54**, 777.
52. Williams, A. P. (1981). In *Amino-Acid Analysis in Clinical Chemistry and Medical Research.* Rattenbury, J. M. and Ambler, R. P. (Eds.). Ellis Horwood, Chichester.
53. Byers, M. (1971). In *Leaf Protein: Its Agronomy, Preparation, Quality and Use.* I.B.P. Handbook No. 20. Pirie, N. W. (Ed.). Blackwell Scientific Publications, Oxford.
54. Walker, H. G., Jr. Kohler, G. O., Kuzmicky, D. D. and Witt, S. C. (1975). In *Protein Nutritional Quality of Foods and Feeds*, Vol. 1, Part 1. Friedman, M. (Ed.). Marcel Dekker Inc., New York.
55. Mason, V. C., Bech-Anderson, S. and Rudemo, M. (1979). *Z. Tierernahr. Futtermittelk.*, **41**, 226.
56. Walker, A. F. (1981). The application of *in vitro* methods to the estimation of protein quality of leaf protein concentrates. Ph.D. Thesis, University of Reading, England.
57. Liu, T. -Y. and Chang, Y. H. (1971). *J. Biol. Chem.*, **246**, 2842.
58. Paddock, G. V., Wilson, G. B. and Wang, A. (1979). *Biochem. Biophys. Res. Commun.*, **87**, 946.
59. Conkerton, E. J. (1974). *J. Agric. Food Chem.*, **22**, 1046.
60. Mitchell, H. H. and Block, R. J. (1946). *J. Biol. Chem.*, **163**, 599.
61. FAO (1957). *Protein Requirements.* FAO Nutr. Stud. No. 16, FAO, Rome.
62. Harper, A. E. (1981). *J. Agric. Food Chem.*, **29**, 429.
63. NAS/NRC (1974). *Recommended Dietary Allowances.* Food and Nutrition Board of NAS/NRC, Washington, D.C.
64. Hsu, H. W., Sutton, N. E., Banjo, M. O., Satterlee, L. D. and Kendrick, J. G. (1978). *Food Technol.*, **32**, 69.
65. Jewell, D. K., Kendrick, J. C. and Satterlee, L. D. (1980). *Nutr. Rep. Int.*, **21**, 25.
66. Carpenter, K. J. (1974). In *Nutrients in Processed Foods — Proteins.* White, P. L. and Fletcher, D. C. (Eds.). Publishing Sciences Group Inc., Acton, Mass.

67. MAURON, J. (1980). In *Food and Health: Science and Technology*. Birch, G. G. and Parker, K. J. (Eds.). Applied Science Publishers, London.
68. EVANS, E. and WITTY, R. (1978). *World Rev. Nutr. Diet.*, **32**, 1.
69. CARPENTER, K. J. (1960). *Biochem. J.*, **77**, 604.
70. CARPENTER, K. J. and WOODHAM, A. A. (1974). *Br. J. Nutr.*, **32**, 647.
71. BOCTOR, A. M. and HARPER, A. E. (1968). *J. Nutr.*, **94**, 289.
72. HURRELL, R. F., CARPENTER, K. J., SINCLAIR, W. J., OTTERBURN, M. S. and ASQUITH, R. S. (1976). *Br. J. Nutr.*, **35**, 383.
73. ROACH, A. G., SANDERSON, P. and WILLIAMS, D. R. (1967). *J. Sci. Food Agric.*, **18**, 274.
74. CARPENTER, K. J. and BOOTH, V. H. (1973). *Nutr. Abstr. Rev.*, **43**, 424.
75. HURRELL, R. F. and CARPENTER, K. J. (1974). *Br. J. Nutr.*, **32**, 589.
76. SANDLER, L. and WARREN, F. L. (1974). *Analyt. Chem.*, **46**, 1870.
77. JONES, G. P. (1974). The use of dye-binding procedures for the evaluation of protein quality. Ph.D. Thesis, University of Reading, England.
78. HURRELL, R. F., LERMAN, P. and CARPENTER, K. J. (1979). *J. Food Sci.*, **44**, 1221.
79. WALKER, A. F. (1979). *Br. J. Nutr.*, **42**, 445.
80. WALKER, A. F. (1979). *Br. J. Nutr.*, **42**, 455.
81. ELLINGER, G. M. and PALMER, R. (1969). *Proc. Nutr. Soc.*, **28**, 42A.
82. LIPTON, S. H. and BODWELL, C. E. (1977). *J. Agric. Food Chem.*, **25**, 1216.
83. FORD, J. E. (1981). In *Protein Quality in Human Nutrition: Assessment and in vitro Estimation*. Avi Publishing Co. Inc., Westport, Conn.
84. FERNELL, W. R. and ROSEN, G. D. (1956). *Br. J. Nutr.*, **10**, 143.
85. SHORROCK, C. and FORD, J. E. (1973). In *Proteins in Human Nutrition*. Porter, J. W. G. and Rolls, B. A. (Eds.). Academic Press, London.
86. LANDERS, R. E. (1975). In *Protein Nutritional Quality of Foods and Feeds*, Vol. 1, Part. 1. Friedman, M. (Ed.). Marcel Dekker Inc., New York.
87. DRYDEN, M. J., KENDRICK, J. G., SATTERLEE, L. D., SCHROEDER, L. J. and BLOCK, R. G. (1977). *J. Food Biochem.*, **1**, 35.
88. SHEPHERD, N. D., TAYLOR, T. G. and WILTON, D. C. (1975). *Proc. Nutr. Soc.*, **34**, 100A.
89. SMITH, E. B. and PENA, P. M. (1977). *J. Food Sci.*, **42**, 674.
90. BAUM, F. and HAENEL, H. (1965). *Nährung*, **9**, 517.
91. WARRANT, R. M. and LABUZA, T. P. (1977). *J. Food Sci.*, **42**, 429.
92. DRYDEN, M. J. and SATTERLEE, L. D. (1978). *J. Food Sci.*, **43**, 650.
93. SUTTON, N. E. (1978). Protein efficiency ratio as estimated by *Tetrahymena thermophila* WH_{14} and its limitations. M.S. Thesis, Univ. of Nebraska, Lincoln, USA.
94. WANG, Y. -Y. D., MILLER, J. and BEUCHAT, L. R. (1980). *Nutr. Rep. Int.*, **21**, 645.
95. FORD, J. E. (1960). *Br. J. Nutr.*, **14**, 485.
96. FORD, J. E. (1962). *Br. J. Nutr.*, **16**, 409.
97. SHORROCK, C. (1972). Studies on the biological availability of amino acids in feedstuffs. Ph.D. Thesis, University of Reading, England.
98. HAWLEY, E. E., MURLIN, J. R., NASSET, E. S. and SAYMANSKI, T. A. (1948). *J. Nutr.*, **36**, 153.

99. WATTS, J. H., BOOKER, L. K., MCAFEE, J. W., WILLIAMS, E. G., WRIGHT, W. G. and JONES, F. (1959). *J. Nutr.*, **67**, 483.
100. BRICKER, M., MITCHELL, H. H. and KINSMAN, G. M. (1945). *J. Nutr.*, **30**, 269.
101. SCRIMSHAW, N. S., TAYLOR, Y. and YOUNG, V. R. (1973). *Am. J. Clin. Nutr.*, **26**, 965.
102. INOUE, G., FUJITA, Y., KISHI, K., YAMAMOTO, S. and NHYAMA, Y. (1974). *Nutr. Rep. Int.*, **10**, 201.
103. MILLWARD, D. J. and GARLICK, P. J. (1972). *Proc. Nutr. Soc.*, **31**, 257.
104. YOUNG, V. R. and SCRIMSHAW, N. S. (1977). In *Evaluation of Proteins for Humans.* Bodwell, C. E. (Ed.). Avi Publishing Co. Inc., Westport, Conn.
105. NRC (1972). *Nutrient Requirement of Laboratory Animals.* No. 10, NAS/NRC, Washington, D.C.
106. HEGSTED, D. M., NEFF, R. and WORCESTER, J. (1968). *J. Agric. Food Chem.*, **16**, 190.
107. SAMONDS, K. and HEGSTED, D. M. (1977). In *Evaluation of Proteins for Humans.* Bodwell, C. E. (Ed.). Avi Publishing Co. Inc., Westport, Conn.
108. MOORE, S. (1963). *J. Biol. Chem.*, **238**, 235.
109. FRIAR, P. M. K. (1978). Adsorption of acid monoazo dyes on anion-exchange materials and proteins. Ph.D. Thesis, University of Reading, England.
110. FRANK, G. and BRAUNITZER, G. (1968). *Hoppe-Seylor's Z. Physiol. Chem.*, **349**, 1456.
111. FAO (1970). *Amino-Acid Content of Foods and Biological Data on Proteins.* FAO Nutr. Stud. No. 24, FAO, Rome.
112. OSER, B. L. (1951). *J. Am. Diet. Assn*, **27**, 396.

INDEX